The Oxford Book of

MODERN SCIENCE WRITING

'A book from which the love of science and the love of language shine.'
Science

'An amalgamation of brilliant reads.'
Amanda Geffner, *New Scientist*

'Beautiful volume…A labour of love.'
Steven Poole, *The Guardian*

'The Oxford Book of Modern Science Writing edited by Richard Dawkins, comes up trumps…It is to be hoped that many will not only read this excellent volume but will then go on to read in their entirety some of the individual works themselves. That is the ultimate success of any anthology.'
Mary Strickland, *Chemistry World*

'Richard Dawkins has assembled a splendid collection of extracts…I can see myself dipping into this excellent bran tub of science writing for years to come, and everything I retrieve will be a source of pleasure and enlightenment.'
Richard Fortey, *The Lancet*

'This is a superb collection…it's a damn good read even if you're only marginally interested in science. I love this book…it's a must-read that will surely make a major contribution to the public understanding of science.'
John Gribbin, *BBC Focus*

'It is a real treasure trove of unexpected pleasures.'
Kenan Malik, *Sunday Telegraph*

'This isn't Dawkins as the centre of attention but as a benign and generous guide to the best science writing, with commentaries from the master…Every reader is likely to make a discovery or two.'
Peter Forbes, *The Independent*

'The book makes for a fascinating browser, but it could also inspire as a bed-time volume, filling the readers' dreams with exploding stars and tiny atoms.'
Peter Ranscombe, *The Scotsman*

'A glorious celebration of literary scientists.'
Harry Richie, *Mail on Sunday*

'A SPARKLING anthology.'
David Sinclair, *The Tribune*

'An excellent collection…A very impressive anthology and one that all science enthusiasts will want to own.'
The Scientific and Medical Network

'It is one of the best introductions to popular science writing and deserves to become a classic in its own right.'
Marc Miquel, Scope

'This is a book one reads over a prolonged period of time, being in essence written for the nightstand or the table next to a comfortable armchair. This book richly deserves to be in any library, public, private, or academic.'
CHOICE

Richard Dawkins was the first holder of the Charles Simonyi Chair of the Public Understanding of Science at Oxford University, and is a Fellow of New College, Oxford. His bestselling books include *The Selfish Gene*; *The God Delusion*; *The Extended Phenotype*; *The Blind Watchmaker*; *River out of Eden*; *Climbing Mount Improbable*; *Unweaving the Rainbow*; and *The Ancestor's Tale*.

Dawkins is a Fellow of both the Royal Society and the Royal Society of Literature. He is the recipient of numerous honours and awards, including the 1987 Royal Society of Literature Award, the 1990 Michael Faraday Award of the Royal Society, the 1994 Nakayama Prize, the 1997 International Cosmos Prize for Achievement in Human Science, the Kistler Prize in 2001, the Shakespeare Prize in 2005, the 2006 Lewis Thomas Prize for Writing about Science, and he was named Author of the Year at the 2007 Galaxy British Books Awards.

The Oxford Book of

MODERN
SCIENCE
WRITING

RICHARD DAWKINS

OXFORD
UNIVERSITY PRESS

OXFORD
UNIVERSITY PRESS

Great Clarendon Street, Oxford OX2 6DP

Oxford University Press is a department of the University of Oxford.
It furthers the University's objective of excellence in research, scholarship,
and education by publishing worldwide in

Oxford New York

Auckland Cape Town Dar es Salaam Hong Kong Karachi
Kuala Lumpur Madrid Melbourne Mexico City Nairobi
New Delhi Shanghai Taipei Toronto

With offices in

Argentina Austria Brazil Chile Czech Republic France Greece
Guatemala Hungary Italy Japan Poland Portugal Singapore
South Korea Switzerland Thailand Turkey Ukraine Vietnam

Oxford is a registered trade mark of Oxford University Press
in the UK and in certain other countries

Published in the United States
by Oxford University Press Inc., New York

Introduction, selection and commentary
© Richard Dawkins 2008

The moral rights of the author have been asserted
Database right Oxford University Press (maker)

First published 2008
First published in paperback 2009

British Library Cataloguing in Publication Data
Data available

Library of Congress Cataloging in Publication Data
Data available

Typeset by SPI Publisher Services, Pondicherry, India
Printed in Great Britain
by Clays Ltd, St Ives plc

ISBN 978–0–19–921680–2 (Hbk.)
ISBN 978–0–19–921681–9 (Pbk.)

1 3 5 7 9 10 8 6 4 2

For Charles Simonyi,
who loves science, loves language,
and understands how to put them together

CONTENTS

PART II

Who Scientists Are

PART III

What Scientists Think

PART IV

What Scientists Delight In

FEATURED WRITERS AND EXTRACTS

Atkins, Peter (1940–) Chemist and writer. Extract from *Creation Revisited*, Penguin, 1994.

Bak, Per (1948–2002) Theoretical physicist. *How Nature Works*, OUP, 1997.

Blakemore, Colin (1944–) Neurobiologist. *Sight Unseen*, BBC books, 1988.

Bonner, John Tyler (1920–) Biologist. *Life Cycles: Reflections of an Evolutionary Biologist*, Princeton University Press, 1993.

Brenner, Sydney (1927–) Biologist and Nobel laureate. 'Theoretical Biology in the Third Millennium', *Phil. Trans. R. Soc. Lond.* B, 354, 1963–1965, 1999.

Bronowski, Jacob (1908–1974) Mathematician and broadcaster. *The Identity of Man*, Prometheus, 2002. First published 1965.

Carson, Rachel (1907–1964) Marine biologist and natural history writer. *The Sea Around Us*, OUP, 1989. First published 1951.

Chandrasekhar, S. (1910–1995) Astrophysicist and Nobel laureate. *Truth and Beauty*, University of Chicago Press, 1987.

Crick, Francis (1916–2004) Molecular biologist and Nobel laureate. *What Mad Pursuit*, Basic Books Inc., 1988, and *Life Itself*, Macdonald and Co., 1981.

Cronin, Helena (1942–) Philosopher of biology. *The Ant and the Peacock*, Cambridge University Press, 1991.

Davies, Paul (1946–) Cosmologist and science writer. *The Goldilocks Enigma*, Allen Lane, 2006.

Dennett, Daniel C. (1942–) Philosopher. *Consciousness Explained*, Penguin, 1993 and *Darwin's Dangerous Idea*, Penguin, 1996.

Deutsch, David (1953–) Physicist. *Fabric of Reality*, Penguin, 1997.

Diamond, Jared (1937–) Evolutionary biologist, biogeographer, and writer. *The Rise and Fall of the Third Chimpanzee*, Vintage, 1992.

Dobzhansky, Theodosius (1900–1975) Geneticist and evolutionary biologist. *Mankind Evolving*, Yale University Press, 1962.

Dyson, Freeman (1923–) Physicist and mathematician. *Disturbing the Universe*, Pan Books, 1981.

Eddington, Sir Arthur (1882–1944) Astrophysicist. *The Expanding Universe*, Cambridge Penguin, 1940.

Edey, Maitland A. (1910–1992) Science writer and conservationist. Donald C. Johanson and Maitland A. Edey, *Lucy: The Beginnings of Humankind*, Penguin, 1990. First published 1981.

Einstein, Albert (1879–1955) Theoretical physicist and Nobel laureate. Reprinted in *Ideas and Opinions*, Wings Books, 1954. 'What is the theory of relativity?' published in *The London Times*, 1919. 'Religion and Science' published in the *New York Times Magazine*, 1930.

Eiseley, Loren (1907–1977) Anthropologist, ecologist, and writer. 'How Flowers Changed the World' and 'Little Men and Flying Saucers', in *The Immense Journey*, Vintage, 1957.

Feynman, Richard P. (1918–1988) Theoretical physicist and Nobel laureate. *The Character of Physical Law*, Penguin, 1992.

Fisher, Sir Ronald (1890–1962) Statistician, evolutionary biologist and geneticist. *The Genetical Theory of Natural Selection*, OUP, 2006.

Ford, Kenneth (1926–) Physicist. John Archibald Wheeler and Kenneth Ford, *Geons, Black Holes and Quantum Foam: a life in physics*, WW Norton, 1998.

Fortey, Richard (1946–) Palaeontologist. *Trilobite!*, Flamingo, 2001. Originally published 2000. *Life: an Unauthorised Biography*, Flamingo, 1998.

Gamow, George (1904–1968) Theoretical physicist and cosmologist. *Mr Tompkins in Paperback*, Cambridge University Press, 1993. *Mr Tomkins in Wonderland* first published 1940.

Gardner, Martin (1914–) Mathematics and science writer. 'Mathematical Games' column, *Scientific American* **223**, October 1970.

Gould, Stephen Jay (1941–2002) Palaeontologist, evolutionary biologist, and writer. 'Worm for a Century and All Seasons', in *Hen's Teeth and Horse's Toes*, W W Norton & Co., 1983.

Greene, Brian (1963–) Theoretical physicist. *The Elegant Universe*, Vintage, 2000.

Gregory, Richard (1923–) Neuropsychologist. *Mirrors in Mind*, W. H. Freeman & Co. Ltd, 1997.

Haldane, J. B. S. (1892–1964) Geneticist and evolutionary biologist. *On Being the Right Size and Other Essays*, Oxford University Press, 1991, formerly published in *Possible Worlds and Other Essays*, Chatto & Windus, 1927. 'Cancer's a Funny Thing', *New Statesman*, 1964.

Hamilton, W. D. (1936–2000) Evolutionary biologist. 'Geometry for the Selfish Herd', *Journal of Theoretical Biology*, **31**, 295–311, reprinted in *Narrow Roads of Gene Land*, Vol. 1. W. H. Freeman, 1996.

Hardin, Garrett (1915–2003) Ecologist. 'The Tragedy of the Commons', *Science*, **162**, No. 3859, 1243–8, 1968.

Hardy, Alister (1896–1985) Marine biologist. *The Open Sea: Its Natural History, Part 1: The World of Plankton*. Collins, 1970. First published 1956.

Hardy, G. H. (1877–1947) Mathematician. *A Mathematician's Apology*, Cambridge University Press, 2007. First published 1940.

Hawking, Stephen (1942–) Theoretical physicist. *A Brief History of Time*, Bantam Books, 1988.

Hofstadter, Douglas R. (1945–) Cognitive scientist, philosopher, polymath, Pulitzer Prize-winning author. *Godel, Escher, Bach*, Basic Books, 1979.

Hogben, Lancelot (1895–1975) Zoologist and statistician. *Mathematics for the Million*, Norton, 1993. Originally published 1937.

Hoyle, Fred (1915–2001) Astronomer and writer. *Man in the Universe*, Columbia University Press, 1966.

Humphrey, Nicholas (1943–) Psychologist. 'One Self', in *The Mind Made Flesh*, Oxford University Press, 2002. First published in *Social Research*, **67**, 32–9, 2000.

Huxley, Julian (1887–1975) Evolutionary biologist. *Essays of a Biologist*, Chatto & Windus, 1929. First published 1923.

Jeans, James (1877–1946) Physicist, astronomer, and mathematician. *The Mysterious Universe*, Cambridge University Press, 1930.

Johanson, Donald C. (1943–) Palaeoanthropologist. Donald C. Johanson and Maitland A. Edey, *Lucy: The Beginnings of Humankind*, Penguin, 1990.

Jones, Steve (1944–) Geneticist and writer. *The Language of Genes*, Flamingo, 2000.

Kingdon, Jonathan (1935–) Biologist, artist, and writer. *Before the Wise Men*, Simon and Schuster, 1993.

Lack, David (1910–1973) Ornithologist. *The Life of the Robin*, Collins, 1965.

Leakey, Richard (1944–) Palaeoanthropologist and conservationist. Richard Leakey and Roger Lewin, *Origins Reconsidered*, Little, Brown, 1992.

Levi, Primo (1919–1987) Chemist and writer. *The Periodic Table*, Penguin Classics, 2000. This translation, by Raymond Rosenthal, first published by Shocken Books, 1984.

Lewin, Roger Anthropologist and science writer. Richard Leakey and Roger Lewin, *Origins Reconsidered*, Little, Brown, 1992.

Maynard Smith, John (1920–2004) Evolutionary biologist and geneticist. *On Evolution*, Edinburgh University Press, 1972.

Mayr, Ernst (1904–2005) Evolutionary biologist. *The Growth of Biological Thought*, Harvard University Press, 1982.

Medawar, Peter B. (1915–1987) Medical scientist and writer, Nobel laureate. 'Science and Literature' from *The Hope of Progress*, Wildwood House, 1974;

'Darwin's Illness', 'The Phenomenon of Man', and the postscript to 'Lucky Jim' from *The Strange Case of the Spotted Mice*, Oxford University Press, 1996; 'D'Arcy Thompson and Growth and Form' from *The Art of the Soluble*, Penguin, 1969.

Oppenheimer, J. Robert (1904–1967) Theoretical physicist. *The Flying Trapeze: three crisis for physicists*, Oxford University Press, 1964.

Penrose, Roger (1931–) Mathematical physicist. *The Emperor's New Mind*, Oxford University Press, 1989.

Perutz, Max (1914–2002) Molecular biologist. 'A Passion for Crystals', in *I Wish I'd Made You Angry Earlier*, Oxford University Press, 1998. First published as an obituary in *The Independent*, 1994.

Pinker, Steven (1954–) Psychologist, cognitive scientist, and writer. *How The Mind Works*, Allen Lane, 1997 and *The Language Instinct*, Allen Lane, 1994.

Rees, Martin (1942–) Cosmologist and writer. *Just Six Numbers*, Phoenix, 2000.

Ridley, Mark (1956–) Zoologist and science writer. *The Explanation of Organic Diversity*, Clarendon Press, 1983.

Ridley, Matt (1958–) Zoologist and science writer. *Genome*, Fourth Estate, 1999.

Sacks, Oliver (1933–) Neurologist and writer. *Uncle Tungsten*, Picador, 2001.

Sagan, Carl (1934–1996) Astronomer and writer. *Pale Blue Dot*, Ballantine, 1997.

Schrodinger, Erwin (1887–1961) Physicist and Nobel laureate. *What is Life?*, Cambridge University Press, 2000. First published 1944.

Shannon, Claude (1916–2001) Electrical engineer and mathematician. Claude Shannon and Warren Weaver, *The Mathematical Theory of Communication*, University of Illinois Press, 1980. Originally published 1949.

Simpson, George Gaylord (1902–1984) Palaeontologist. *Meaning of Evolution*, Yale University Press, 1963. First published 1949.

Smolin, Lee (1955–) Theoretical physicist. *The Life of the Cosmos*, Phoenix, 1997.

Snow, C. P. (1905–1980) Physicist and novelist. G. H. Hardy's *A Mathematician's Apology* (Foreword), Cambridge University Press, 1967.

Stannard, Russell (1931–) Physicist and writer. *The Time and Space of Uncle Albert*, Faber and Faber, 1989.

Stewart, Ian (1945–) Mathematician and writer. *From Here to Infinity*, Oxford University Press, 1996. First published as *The Problems of Mathematics*, 1987.

Thomas, Lewis (1913–1993) Physician and essayist. *Late Night Thoughts*, Oxford University Press, 1985. First published 1980.

Thompson, D'Arcy Wentworth (1860–1948) Biologist and classicist. *On Growth and Form*, Cambridge University Press, 2004. Original complete work published 1917.

Tinbergen, Niko (1907–1988) Ethologist, ornithologist, and Nobel laureate. *Curious Naturalists*, Country Life, 1958.

Trivers, Robert (1943–) Evolutionary biologist. *Social Evolution*, Benjamin-Cummings Publishing Co., 1985.

Turing, Alan (1912–1954) Mathematician. *Computing Machinery and Intelligence, Mind*, **59**, 1950.

Watson, James (1928–) Molecular biologist and Nobel laureate. *Avoid Boring People*, Oxford University Press, 2007.

Weaver, Warren (1894–1978) Mathematician and scientist. Claude Shannon and Warren Weaver, *The Mathematical Theory of Communication*, University of Illinois Press, 1980. Originally published 1949.

Weinberg, Steven (1933–) Theoretical physicist and Nobel laureate. *Dreams of a Final Theory*, Vintage, 1993.

Wheeler, John Archibald (1911–) Theoretical physicist. John Archibald Wheeler and Kenneth Ford, *Geons, Black Holes and Quantum Foam*, WW Norton, 1998.

Williams, George C. (1926–) Evolutionary biologist. *Adaptation and Natural Selection*, Princeton University Press, 1966.

Wilson, Edward O. (1929–) Biologist. *The Diversity of Life*, Penguin, 2001.

Wolpert, Lewis (1929–) Developmental biologist and writer. *The Unnatural Nature of Science*, Faber and Faber, 1992.

INTRODUCTION

My introductory remarks are distributed through the book itself, so I shall here limit myself mostly to acknowledgements. The idea for an anthology of modern science writing was put to me by Latha Menon of Oxford University Press, and it was a pleasure to work with her on it. She and I had previously collaborated on a collection of my own occasional writings, and we slipped effortlessly back into the same synoptic vein as before. We disagreed only over whether or not to include anything from my own books. I won, and we didn't.

This is a collection of good writing by professional scientists, not excursions into science by professional writers. Another difference from John Carey's admirable *Faber Book of Science* is that we go back only one century. Within that century, no attempt was made to arrange the pieces chronologically. Instead, the selections fall roughly into four themes, although some of the entries could have fitted into more than one of these divisions. My biggest regret concerns the number of excellent scientists that I have had to leave out, for reasons of space. I would apologize to them, did I not suspect that my own pain at their omission is greater than theirs. The collection is limited to the English language and, with very few exceptions, I have omitted translations from books originally composed in other languages.

My wife, Lalla Ward, has again lent her finely tuned ear for the English language, together with her unfailing encouragement. I remain deeply grateful to her.

I have long wanted to dedicate a book to Charles Simonyi, but I was anxious to be clear that it was a dedication to him as an individual and friend, rather than as the munificent benefactor of the Oxford professorship in Public Understanding of Science that I hold. Now, in the year of my retirement, it finally seems appropriate to offer this volume to him as a personal friend, while at the same time conveying Oxford's gratitude to a major benefactor through a book published by the University Press. Charles Simonyi is a sort of combination of International Renaissance Man, Playboy of the Scientific World, Test Pilot of the Intellect, and Space-age Orbiter of the Mind as well as of the Planet. Although most of the words in an anthology belong to others, I hope that my love of science and of writing, which Charles shares and which he generously chose to encourage in me, will shine through both my selections and my commentary, and give him pleasure. ■

Richard Dawkins
Oxford, September 2007

PART I

WHAT
SCIENTISTS
STUDY

James Jeans

from THE MYSTERIOUS UNIVERSE

■ Our ability to understand the universe and our position in it is one of the glories of the human species. Our ability to link mind to mind by language, and especially to transmit our thoughts across the centuries is another. Science and literature, then, are the two achievements of *Homo sapiens* that most convincingly justify the specific name. In attempting, however inadequately, to bring the two together, this book can be seen as a celebration of humanity. It is only superficially paradoxical to begin our celebration by cutting humanity down to size, and no science puts us in our place better than astronomy. I begin with a fragment from James Jeans's 1930 book, *The Mysterious Universe*, which is a fine example of the humbling prose poetry that the stars so intoxicatingly inspire. ■

Standing on our microscopic fragment of a grain of sand, we attempt to discover the nature and purpose of the universe which surrounds our home in space and time. Our first impression is something akin to terror. We find the universe terrifying because of its vast meaningless distances, terrifying because of its inconceivably long vistas of time which dwarf human history to the twinkling of an eye, terrifying because of our extreme loneliness, and because of the material insignificance of our home in space—a millionth part of a grain of sand out of all the sea-sand in the world. But above all else, we find the universe terrifying because it appears to be indifferent to life like our own; emotion, ambition and achievement, art and religion all seem equally foreign to its plan. Perhaps indeed we ought to say it appears to be actively hostile to life like our own. For the most part, empty space is so cold that all life in it would be frozen; most of the matter in space is so hot as to make life on it impossible; space is traversed, and astronomical bodies continually bombarded, by radiation of a variety of kinds, much of which is probably inimical to, or even destructive of, life.

Into such a universe we have stumbled,

least as the result of what may properly be

use of such a word need not imply any sur

accidents will happen, and if the universe g

conceivable accident is likely to happen in

who said that six monkeys, set to strum u

for millions of millions of years, would b

the books in the British Museum. If we ex

particular monkey had typed, and found t

strumming, to type a Shakespeare sonnet

occurrence as a remarkable accident, but

millions of pages the monkeys had turned

we might be sure of finding a Shakespear

them, the product of the blind play of cha

Martin Re

from JUST SIX N

As Astronomer Royal and Preside

Rees, too, is no stranger to the romance

approach to putting us in our place in

ouraborus to situate us exactly in the mid

of magnitudes ranging from the astron

revert to this later in the book, when I d

by the evolved human mind as we try to

science far from the middle ground in w

The first extract comes from Rees's 19

extract from the same book explains its

made amazing strides towards explaini

our ignorance back into the first fracti

But our explanations of the deep probl

James Jeans

from THE MYSTERIOUS UNIVERSE

■ Our ability to understand the universe and our position in it is one of the glories of the human species. Our ability to link mind to mind by language, and especially to transmit our thoughts across the centuries is another. Science and literature, then, are the two achievements of *Homo sapiens* that most convincingly justify the specific name. In attempting, however inadequately, to bring the two together, this book can be seen as a celebration of humanity. It is only superficially paradoxical to begin our celebration by cutting humanity down to size, and no science puts us in our place better than astronomy. I begin with a fragment from James Jeans's 1930 book, *The Mysterious Universe*, which is a fine example of the humbling prose poetry that the stars so intoxicatingly inspire. ■

Standing on our microscopic fragment of a grain of sand, we attempt to discover the nature and purpose of the universe which surrounds our home in space and time. Our first impression is something akin to terror. We find the universe terrifying because of its vast meaningless distances, terrifying because of its inconceivably long vistas of time which dwarf human history to the twinkling of an eye, terrifying because of our extreme loneliness, and because of the material insignificance of our home in space—a millionth part of a grain of sand out of all the sea-sand in the world. But above all else, we find the universe terrifying because it appears to be indifferent to life like our own; emotion, ambition and achievement, art and religion all seem equally foreign to its plan. Perhaps indeed we ought to say it appears to be actively hostile to life like our own. For the most part, empty space is so cold that all life in it would be frozen; most of the matter in space is so hot as to make life on it impossible; space is traversed, and astronomical bodies continually bombarded, by radiation of a variety of kinds, much of which is probably inimical to, or even destructive of, life.

Into such a universe we have stumbled, if not exactly by mistake, at least as the result of what may properly be described as an accident. The use of such a word need not imply any surprise that our earth exists, for accidents will happen, and if the universe goes on for long enough, every conceivable accident is likely to happen in time. It was, I think, Huxley who said that six monkeys, set to strum unintelligently on typewriters for millions of millions of years, would be bound in time to write all the books in the British Museum. If we examined the last page which a particular monkey had typed, and found that it had chanced, in its blind strumming, to type a Shakespeare sonnet, we should rightly regard the occurrence as a remarkable accident, but if we looked through all the millions of pages the monkeys had turned off in untold millions of years, we might be sure of finding a Shakespeare sonnet somewhere amongst them, the product of the blind play of chance.

Martin Rees

from JUST SIX NUMBERS

As Astronomer Royal and President of the Royal Society, Martin Rees, too, is no stranger to the romance of the stars and of science. His approach to putting us in our place invokes the mythical symbol of the *ouraborus* to situate us exactly in the middle of the (logarithmic) spectrum of magnitudes ranging from the astronomical to the sub-atomic. I shall revert to this later in the book, when I discuss the difficulties experienced by the evolved human mind as we try to understand the extreme realms of science far from the middle ground in which our ancestors survived.

The first extract comes from Rees's 1999 book *Just Six Numbers*. A second extract from the same book explains its central theme. Modern physics has made amazing strides towards explaining the universe, heroically driving our ignorance back into the first fraction of a second after the Big Bang. But our explanations of the deep problems of existence rely on some half dozen numbers, the fundamental constants of physics, whose values we can measure but cannot derive from existing theories. They are just there; and many physicists, including Rees himself (though not, for example, Victor Stenger, a physicist for whom I also have a very high regard) believe that their precise values are crucial to the existence of a universe capable of producing biological evolution of some kind. Rees takes each of the six constants in turn, and the one I have chosen for this anthology is *N*, the ratio between the strength of the electrical force that holds atoms together and the gravitational force that holds the universe together. ■

Large Numbers and Diverse Scales

We are each made up of between 10^{28} and 10^{29} atoms. This 'human scale' is, in a numerical sense, poised midway between the masses of atoms and stars. It would take roughly as many human bodies to make up the mass of the Sun as there are atoms in each of us. But our Sun is just an ordinary star in the galaxy that contains a hundred billion stars altogether. There are at least as many galaxies in our observable universe as there are stars in a galaxy. More than 10^{78} atoms lie within range of our telescope.

Living organisms are configured into layer upon layer of complex structure. Atoms are assembled into complex molecules; these react, via complex pathways in every cell, and indirectly lead to the entire interconnected structure that makes up a tree, an insect or a human. We straddle the cosmos and the microworld—intermediate in size between the Sun, at a billion metres in diameter, and a molecule at a billionth of a metre. It is actually no coincidence that nature attains its maximum complexity on this intermediate scale: anything larger, if it were on a habitable planet, would be vulnerable to breakage or crushing by gravity.

We are used to the idea that we are moulded by the microworld: we are vulnerable to viruses a millionth of a metre in length, and the minute DNA double-helix molecule encodes our total genetic heritage. And it's just as obvious that we depend on the Sun and its power. But what about the still vaster scales? Even the nearest stars are millions of times further away than the Sun, and the known cosmos extends a billion times further still. Can we understand why there is so much beyond our Solar System? In this book I shall describe several ways in which we are linked to the stars, arguing that we cannot understand our origins without the cosmic context.

The intimate connections between the 'inner space' of the subatomic world and the 'outer space' of the cosmos are illustrated by the picture in Figure 1—an *ouraborus*, described by *Encyclopaedia Britannica* as the 'emblematic serpent of ancient Egypt and Greece, represented with its tail in its mouth continually devouring itself and being reborn from itself... [It] expresses the unity of all things, material and spiritual, which never disappear but perpetually change form in an eternal cycle of destruction and re-creation'.

On the left in the illustration are the atoms and subatomic particles; this is the 'quantum world'. On the right are planets, stars and galaxies. This book will highlight some remarkable interconnections between the microscales on the left and the macroworld on the right. Our everyday world is determined by atoms and how they combine together into molecules, minerals

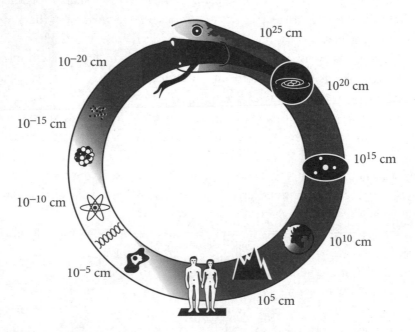

Figure 1. The *ouraborus*. There are links between the microworld of particles, nuclei and atoms (left) and the cosmos (right).

and living cells. The way stars shine depends on the nuclei within those atoms. Galaxies may be held together by the gravity of a huge swarm of subnuclear particles. Symbolized 'gastronomically' at the top, is the ultimate synthesis that still eludes us—between the cosmos and the quantum.

Lengths spanning sixty powers of ten are depicted in the *ouraborus*. Such an enormous range is actually a prerequisite for an 'interesting' universe. A universe that didn't involve large numbers could never evolve a complex hierarchy of structures: it would be dull, and certainly not habitable. And there must be long timespans as well. Processes in an atom may take a millionth of a billionth of a second to be completed; within the central nucleus of each atom, events are even faster. The complex processes that transform an embryo into blood, bone and flesh involve a succession of cell divisions, coupled with differentiation, each involving thousands of intricately orchestrated regroupings and replications of molecules; this activity never ceases as long as we eat and breathe. And our life is just one generation in humankind's evolution, an episode that is itself just one stage in the emergence of the totality of life.

The tremendous timespans involved in evolution offer a new perspective on the question 'Why is our universe so big?' The emergence of human life here on Earth has taken 4.5 billion years. Even before our Sun and its planets could form, earlier stars must have transmuted pristine hydrogen into carbon, oxygen and the other atoms of the periodic table. This has taken about ten billion years. The size of the observable universe is, roughly, the distance travelled by light since the Big Bang, and so the present visible universe must be around ten billion light-years across.

This is a startling conclusion. The very hugeness of our universe, which seems at first to signify how unimportant we are in the cosmic scheme, is actually entailed by our existence! This is not to say that there couldn't have been a smaller universe, only that we could not have existed in it. The expanse of cosmic space is not an extravagant superfluity; it's a consequence of the prolonged chain of events, extending back before our Solar System formed, that preceded our arrival on the scene

This may seem a regression to an ancient 'anthropocentric' perspective—something that was shattered by Copernicus's revelation that the Earth moves around the Sun rather than vice versa. But we shouldn't take Copernican modesty (sometimes called the 'principle of mediocrity') too far. Creatures like us require special conditions to have evolved, so our perspective is bound to be in some sense atypical. The vastness of our universe shouldn't surprise us, even though we may still seek a deeper explanation for its distinctive features.

[...]

The Value of N and Why it is So Large

Despite its importance for us, for our biosphere, and for the cosmos, gravity is actually *amazingly feeble* compared with the other forces that affect atoms. Electric charges of opposite 'sign' attract each other: a hydrogen atom consists of a positively charged proton, with a single (negative) electron trapped in orbit around it. Two protons would, according to Newton's laws, attract each other gravitationally, as well as exerting an electrical force of repulsion on one another. Both these forces depend on distance in the same way (both follow an 'inverse square' law), and so their relative strength is measured by an important number, N, which is the same irrespective of how widely separated the protons are. When two hydrogen atoms are bound together in a molecule, the electric force between the protons is neutralized by the two electrons. The gravitational attraction between the protons is thirty-six powers of ten feebler than the electrical forces, and quite unmeasurable. Gravity can safely be ignored by chemists when they study how groups of atoms bond together to form molecules.

How, then, can gravity nonetheless be dominant, pinning us to the ground and holding the moon and planets in their courses? It's because gravity is *always an attraction*: if you double a mass, then you double the gravitational pull it exerts. On the other hand, electric charges can repel each other as well as attract; they can be either positive or negative. Two charges only exert twice the force of one if they are of the same 'sign'. But any everyday object is made up of huge numbers of atoms (each made up of a positively charged nucleus surrounded by negative electrons), and the positive and negative charges almost exactly cancel out. Even when we are 'charged up' so that our hair stands on end, the imbalance is less than one charge in a billion billion. But everything has the same sign of gravitational 'charge', and so gravity 'gains' relative to electrical forces in larger objects. The balance of electric forces is only slightly disturbed when a solid is compressed or stretched. An apple falls only when the combined gravity of all the atoms in the Earth can defeat the electrical stresses in the stalk holding it to the tree. Gravity is important to us because we live on the heavy Earth.

We can quantify this. In Chapter 1, we envisaged a set of pictures, each being viewed from ten times as far as the last. Imagine now a set of differently sized spheres, containing respectively 10, 100, 1000, ... atoms, in other words each ten times heavier than the one before. The eighteenth would be as big as a grain of sand, the twenty-ninth the size of a human, and the fortieth that of a largish asteroid. For each thousand-fold increase in mass, the volume also goes up a thousand times (if the spheres are equally dense) but the radius goes up only by ten times. The importance of the sphere's own gravity, measured by how much energy it takes to remove an atom from its gravitational pull, depends on mass divided by radius, and so goes up a factor of a hundred. Gravity starts off, on the atomic scale, with a handicap of thirty-six powers of ten; but it gains two powers of ten (in other words 100) for every three powers (factors of 1000) in mass. So gravity will have caught up for the fifty-fourth object ($54 = 36 \times 3/2$), which has about Jupiter's mass. In any still heavier lump more massive than Jupiter, gravity is so strong that it overwhelms the forces that hold solids together.

Sand grains and sugar lumps are, like us, affected by the gravity of the massive Earth. But their *self-gravity*—the gravitational pull that their constituent atoms exert on each other, rather than on the entire Earth—is negligible. Self-gravity is not important in asteroids, nor in Mars's two small potato-shaped moons, Phobos and Deimos. But bodies as large as planets (and even our own large Moon) are not rigid enough to maintain an irregular shape: gravity makes them nearly round. And masses above that of Jupiter get crushed by their own gravity to extraordinary densities (unless the centre gets hot enough to supply a balancing pressure, which is what happens in the Sun and other stars like it). It is because gravity is so weak that a typical star like the Sun is so massive. In any lesser aggregate, gravity could not compete with the pressure, nor squeeze the material hot and dense enough to make it shine.

The Sun contains about a thousand times more mass than Jupiter. If it were cold, gravity would squeeze it a million times denser than an ordinary solid: it would be a 'white dwarf' about the size of the Earth but 330,000 times more massive. But the Sun's core actually has a temperature of fifteen million degrees—thousands of times hotter than its glowing surface, and the pressure of this immensely hot gas 'puffs up' the Sun and holds it in equilibrium.

The English astrophysicist Arthur Eddington was among the first to understand the physical nature of stars. He speculated about how much we could learn about them just by theorizing, if we lived on a perpetually cloud-bound planet. We couldn't, of course, guess how many there are, but simple reasoning along the lines I've just outlined could tell us how big they would have to be, and it isn't too difficult to extend the argument further, and work out how brightly such objects could shine. Eddington concluded that: 'When we draw aside the veil of clouds beneath which our physicist is working and let him look up at the sky, there he will find a thousand million globes of gas, nearly all with [these] masses.'

Gravitation is feebler than the forces governing the microworld by the number N, about 10^{36}. What would happen if it weren't quite so weak? Imagine, for instance, a universe where gravity was 'only' 10^{30} rather than 10^{36} feebler than electric forces. Atoms and molecules would behave just as in our actual universe, but objects would not need to be so large before gravity became competitive with the other forces. The number of atoms needed to make a star (a gravitationally bound fusion reactor) would be a billion times less in this imagined universe. Planet masses would also be scaled down by a billion. Irrespective of whether these planets could retain steady orbits, the strength of gravity would stunt the evolutionary potential on them. In an imaginary strong-gravity world, even insects would need thick legs to support them, and no animals could get much larger. Gravity would crush anything as large as ourselves.

Galaxies would form much more quickly in such a universe, and would be miniaturized. Instead of the stars being widely dispersed, they would be so densely packed that close encounters would be frequent. This would in itself preclude stable planetary systems, because the orbits would be disturbed by passing stars—something that (fortunately for our Earth) is unlikely to happen in our own Solar System.

But what would preclude a complex ecosystem even more would be the limited time available for development. Heat would leak more quickly from these 'mini-stars': in this hypothetical strong-gravity world, stellar lifetimes would be a million times shorter. Instead of living for ten billion years, a typical star would live for about 10,000 years. A mini-Sun would burn faster, and would have exhausted its energy before even the first steps in organic evolution had got under way. Conditions for complex evolution would undoubtedly be less favourable if (leaving everything else unchanged) gravity were stronger. There wouldn't be such a huge gulf as there is in our actual universe between the immense timespans of astronomical processes and the basic microphysical timescales for physical or chemical reactions. The converse, however, is that an even *weaker* gravity could allow even more elaborate and longer-lived structures to develop.

Gravity is the organizing force for the cosmos...[It] is crucial in allowing structure to unfold from a Big Bang that was initially almost featureless. But it is only because it is weak compared with other forces that large and long-lived structures can exist. Paradoxically, the weaker gravity is (provided that it isn't actually zero), the grander and more complex can be its consequences. We have no theory that tells us the value of N. All we know is that nothing as complex as humankind could have emerged if N were much less than 1,000,000,000,000,000,000,000, 000,000,000,000,000.

Peter Atkins

from CREATION REVISITED

The chemist Peter Atkins writes substantial textbooks, which American university bookshops order by the cubic yard. He is also, in my opinion, one of the finest living writers of scientific literature, a master of scientific wit ('Thermodynamicists get very excited when nothing happens') and lyrical prose poetry extolling the wonders of science and the scientific world view. His literary flair is most hauntingly demonstrated in *The Creation* (second edition *Creation Revisited*):

> When we have dealt with the values of the fundamental constants by see-
> ing that they are unavoidably so, and have dismissed them as irrelevant, we
> shall have arrived at complete understanding. Fundamental science then
> can rest. We are almost there. Complete knowledge is just within our grasp.
> Comprehension is moving across the face of the Earth, like the sunrise.

The extract from the same delightful book that I have chosen expounds
one of the central ideas of all science—C. P. Snow's litmus test of scientific culture—the Second Law of Thermodynamics. Atkins shows how the
universal downhill degradation towards disorder can be harnessed locally to
drive processes uphill (the principle of the ram pump) and build up order—
including life and everything that makes it worth having. Notice, by the way,
that Atkins uses the word 'chaos' in its normal sense of disorder, which is
rather different from a special technical sense popularized as the 'butterfly
effect' (one flap of a butterfly's wing could in theory initiate a chain of events
that leads to a hurricane). Important and interesting as that technical sense
of 'chaos' undoubtedly is, I deplore the hijacking of the word. ■

––––––––––

Why Things Change

Change takes a variety of forms. There is simple change, as when a
bouncing ball comes to rest, or when ice melts. There is more complex
change, as in digestion, growth, reproduction, and death. There is also
what appears to be excessively subtle change, as in the formation of
opinions and the creation and rejection of ideas. Though diverse in its
manifestations, change does in fact have a common source. Like everything fundamental, that source is perfectly simple.

Organized change, the contriving of some end, such as a pot, a crop,
or an opinion, is powered by the same events that stop balls bouncing
and melt ice. All change, I shall argue, arises from an underlying collapse
into chaos. We shall see that what may appear to us to be motive and
purpose is in fact ultimately motiveless, purposeless decay. Aspirations,
and their achievement, feed on decay.

The deep structure of change is decay. What decays is not the quantity
but the *quality* of energy. I shall explain what is meant by high quality energy, but for the present think of it as energy that is localized, and potent
to effect change. In the course of causing change it spreads, becomes

chaotically distributed like a fallen house of cards, and loses its initial potency. Energy's quality, but not its quantity, decays as it spreads in chaos.

Harnessing the decay results not only in civilizations but in all the events in the world and the universe beyond. It accounts for all discernible change, both animate and inanimate. The quality of energy is like a slowly unwinding spring. The quality spontaneously declines and the spring of the universe unwinds. The quality spontaneously degrades, and the spontaneity of the degradation drives the interdependent processes webbed around and within us, as through the interlocked gear wheels of a sophisticated machine. Such is the complexity of the interlocking that here and there chaos may temporarily recede and quality flare up, as when cathedrals are built and symphonies are performed. But these are temporary and local deceits, for deeper in the world the spring inescapably unwinds. Everything is driven by decay. Everything is driven by motiveless, purposeless decay.

As we have said, by 'quality' of energy is meant the extent of its dispersal. High-quality, useful energy, is localized energy. Low quality, wasted energy, is chaotically diffuse energy. Things can get done when energy is localized; but energy loses its potency to motivate change when it has become dispersed. The degradation of quality is chaotic dispersal.

I shall now argue that such dispersal is ultimately natural, motiveless, and purposeless. It occurs naturally and spontaneously, and when it occurs it causes change. When it is precipitate it destroys. When it is geared through chains of events it can produce civilizations.

The naturalness of the tendency of energy to spread can be appreciated by thinking of a crowd of atoms jostling. Localized energy, energy in a circumscribed region, corresponds to vigorous motion in a corner of the crowd. As the atoms jostle, they hand on their energy and induce their neighbours to jostle too, and soon the jostling disperses like the order of a shuffled pack. There is very little chance that the original corner of the crowd will ever again be found jostled back into its original activity with all the rest at rest. Random, motiveless jostling has resulted in irreversible change.

This natural tendency to disperse accounts for simple processes like the cooling of hot metal. The energy of the block, an energy captured in the vigorous vibrations of its atoms, is jostled into its surroundings. The individual jostlings may result in energy being passed in either direction; but there are so many more atoms in the world outside than in the block itself, that it is much more probable that at all later times the energy of the block will be found (or lost) dispersed.

Illusions of purpose are captured by the model. We may think that there are reasons why one change occurs and not another. We may think that there are reasons for specific changes in the location of energy (such as a change of structure, as in the opening of a flower); but at root, all there is, is degradation by dispersal.

Suppose that in some region there are many more places for energy to accumulate than elsewhere. Then jostling and random leaping results in its heaping there. If the energy began in a heap initially organized elsewhere, it will be found later in a heap in the region where the platforms are most dense. A casual observer will wonder why the energy chose to go there, conclude there must have been a purpose, and try to find it. We, however, can see that achieving being there should not be confused with choosing to go there.

Changes of location, of state, of composition, and of opinion are all at root dispersal. But if that dispersal spreads energy into regions where it can be located densely, it gives the illusion of specific change rather than mere spreading. At the deepest level, purpose vanishes and is replaced by the consequences of having the opportunity to explore at random, discovering dense locations, and lingering there until new opportunities for exploration arise.

Events are the manifestations of overriding probabilities. All the events of nature, from the bouncing of balls to the conceiving of gods, are aspects and elaborations of this simple idea. But we should not let pass without emphasis the word probability. The energy just might by chance jostle back into its original heap, and a structure reform. The energy just might, by chance, jostle its way back into the block from the world at large, and an observer see a cool block spontaneously becoming hot or a house of cards reforming. These possibilities are such remote chances that we dismiss them as wholly improbable. Yet, while improbable, they are not impossible.

The ultimate simplicity underlying the tendency to change is more effectively shrouded in some processes than in others. While cooling is easy to explain as natural, jostling dispersal, the processes of evolution, free will, political ambition, and warfare have their intrinsic simplicity buried more deeply. Nevertheless, even though it may be concealed, the spring of all creation is decay, and every action is a more or less distant consequence of the natural tendency to corruption.

The tendency of energy to chaos is transformed into love or war through the agency of chemical reactions. All actions are chains of reactions. From thinking to doing, in simply thinking, or in responding, the mechanism in action is chemical reaction.

At its most rudimentary, a chemical reaction is a rearrangement of atoms. Atoms in one arrangement constitute one species of molecule, and atoms in another, perhaps with additions or deletions, constitute another. In some reactions a molecule merely changes its shape; in some, a molecule adopts the atoms provided by another, incorporates them, and attains a more complex structure. In others, a complex molecule is itself eaten, either wholly or in part, and becomes the source of atoms for another.

Molecules have no inclination to react, and none to remain unreacted. There is, of course, no such thing as motive and purpose at this level of behaviour. Why, then, do reactions occur? At this level too, therefore, there can be no motive or purpose in love or war. Why then do they occur?

A reaction tends to occur if in the process energy is degraded into a more dispersed, more chaotic form. Every arrangement of atoms, every molecule, is constantly subject to the tendency to lose energy as jostling carries it away to the surroundings. If a cluster of atoms happens by chance to wander into an arrangement that corresponds to a new molecule, that transient arrangement may suddenly be frozen into permanence as the energy released leaps away. Chemical reactions are transformations by misadventure.

Atoms are only loosely structured into molecules, and explorations of rearrangements resulting in reactions are commonplace. That is one reason why consciousness has already emerged from the inanimate matter of the original creation. If atoms had been as strongly bound as nuclei, the initial primitive form of matter would have been locked into permanence, and the universe would have died before it awoke.

The frailty of molecules, though, raises questions. Why has the universe not already collapsed into unreactive slime? If molecules were free to react each time they touched a neighbour, the potential of the world for change would have been realized long ago. Events would have taken place so haphazardly and rapidly that the rich attributes of the world, like life and its own self-awareness, would not have had time to grow.

The emergence of consciousness, like the unfolding of a leaf, relies upon restraint. Richness, the richness of the perceived world and the richness of the imagined worlds of literature and art—the human spirit—is the consequence of controlled, not precipitate, collapse.

Helena Cronin

from THE ANT AND THE PEACOCK

■ We now switch from physical science to my own subject of biology. Helena Cronin's beautifully written *The Ant and the Peacock* is mostly about two special problems that arose out of Darwin's work, altruism and sexual selection. But the book begins with as elegant a word picture as you'll find of the central idea of biology itself, Darwinian evolution. ■

We are walking archives of ancestral wisdom. Our bodies and minds are live monuments to our forebears' rare successes. This Darwin has taught us. The human eye, the brain, our instincts, are legacies of natural selection's victories, embodiments of the cumulative experience of the past. And this biological inheritance has enabled us to build a new inheritance: a cultural ascent, the collective endowment of generations. Science is part of this legacy, and this book is about one of its foremost achievements: Darwinian theory itself.

[...]

A World Without Darwin

Imagine a world without Darwin. Imagine a world in which Charles Darwin and Alfred Russel Wallace had not transformed our understanding of living things. What, that is now comprehensible to us, would become baffling and puzzling? What would we see as in urgent need of explanation?

The answer is: practically everything about living things—about all of life on earth and for the whole of its history (and, probably, as we'll see, about life elsewhere, too). But there are two aspects of organisms that had baffled and puzzled people more than any others before Darwin and Wallace came up with their triumphant and elegant solution in the 1850s.

The first is design. Wasps and leopards and orchids and humans and slime moulds have a designed appearance about them; and so do eyes and kidneys and wings and pollen sacs; and so do colonies of ants, and flowers attracting bees to pollinate them, and a mother hen caring for her chicks. All this is in sharp contrast to rocks and stars and atoms and fire. Living things are beautifully and intricately adapted, and in myriad ways, to their inorganic surroundings, to other living things (not least to those most like themselves), and as superbly functioning wholes. They have an air of purpose about them, a highly organised complexity, a precision and efficiency. Darwin aptly referred to it as 'that perfection of structure and co-adaptation which most justly excites our admiration'. How has it come about?

The second puzzle is 'likeness in diversity'—the strikingly hierarchical relationships that can be found throughout the organic world, the differences and yet obvious similarities among groups of organisms, above all the links that bind the serried multitudes of species. By the mid-nineteenth century, these fundamental patterns had emerged from a range of biological disciplines. The fossil record was witness to continuity in time; geographical distribution to continuity in space; classification systems were built on what was called unity of type; morphology and embryology (particularly comparative studies) on so-called mutual affinities; and all these subjects revealed a remarkable abundance of further regularities and ever-more diversity. How could these relationships be accounted for? And whence such profligate speciation?

In the light of Darwinian theory, the answers to both questions, and to a host of other questions about the organic world, fall into place. Darwin and Wallace assumed that living things had evolved. Their problem was to find the mechanism by which this evolution had occurred, a mechanism that could account for both adaptation and diversity. Natural selection was their solution. Individuals vary and some of their variations are heritable. These heritable variations arise randomly—that is, independently of their effects on the survival and reproduction of the organism. But they are perpetuated differentially, depending on the adaptive advantage they confer.

Thus, over time, populations will come to consist of the better adapted organisms. And, as circumstances change, different adaptations become advantageous, gradually giving rise to divergent forms of life.

The key to all of this—to how natural selection is able to produce its wondrous results—is the power of many, many small but cumulative changes. Natural selection cannot jump from the primaeval soup to orchids and ants all in one go, at a single stroke. But it can get there through millions of small changes, each not very different from what went before but amounting over very long periods of time to a dramatic transformation. These changes arise randomly—without relation to whether they'll be good, bad or indifferent. So if they happen to be of advantage that's just a matter of chance. But it's not a grossly improbable chance, because the change is very small, from an organism that's not much like an exquisitely fashioned orchid to one that's ever-so-slightly more like it. So what would otherwise be a vast dollop of luck is smeared out into acceptably probable portions. And natural selection not only seizes on each of these chance advantages but also preserves them cumulatively, conserving them one after another throughout a vast series, until they gradually build up into the intricacy and diversity of adaptation that can move us to awed admiration. Natural selection's power, then, lies in randomly generated diversity that is pulled into line and shaped over vast periods of time by a selective force that is both opportunistic and conserving.

R. A. Fisher

from THE GENETICAL THEORY OF NATURAL SELECTION

My overwhelming impression on opening any page of Darwin is of being ushered into the presence of a great intellect. I feel the same sense of reverential hush when I read Darwin's great twentieth-century successor R. A. Fisher. I think it is right to include Fisher in this collection, even though his writing is more difficult than Darwin's, especially to non-mathematicians (in which large category Darwin would have been the first to place himself). The opening pages of Fisher's *The Genetical Theory of Natural Selection* lack the explicit mathematics of later parts of the book, but you can tell that the biologist in whose presence we find ourselves was a mathematician first. Fisher was one of the three great founders of population, mathematical, and evolutionary genetics, one of the half dozen or so founders of the neo-Darwinian Modern Synthesis, and arguably the founder of modern statistics. Alas, I never met him, but as an undergraduate I once encountered the eccentric Oxford geneticist E. B. Ford escorting an old gentleman with a very white beard and very thick glasses through the Museum, and I like to think that this must have been Ford's mentor and hero, Sir Ronald Fisher. ▪

Difficulties Felt by Darwin

The argument based on blending inheritance and its logical conse-quences, though it certainly represents the general trend of Darwin's thought upon inheritance and variation, for some years after he com-menced pondering on the theory of Natural Selection, did not satisfy him completely. Reversion he recognized as a fact which stood out-side his scheme of inheritance, and that he was not altogether satis-fied to regard it as an independent principle is shown by his letter to Huxley already quoted. By 1857 he was in fact on the verge of devising a scheme of inheritance which should include reversion as one of its consequences. The variability of domesticated races, too, presented a difficulty which, characteristically, did not escape him. He notes (pp. 77, 78, *Foundations*) in 1844 that the most anciently domesticated ani-mals and plants are not less variable, but, if anything more so, than those more recently domesticated; and argues that since the supply of food could not have been becoming much more abundant progres-sively at all stages of a long history of domestication, this factor cannot alone account for the great variability which still persists. The passage runs as follows:

If it be an excess of food, compared with that which the being obtained in its natural state, the effects continue for an improbably long time; during how many ages has wheat been cultivated, and cattle and sheep reclaimed, and we cannot suppose their *amount* of food has gone on increasing, nevertheless these are amongst the most variable of our domestic productions.

This difficulty offers itself also to the second supposed cause of variability, namely changed conditions, though here it may be argued that the conditions of cultivation or nurture of domesticated species have always been changing more or less rapidly. From a passage in the *Variation of Animals and Plants* (p. 301), which runs:

> Moreover, it does not appear that a change of climate, whether more or less genial, is one of the most potent causes of variability; for in regard to plants Alph. De Candolle, in his *Geographie Botanique*, repeatedly shows that the native country of a plant, where in most cases it has been longest cultivated, is that where it has yielded the greatest number of varieties.

it appears that Darwin satisfied himself that the countries in which animals or plants were first domesticated, were at least as prolific of new varieties as the countries into which they had been imported, and it is natural to presume that his inquiries under this head were in search of evidence bearing upon the effects of changed conditions. It is not clear that this difficulty was ever completely resolved in Darwin's mind, but it is clear from many passages that he saw the necessity of supplementing the original argument by postulating that the causes of variation which act upon the reproductive system must be capable of acting in a delayed and cumulative manner so that variation might still be continued for many subsequent generations.

Particulate Inheritance

It is a remarkable fact that had any thinker in the middle of the nineteenth century undertaken, as a piece of abstract and theoretical analysis, the task of constructing a particulate theory of inheritance, he would have been led, on the basis of a few very simple assumptions, to produce a system identical with the modern scheme of Mendelian or factorial inheritance. The admitted non-inheritance of scars and mutilations would have prepared him to conceive of the hereditary nature of an organism as something nonetheless definite because possibly represented inexactly by its visible appearance. Had he assumed that this hereditary nature was completely determined by the aggregate of the hereditary particles (genes), which enter into its composition, and at the same time assumed that organisms of certain possible types of hereditary composition were capable of breeding true, he would certainly have inferred that each organism must receive a definite portion of its genes from each parent, and that consequently it must transmit only a corresponding portion to each of its offspring. The simplification that, apart from sex and possibly other characters related in their inheritance to sex, the contributions of the two parents were equal, would not have been confidently assumed without the evidence of reciprocal crosses; but our imaginary theorist, having won so far, would scarcely have failed to imagine a conceptual framework in which each gene had its proper place or locus, which could be occupied alternatively, had the parentage been different, by a gene of a different kind. Those organisms (homozygotes) which received like genes, in any pair of corresponding loci, from their two parents, would necessarily hand on genes of this kind to all of their offspring alike; whereas those (heterozygotes) which received from their two parents genes of different kinds, and would be, in respect of the locus in question, crossbred, would have, in respect of any particular offspring, an equal chance of transmitting either kind. The heterozygote when mated to either kind of homozygote would produce both heterozygotes and homozygotes in a ratio which, with increasing numbers of offspring, must tend to equality, while if two heterozygotes were mated, each homozygous form would be expected to appear in a quarter of the offspring, the remaining half being heterozygous. It thus appears that, apart from dominance and linkage, including sex linkage, all the main characteristics of the Mendelian system flow from assumptions of particulate inheritance of the simplest character, and could have been deduced *a priori* had any one conceived it possible that the laws of inheritance could really be simple and definite.

The segregation of single pairs of genes, that is of single factors, was demonstrated by Mendel in his paper of 1865. In addition Mendel demonstrated in his material the fact of dominance, namely that the heterozygote was not intermediate in appearance, but was almost or quite indistinguishable from one of the homozygous forms. The fact of dominance, though of the greatest theoretical interest, is not an essential feature of the factorial system, and in several important cases is lacking altogether. Mendel also demonstrated what a theorist could scarcely have ventured to postulate, that the different factors examined by him in combination, segregated in the simplest possible manner, namely independently. It was not till after the rediscovery of Mendel's laws at the end of the century that cases of linkage were discovered, in which, for factors in the same linkage group, the pair of genes received from the same parent are more often than not handed on together to the same child. The conceptual framework of loci must therefore be conceived as made of several parts, and these are now identified, on evidence which appears to be singularly complete, with the dark-staining bodies or chromosomes which are to be seen in the nuclei of cells at certain stages of cell division.

The mechanism of particulate inheritance is evidently suitable for reproducing the phenomenon of reversion, in which an individual resembles a grandparent or more remote ancestor, in some respect in which it differs from its parents; for the ancestral gene combination may by chance be reproduced. This takes its simplest form when dominance occurs, for every union of two heterozygotes will then produce among the offspring some recessives, differing in appearance from their parents, but probably resembling some grandparent or ancestor.

Theodosius Dobzhansky

from MANKIND EVOLVING

■ One of Fisher's co-founders of the neo-Darwinian Modern Synthesis was the Russian American geneticist Theodosius Dobzhansky (1900–75). Less mathematical than Fisher, he was a fine researcher, and the author of one of the most influential of the founding texts of the Synthesis, *Genetics and the Origin of Species*. The passage I have chosen is from one of his later books, *Mankind Evolving* (1962), which influenced me when I was an undergraduate and heard him give a lecture at Oxford as the guest of E. B. Ford. Dobzhansky's chapter is a lucid exposition of how genes interact with environment in the determination of the variation among human individuals. The stress, importantly, is on the word variation. 'Nature versus nurture' is a topic that frequently inspires rather boring writing. Dobzhansky's chapter is an honourable exception. ▓

Equal but Dissimilar

> My idea of society is that while we are born equal, meaning that we have a right to equal opportunity, all have not the same capacity.
>
> MAHATMA GANDHI

'I have made the four winds that every man might breathe there-of like his brother during his time. I have made every man like his brother, and I have forbidden that they do evil; it was their hearts which undid that which I had said.' This utterance, ascribed to the Egyptian god Re, ante-dates by some four and a half thousand years the Declaration of Independence, which states: 'We hold these truths to be self-evident, that all men are created equal.' But, surely, Re as well as Thomas Jefferson knew that brothers very often look and act unlike. Brothers, though dissimilar, are yet equal in their rights to share in the patrimony of their fathers.

A newborn infant is not a blank page; however, his genes do not seal his fate. His reactions to the world around him will differ in many ways from those of other infants, including his brothers. My genes have indeed determined what I am, but only in the sense that, given the succession of environments and experiences that were mine, a carrier of a different set of genes might have become unlike myself.

It is sometimes said that the genes determine the limits up to which, but not beyond which, a person's development may advance. This only con-fuses the issue. There is no way to predict all the phenotypes that a given genotype might yield in every one of the infinity of possible environments. Environments are infinitely diversified, and in the future there will exist environments that do not exist now. The infant now promenading in his perambulator under my window may become many things. To be sure, he is not likely to grow eight feet tall, but we do not know how to obtain the evidence needed to determine how tall he may grow in some environments that may be contrived to stimulate growth. It is an illusion that there is something fundamental or intrinsic about limits, particularly upper limits. Every statistician knows that limits are elusive and hard to determine, most of all when the environmental conditions are not specified.

Even at the risk of belaboring the obvious, let it be repeated that heredity cannot be called the 'dice of destiny'. Variations in body build, in physiology, and in mental traits are in part genetically conditioned, but this does not make education and social improvements any less necessary, or the hopes of benefits to be derived from these improvements any less well founded. What genetic conditioning does mean is that there is no single human nature, only human natures with different requirements for optimal growth and self-realization. The evidence of genetic conditioning of human traits, especially mental traits, must be examined with the greatest care.

FAMILY RESEMBLANCES

Heredity is said to cause the resemblance between children and their parents. This definition is good as far as it goes, but it does not go far enough. Mendel found that some of the progeny of two dominant heterozygous parents will differ from them because of homozygosis for recessive genes. Heredity may thus make children different from their parents. Hence heredity is better described as the transmission of self-reproducing entities, genes. The oldest and simplest method of studying heredity is, nevertheless, to observe resemblances and differences within and between families. Galton was the pioneer of systematic studies of this sort.

In *Hereditary Genius* Galton (1869) studied 300 families which produced one or more eminent men, the eminence being defined as attainment of a position of influence or renown, such as was achieved by about one person in 4,000, or 0.025 per cent, in the English population. Galton's eminent men were statesmen, judges, military commanders, church dignitaries, and famous writers and scientists. Table 1 shows that the incidence of eminence is higher in relatives of eminent men than in the general population, and that close relatives of eminent men are more likely to achieve eminence than more remote relatives.

Table 1. Numbers of eminent male relatives per 100 eminent men in 19th-century England *(after Galton)*

Relation	Number eminent	Relation	Number eminent
Father	31	Grandson	14
Brother	41	First cousin	13
Son	48	Great grandfather	3
Uncle	18	Great uncle	5
Nephew	22	Great grandson	3
Grandfather	17	Great nephew	10

Clearly, eminence 'runs in families'. But does it follow that possession of a certain genetic endowment is necessary or sufficient to attain eminence? Surely having influential relatives is helpful, even in societies with class barriers less rigid than those of nineteenth-century England. Galton was not oblivious of this possibility. But he dismissed it because he defined the genetic endowment of eminent men as that which, 'when left to itself, will, urged by an inherent stimulus, climb the path that leads to eminence, and has strength to reach the summit—one which, if hindered or thwarted will fret and strive until the hindrance is overcome, and it is again free to follow its labor-loving instinct.'

This sounds perilously close to circular reasoning. Galton's clinching argument seems whimsical at present, but it was taken seriously a century ago. Class barriers are less rigid in the United States than in England; it should be easier to achieve eminence in the former than in the latter country; one might accordingly expect that the United States will produce more men of genius than England; in point of fact, the opposite is true; therefore, according to Galton, to become eminent one must inherit genes that are more frequent in the English than in the American population.

His observations, though not his conclusions, have been repeatedly confirmed in different countries by investigators studying the pedigrees and descendants of persons of varying degrees of eminence, from the indisputable eminence of geniuses such as Darwin or J. S. Bach to the relatively puny eminence of the persons 'starred' in *American Men of Science* or included in various 'Who's Who' directories.

Although it has been found nearly everywhere that eminence runs in families, it is just at this point that one must proceed with the greatest caution. The environmental bias in the data is patent—other things being equal, a son may find his way smoothest if he follows the calling in which his father excelled. But to reject the data as throwing no light at all on genetic conditioning is unreasonable: notable development of certain special abilities does occur in persons who are possessors of special genetic endowments. The data on inheritance of musical talent are particularly abundant and convincing.

Among the fifty-four known male ancestors, relatives, and descendants of J. S. Bach, forty-six were professional musicians, and among these seventeen were composers of varying degrees of distinction. Granted that in many parts of the world it is customary for members of a family to seek their livelihood in the same profession; granted also that growing up in a family of musicians is propitious for becoming a musician; it is still quite unlikely that the genetic equipment of the Bachs had nothing to do with their musicianship. The recurrence of marked musical ability among the relatives of great musicians is so general a rule that exceptions are worthy of notice. No musical talent is known among the 136 ancestors and relatives of Schumann. Although the composer was married to a virtuoso pianist, none of their eight children possessed great musical ability. Such exceptions do not disprove that musicianship is genetically conditioned; they only show that its genetic basis is complex.

[...]

NATURE AND NURTURE IN CONDOMINIUM
In 1924, with nature-versus-nurture polemics close to their peak, J. B. Watson, the leader of the school of behaviorism in psychology and one of the staunchest partisans of the nurture hypothesis, wrote the following fighting lines:

> Give me a dozen healthy infants, well formed, and my own special world to bring them up in, and I'll guarantee to take any one at random and train him to become any type of specialist I might select—doctor, lawyer, artist, merchant-chief, and yes, even beggar and thief, regardless of his talents, penchants, tendencies, abilities, vocations, and race of his ancestors.

Watson's challenge was pure rhetoric—nobody has made an experiment according to his specifications. Now, more than a third of a century after Watson, we can deal more easily with his verbiage. Notice that the experiment would have to be done on healthy and well-formed individuals; this at once makes a considerable portion of mankind ineligible, since much poor health and malformation are plainly genetically conditioned. Is not normality here defined as developmental plasticity and educability? Notice further that Watson would have trained his normal subjects regardless of their talents, penchants, abilities, etc. But what are these things? If they are products of upbringing, they need not be mentioned in this context at all; if they have, at least in part, a genetic basis, then it is probably easier to train some persons to be doctors and others artists or merchant chiefs, etc.

Many, perhaps most, human infants could be trained, either as lawyers, or beggars, or thieves, etc., by suitably manipulating their environment. But this does not contradict the existence of genetic diversity, so that in a given environment some persons will probably become lawyers and others thieves. Or, to put it another way, different environments may be needed to make lawyers or thieves of different individuals. Or, again, those who are in fact lawyers could perhaps have become thieves, and the actual thieves could have become lawyers, if the circumstances of their lives had been different. In short, nature is not sovereign over some traits and potentialities and nurture over others; they share all traits in condominium.

G. C. Williams

from ADAPTATION AND NATURAL SELECTION

As Fisher was perhaps the first to realize clearly, evolution, at bottom, consists of the changing frequencies of genes in gene pools. It was left to the distinguished American biologist George C. Williams to apply the same insight clearly to adaptation, the tendency of living organisms to look as though they were designed for a purpose. Before Williams's 1966 book *Adaptation and Natural Selection*, the *cui bono* question (for whose benefit do adaptations evolve?) was likely to be answered by some vague nonsense about 'the good of the species'. Darwin, who knew better, would have said 'for the good of the individual in the struggle for survival and reproduction against rival members of the same species'. It was Williams, especially in the passage extracted here, who powerfully emphasized that 'for the good of the gene' was the answer that most naturally flowed from neo Darwinism. Adaptations are 'for the benefit of' the genes responsible for the difference between individuals that possess them and individuals that don't. This is the central idea to which I later gave the title The Selfish Gene—although as it happened my original influence came from W. D. Hamilton (see below) rather than Williams. The extract reproduced here is Williams's vivid and persuasive way of expressing the argument. ▪

The essence of the genetical theory of natural selection is a statistical bias in the relative rates of survival of alternatives (genes, individuals, etc.). The effectiveness of such bias in producing adaptation is contingent on the maintenance of certain quantitative relationships among the operative factors. One necessary condition is that the selected entity must have a high degree of permanence and a low rate of endogenous change, relative to the degree of bias (differences in selection coefficients). Permanence implies reproduction with a potential geometric increase.

Acceptance of this theory necessitates the immediate rejection of the importance of certain kinds of selection. The natural selection of phenotypes cannot in itself produce cumulative change, because phenotypes are extremely temporary manifestations. They are the result of an interaction between genotype and environment that produces what we recognize as an individual. Such an individual consists of genotypic information and information recorded since conception. Socrates consisted of the genes his parents gave him, the experiences they and his environment later provided, and a growth and development mediated by numerous meals. For all I know, he may have been very successful in the evolutionary sense of leaving numerous offspring. His phenotype, nevertheless, was utterly destroyed by the hemlock and has never since been duplicated. If the hemlock had not killed him, something else soon would have. So however natural selection may have been acting on Greek phenotypes in the fourth century BC, it did not of itself produce any cumulative effect.

The same argument also holds for genotypes. With Socrates' death, not only did his phenotype disappear, but also his genotype. Only in species that can maintain unlimited clonal reproduction is it theoretically possible for the selection of genotypes to be an important evolutionary factor. This possibility is not likely to be realized very often, because only rarely would individual clones persist for the immensities of time that are important in evolution. The loss of Socrates' genotype is not assuaged by any consideration of how prolifically he may have reproduced. Socrates' genes may be with us yet, but not his genotype, because meiosis and recombination destroy genotypes as surely as death.

It is only the meiotically dissociated fragments of the genotype that are transmitted in sexual reproduction, and these fragments are further fragmented by meiosis in the next generation. If there is an ultimate indivisible fragment it is, by definition, 'the gene' that is treated in the abstract discussions of population genetics. Various kinds of suppression of recombination may cause a major chromosomal segment or even a whole chromosome to be transmitted entire for many generations in certain lines of descent. In such cases the segment or chromosome behaves in a way that approximates the population genetics of a single gene. In this book I use the term *gene* to mean 'that which segregates and recombines with appreciable frequency'. Such genes are potentially immortal, in the sense of there being no physiological limit to their survival, because of their potentially reproducing fast enough to compensate for their destruction by external agents.

Francis Crick

from LIFE ITSELF

■ The salient feature of Mendelian genetics, the one that equips it to undergird the neo-Darwinian synthesis, is that it is digital. A given gene (Gregor Mendel himself didn't use the word) either passes to a given off-spring (grand-offspring etc.) or it does not. There are no half measures, and genes never blend with one another. Heredity is all-or-none. That's digital. But what neither Mendel nor anyone else before 1953 knew was that genes themselves are digital, within themselves. A gene is a sequence of code letters, drawn from an alphabet of precisely four letters, and the genetic code is universal throughout all known living things. Life is the execution of programs written using a small digital alphabet in a single, universal machine language. This realization was the hammer blow that knocked the last nail into the coffin of vitalism and, by extension, of dualism. The hammer was wielded, with undisguised youthful relish, by James Watson and Francis Crick. Their famous one-page paper in *Nature* of 1953 concludes with what may be the greatest piece of calculated understatement ever: 'It has not escaped our notice that the specific pairing we have postulated immediately suggests a possible copying mechanism for the genetic material.'

The quality of mind that enabled Watson and Crick to race ahead of their laboratory-based rival Rosalind Franklin is well demonstrated by that sentence, and it is shown again in the extract I have chosen from Crick's book *Life Itself: Its Origin and Nature* (1981). Watson and Crick were not only concerned with finding out how things actually are—although that was of course their ultimate goal. They also kept in mind the way DNA ought to be if it was to do its job as the genetic molecule, and this gave them a short cut, which was ignored by the painstaking Rosalind Franklin. Watson said something similar in *Avoid Boring People* (see below): 'Knowing why is more important than learning what.' Crick exemplified this again and again in after years, as we learn from Matt Ridley's biography. At times it led him astray, as when he was seeking the genetic code and was temporarily seduced by a brilliantly neat idea about an ideally economical one. Nature, it turned out, was less elegant than Crick's mind, but it is in general true that deep cogitation on the way nature ought to be constitutes a good prelude to the eventual investigation of the way it actually is. Only a prelude, however: the ultimate test of an idea is not its elegance but how well it explains reality. ■

Nucleic Acids and Molecular Replication

Now that we have described the requirements for a living system in rather abstract terms, we must examine more closely how the various processes are carried out in the organisms we find all around us. As we have seen, the absolutely central requirement is for some rather precise method of replication and, in particular, for copying a long linear macromolecule put together from a standard set of subunits. On earth this role is played by one or the other of the two great families of nucleic acids, the DNA family and the RNA family. The general plan of these molecules is extremely simple, so simple indeed that it strongly suggests that they go right back to the very beginning of life.

DNA and RNA are rather similar—molecular cousins, you might say— so let us describe DNA first and then how RNA differs from it. One chain of DNA consists of a uniform backbone, the sequence of atoms repeating over and over again, with a side-group joined on at every repeat. Chemically the backbone goes...phosphate-sugar phosphate-sugar...etc., repeating many thousands or even millions of times. The sugar is not the sugar you have on your breakfast table but a smaller one called deoxyribose—that is, ribose with one 'oxy' group missing (hence the name DNA, standing for DeoxyriboNucleic Acid—'nucleic' because it is found in the nucleus of higher cells, and 'acid' because of the phosphate groups, each of which in normal conditions carries a negative charge). Each sugar has a side-group joined to it. The side-groups differ, but there are only four main types of them. These four side-groups of DNA (for technical reasons called *bases*) are conveniently denoted by their initial letters, A, G, T and C (standing for *A*denine, *G*uanine, *T*hymine and *C*ytosine, respectively). Because of their exact size and shape and the nature of the chemical constituents, A will pair neatly with T, G with C. (A and G are big, T and C are smaller, so each pair consists of one big one with one smaller one.)

Both DNA and RNA rather easily form two-chain structures, in which the two chains lie together, side by side, twisted around one another to form a double helix and linked together by their bases. At each level there is a base-pair, formed between a base on one chain paired (using the pairing rules) with a base on the other. The bonds holding these pairs together are individually rather weak, though collectively

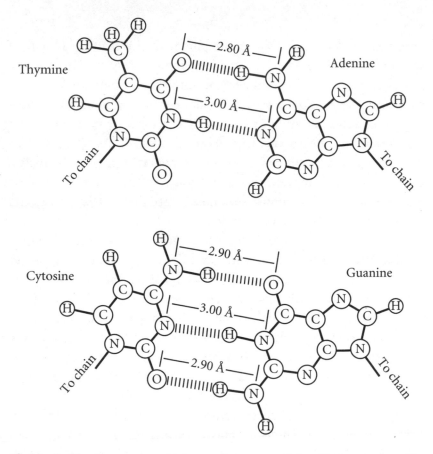

Figure 2. The base-pairs which are the secret of the DNA structure. The bases are held together by weak hydrogen bonds, shown by the interrupted lines. Thymine always pairs with adenine; cytosine with guanine.

they make a double helix reasonably stable. But if the structure is heated the increased thermal agitation will jostle the chains apart, so that they separate and float away from each other in the surrounding water.

The genetic message is conveyed by the exact base-sequence along one chain. Given this sequence, then the sequence of its complementary companion can be read off, using the base-pairing rules (A with T, G with C). The genetic information is recorded twice, once on each chain. This can be useful if one chain is damaged, since it can be repaired using the information—the base-sequence—of the other chain.

There is one unexpected peculiarity. In the usual double helix the two backbones of the two chains are not approximately parallel but antiparallel. If the sequence of the atoms in one backbone runs up, that in the other runs down. This does cause certain complications, but not as much as one might expect. At bottom it springs from the type of symmetry possessed by the double helix. This is produced by the pseudosymmetry of the base-pairing. It happens to be the convenient way for these particular chemicals to fit neatly together.

It is easy to see that a molecule of this type, consisting of a pair of chains whose irregular elements (the bases) fit together, is ideal for molecular replication, especially since the two chains can be rather easily separated from each other by mild methods. This is because the bonds *within* each chain, holding each chain together, are strong chemical bonds, fairly immune to normal thermal battering, whereas the two chains cling to each other by rather weak bonds so that they can be prized apart without too much difficulty and without breaking the individual backbones. The two chains of DNA are like two lovers, held tightly together in an intimate embrace, but separable because however closely they fit together each has a unity which is stronger than the bonds which unite them.

Because they fit together so precisely, each chain can be regarded as a mold for the other one. Conceptually the basic replication mechanism is very straightforward. The two chains are separated. Each chain then acts as a template for the assembly of a new companion chain, using as raw material a supply of four standard components. When this operation has been completed we shall have two pairs of chains instead of one, and since to do a neat job the assembly must obey the base-pairing rules (A with T, G with C), the base-sequences will have been copied exactly. We shall end up with two double helices where we only had one before. Each daughter double helix will consist of one old chain and one newly synthesized chain fitting closely together, and more important, the base-sequence of these two daughters will be identical to that of the original parental DNA.

The basic idea could hardly be simpler. The only rather unexpected feature is that the two chains are not identical but complementary. One could conceive an even simpler mechanism in which like paired with like, so that the two paired chains were identical, but the nature of chemical interactions makes it somewhat easier for complementary molecules, rather than identical ones, to fit together.

How does such a process compare with the grosser copying mechanisms commonly used today? A line of type, made up for printing, consists (or used to consist) of a set of standard symbols arranged in a line or a series of lines. Each letter from the font has a standard part, the same for all letters, which fits into the grooves which hold the type in place, and a part which is characteristic of each letter. After that the resemblance ceases. There is nothing in DNA replication which corresponds to the ink. The letters printed on the page are the mirror images of the typeface, not the complement (which would stick out when the typeface went in), and, most important, the resulting line of print cannot then be put back into the same machine to reproduce the typeface. Printing presses produce many thousands of copies of newspapers, but newspapers are not copied back into type.

DNA replication is not like that. For natural selection to work it is essential that the copy can itself be copied. DNA replication is more like production of a piece of sculpture from a mold, since if it is sufficiently simple the sculpture can itself be used to produce a further mold. The main difference is that a strand of DNA is built from just four standard pieces. This is obviously not true of most pieces of sculpture.

If we examine the process of DNA replication, we see that there are a number of basic requirements. If we start with a double helix, the two chains must be separated in some way. There must be available a supply of the four components. Each of these consists of the relevant piece of the backbone—one sugar molecule joined to one phosphate—plus one of the four bases attached to the sugar. Such a tripartite molecule is called a nucleotide. In practice these precursors have not just one phosphate but three in a row, the other two being split off in the process of polymerization, thus providing the energy to drive the synthesis in the desired direction. Though one can conceive of the process proceeding without extra components, in an evolved system we would expect to find at least one enzyme (a protein with catalytic activity, that is) which would accelerate the synthesis and make it more accurate.

Such are the requirements in outline. When a real replicative system is examined it is found to be considerably more elaborate. To begin with, the two chains are not first completely separated before synthesis starts.

Synthesis of the new chains proceeds during the process of separation, so that some parts of the double helix have been replicated before other more distant parts have been separated. There are special proteins whose job it is to unwind the double helix, together with others which can put nicks in the backbone, to allow one chain to rotate around the other, and then join up the broken chain again. Since the two chains of the double helix run in opposite directions, and since, chemically speaking, the synthesis goes in only one direction, we find that synthesis is directed forward on one chain and backward on the other, so the mechanism has to allow for this complication. Moreover, a new fragment of a DNA chain is usually started as a small length of RNA, to which a longer piece of DNA is then joined. There are additional proteins which then cut out this RNA primer and replace it with an equivalent bit of DNA chain and then join everything together without a break. To synthesize one particular small virus made of DNA we know that almost twenty distinct proteins are required, some to do one job, some to do another. This is very characteristic of biological processes. The underlying mechanism may be simple, but if the process is biologically important, then, in the long course of evolution, natural selection will have improved it and embroidered it, so that it can work both faster and more accurately. It is because of this baroque elaboration that biological mechanisms are often so difficult to unravel.

Matt Ridley

from GENOME

Matt Ridley is the author of a string of excellent books, including the official biography of Francis Crick just mentioned. Ridley's clarity as a writer stems from his deeply intelligent understanding of his subject matter, coupled with a gift for vividly arresting imagery. *Genome* (1999) is based upon the conceit of allotting each of 23 chapters to one of our 23 chromosomes (rather in the manner of Primo Levi's *The Periodic Table*, where each chapter is named after an element). Instead of the chapter being an exhaustive treatment of what is known of that chromosome (nowadays rather a lot), Ridley picks out a more general lesson, which happens to be conveniently tied to that chromosome. It is a style that I adopted in *The Ancestor's Tale*, where I tied a series of general lessons about evolution to corresponding Chaucerian Tales, 'told by' particular animals on a 'pilgrimage'. Ridley's 'Chromosome 1' is an opportunity to muse on life itself (Crick's title, and Ridley covers some of the same ground), especially the centrality of 'information' in the modern technical sense of computers and telecommunications, concluding with an allusion to information's mathematical affinity to the 'entropy' of the thermodynamicists. ■

Chromosome 1: Life

In the beginning was the word. The word proselytized the sea with its message, copying itself unceasingly and forever. The word discovered how to rearrange chemicals so as to capture little eddies in the stream of entropy and make them live. The word transformed the land surface of the planet from a dusty hell to a verdant paradise. The word eventually blossomed and became sufficiently ingenious to build a porridgy contraption called a human brain that could discover and be aware of the word itself.

My porridgy contraption boggles every time I think this thought. In four thousand million years of earth history, I am lucky enough to be alive today. In five million species, I was fortunate enough to be born a conscious human being. Among six thousand million people on the planet, I was privileged enough to be born in the country where the word was discovered. In all of the earth's history, biology and geography, I was born just five years after the moment when, and just two hundred miles from the place where, two members of my own species discovered the structure of DNA and hence uncovered the greatest, simplest and most surprising secret in the universe. Mock my zeal if you wish; consider me a ridiculous materialist for investing such enthusiasm in an acronym. But follow me on a journey back to the very origin of life, and I hope I can convince you of the immense fascination of the word.

'As the earth and ocean were probably peopled with vegetable productions long before the existence of animals; and many families of these animals long before other families of them, shall we conjecture that one and the same kind of living filaments is and has been the cause of all organic life?' asked the polymathic poet and physician Erasmus Darwin in 1794. It was a startling guess for the time, not only in its bold conjecture that all organic life shared the same origin, sixty-five years before his grandson Charles's book on the topic, but for its weird use of the word 'filaments'. The secret of life is indeed a thread.

Yet how can a filament make something live? Life is a slippery thing to define, but it consists of two very different skills: the ability to replicate, and the ability to create order. Living things produce approximate copies of themselves: rabbits produce rabbits, dandelions make dandelions. But rabbits do more than that. They eat grass, transform it into rabbit flesh and somehow build bodies of order and complexity from the random chaos of the world. They do not defy the second law of thermodynamics, which says that in a closed system everything tends from order towards disorder, because rabbits are not closed systems. Rabbits build packets of order and complexity called bodies but at the cost of expending large amounts of energy. In Erwin Schrödinger's phrase, living creatures 'drink orderliness' from the environment.

The key to both of these features of life is information. The ability to replicate is made possible by the existence of a recipe, the information that is needed to create a new body. A rabbit's egg carries the instructions for assembling a new rabbit. But the ability to create order through metabolism also depends on information—the instructions for building and maintaining the equipment that creates the order. An adult rabbit, with its ability to both reproduce and metabolize, is prefigured and presupposed in its living filaments in the same way that a cake is prefigured and presupposed in its recipe. This is an idea that goes right back to Aristotle, who said that the 'concept' of a chicken is implicit in an egg, or that an acorn was literally 'informed' by the plan of an oak tree. When Aristotle's dim perception of information theory, buried under generations of chemistry and physics, re-emerged amid the discoveries of modern genetics, Max Delbruck joked that the Greek sage should be given a posthumous Nobel prize for the discovery of DNA.

The filament of DNA is information, a message written in a code of chemicals, one chemical for each letter. It is almost too good to be true, but the code turns out to be written in a way that we can understand. Just like written English, the genetic code is a linear language, written in a straight line. Just like written English, it is digital, in that every letter bears the same importance. Moreover, the language of DNA is considerably simpler than English, since it has an alphabet of only four letters, conventionally known as A, C, G and T.

Now that we know that genes are coded recipes, it is hard to recall how few people even guessed such a possibility. For the first half of the twentieth century, one question reverberated unanswered through biology: what is a gene? It seemed almost impossibly mysterious. Go back not to 1953, the year of the discovery of DNA's symmetrical structure, but ten years further, to 1943. Those who will do most to crack the mystery, a whole decade later, are working on other things in 1943. Francis Crick is working on the design of naval mines near Portsmouth. At the same time James Watson is just enrolling as an undergraduate at the precocious age of fifteen at the University of Chicago; he is determined to devote his life to ornithology. Maurice Wilkins is helping to design the atom bomb in the United States. Rosalind Franklin is studying the structure of coal for the British government.

In Auschwitz in 1943, Josef Mengele is torturing twins to death in a grotesque parody of scientific inquiry. Mengele is trying to understand heredity, but his eugenics proves not to be the path to enlightenment. Mengele's results will be useless to future scientists.

In Dublin in 1943, a refugee from Mengele and his ilk, the great physicist Erwin Schrödinger is embarking on a series of lectures at Trinity College entitled 'What is life?' He is trying to define a problem. He knows that chromosomes contain the secret of life, but he cannot understand how: 'It is these chromosomes...that contain in some kind of code-script the entire pattern of the individual's future development and of its functioning in the mature state.' The gene, he says, is too small to be anything other than a large molecule, an insight that will inspire a generation of scientists, including Crick, Watson, Wilkins and Franklin, to tackle what suddenly seems like a tractable problem. Having thus come tantalisingly close to the answer, though, Schrödinger veers off track. He thinks that the secret of this molecule's ability to carry heredity lies in his beloved quantum theory, and is pursuing that obsession down what will prove to be a blind alley. The secret of life has nothing to do with quantum states. The answer will not come from physics.

In New York in 1943, a sixty-six-year-old Canadian scientist, Oswald Avery, is putting the finishing touches to an experiment that will decisively identify DNA as the chemical manifestation of heredity. He has proved in a series of ingenious experiments that a pneumonia bacterium can be transformed from a harmless to a virulent strain merely by absorbing a simple chemical solution. By 1943, Avery has concluded that the transforming substance, once purified, is DNA. But he will couch his conclusions in such cautious language for publication that few will take notice until much later. In a letter to his brother Roy written in May 1943, Avery is only slightly less cautious:

> If we are right, and of course that's not yet proven, then it means that nucleic acids [DNA] are not merely structurally important but functionally active substances in determining the biochemical activities and specific characteristics of cells—and that by means of a known chemical substance it is possible to induce predictable and hereditary changes in cells. That is something that has long been the dream of geneticists.

Avery is almost there, but he is still thinking along chemical lines. 'All life is chemistry', said Jan Baptist van Helmont in 1648, guessing. 'At least some life is chemistry', said Friedrich Wöhler in 1828 after synthesising urea from ammonium chloride and silver cyanide, thus breaking the hitherto sacrosanct divide between the chemical and biological worlds: urea was something that only living things had produced before. That life is chemistry is true but boring, like saying that football is physics. Life, to a rough approximation, consists of the chemistry of three atoms, hydrogen, carbon and oxygen, which between them make up ninety-eight per cent of all atoms in living beings. But it is the emergent properties of life—such as heritability—not the constituent parts that are interesting. Avery cannot conceive what it is about DNA that enables it to hold the secret of heritable properties. The answer will not come from chemistry.

In Bletchley, in Britain, in 1943, in total secrecy, a brilliant mathematician, Alan Turing, is seeing his most incisive insight turned into physical reality. Turing has argued that numbers can compute numbers. To crack the Lorentz encoding machines of the German forces, a computer called Colossus has been built based on Turing's principles: it is a universal machine with a modifiable stored program. Nobody realises it at the time, least of all Turing, but he is probably closer to the mystery of life than anybody else. Heredity is a modifiable stored program; metabolism is a universal machine. The recipe that links them is a code, an abstract message that can be embodied in a chemical, physical or even immaterial form. Its secret is that it can cause itself to be replicated. Anything that can use the resources of the world to get copies of itself made is alive; the most likely form for such a thing to take is a digital message—a number, a script or a word.

In New Jersey in 1943, a quiet, reclusive scholar named Claude Shannon is ruminating about an idea he had first had at Princeton a few years earlier. Shannon's idea is that information and entropy are opposite faces of the same coin and that both have an intimate link with energy. The less entropy a system has, the more information it contains. A steam engine parcels out entropy to generate energy because of the information injected into it by its designer. So does a human body. Aristotle's information theory meets Newton's physics in Shannon's brain. Like Turing, Shannon has no thoughts about biology. But his insight is of more relevance to the question of what is life than a mountain of chemistry and physics. Life, too, is digital information written in DNA.

In the beginning was the word. The word was not DNA. That came afterwards, when life was already established, and when it had divided the labour between two separate activities: chemical work and information storage, metabolism and replication. But DNA contains a record of the word, faithfully transmitted through all subsequent aeons to the astonishing present.

Sydney Brenner

'THEORETICAL BIOLOGY IN THE THIRD MILLENNIUM'

■ One of the things Matt Ridley makes clear in his biography is that Francis Crick's genius thrived on collaboration and conversation. He would talk and talk about science with intelligent colleagues whose expertise complemented his own. After Watson left Cambridge to return to America, Crick teamed up with the effervescently brilliant geneticist Sydney Brenner, recently arrived from South Africa via Oxford, to overtake the next big milestone in the molecular biology journey, the genetic code itself. Using stunningly clever experiments with viral parasites of bacteria, Brenner, Crick and their colleagues demonstrated that it had to be a triplet code. After this, others came along and showed exactly which three letter word in the DNA lexicon corresponded to which amino acid in a protein chain.

Brenner later founded a whole major field of experimental biology, using the tiny nematode worm *Caenorhabditis elegans* (misleadingly called 'the' nematode or even 'the worm' by molecular biologists, as though there aren't any others). Brenner chose this animal for very particular reasons, initiated a globally successful research program on it, and eventually (far later than most observers thought just) was rewarded with the Nobel Prize. He is noted for his cutting wit, displayed in his lectures (he is one of the best lecturers I have ever heard) and in the series of regular columns he wrote in the journal *Current Biology* under the pseudonym 'Uncle Syd'. Regrettably, he has never published a book, and the following paper, written for the millennium, gives some idea of what we are missing by way of thoughtful and provocative reflections on science in general and biology in particular. ■

Like begets like is the fundamental law of biology and probably the oldest piece of genetic knowledge. During the 20th century—the last of this millennium—our understanding of inheritance has undergone several revolutionary changes; first with the rediscovery of Mendel's laws in 1901, through the DNA double helix of Watson and Crick, and culminating, in the last decade, in DNA sequencing of genomes. Genetics changed from a subject concerned simply with the segregation of characters in crosses to the direct analysis of the genes. This has led us to the insight that organisms are unique, complex systems in the natural world, which contain internal description of their structure, function, development and history encoded in the DNA sequences of their genes.

Parallel advances in biochemistry have provided us with detailed knowledge of how energy is converted to chemical bonds and chemical bonds to energy, and how the elementary chemical components of living cells are synthesized. We have come to understand the mechanisms of information transfer from genes to proteins. We know that the information is copied into messenger RNA, that this RNA is translated in ribosomes and that the code-script is read in triplets by transfer RNAs, each carrying one of the 20 amino acids. We know the special signals for starting and stopping the polypeptide chain and the code for each amino acid. The genetic code is universal, with some minor exceptions to this rule in a few organisms and organelles.

Several major technical advances occurred in the mid-1970s. These were the invention of DNA molecular cloning, and methods for sequencing DNA molecules and synthesizing oligonucleotides. These techniques allowed geneticists to clone their genes and characterize them directly, and gave biochemists access to large amounts of the proteins they were studying. In principle, the sequence of amino acids can be read from the DNA sequence, although the presence of introns found in the genomes of higher organisms may cause some difficulties. In any event, sequencing DNA became the preferred way of finding the amino-acid sequences of proteins, the direct determination of which had previously been a long and laborious process.

It was an essential feature of Crick's sequence hypothesis that the information contained in the amino-acid sequence was sufficient to determine how the chain folds to give the three-dimensional (3D) structure of globular proteins. For many proteins, this process occurs spontaneously, but in a large number of cases, special proteins called chaperonins are used to facilitate the folding of the molecules. Advances in X-ray crystallography, electron microscopy and nuclear magnetic resonance methods allowed us to determine the structures of large numbers of protein molecules and even complex protein assemblies, but the problem of going from the one-dimensional polypeptide to the folded, active structure remains unsolved and may even be insoluble.

These new methods came as a godsend to those studying the genetics of organisms. Cloning the mutated gene gave us a direct approach to the protein product of the gene and, as knowledge increased, to an insight of how it might function and thereby contribute to the observed phenotype. They liberated experimental genetics from the tyranny of breeding cycles and provided new approaches, particularly to human genetics, which had hitherto been intractable. They enabled us to move genes from one organism to another and allowed us to analyse the function of human genes in yeast cells, and to study how fish genes behave in mice.

An important feature of living organisms is the regulation of their functions. At the genetic level, Jacob and Monod showed that there were proteins that recognized segments of DNA and turned the adjacent genes off. Repression was originally thought to be the only mode of control, but we now know that there are many regulatory proteins that act positively. In higher metazoa, there are large numbers of controlling genes, which specify the times and locations of expression of the many genes acting in development and in adaptive responses in the cells of the adult. Different cells contain different subsets of a panoply of receptors embedded in their membranes, which serve to transmit signals delivered to the outside of the cell to the inside. The signal-transduction machinery, a complicated set of interacting proteins, converts these signals into chemical currencies, which are used to control a multitude of cellular functions including growth, movement, division, secretion and differentiation. In multicellular organisms, increased complexity has been achieved not by the invention of new genes but simply by the regulation of gene expression. This reaches its apotheosis in the central nervous system of advanced animals in which the same repertoire of molecular entities is used to generate complex cellular networks.

Finally, and unexpectedly, contemporary cells were found to contain RNA molecules that display catalytic functions. These are likely to be RNA relics, survivors from very early evolution before living systems used proteins. The discovery of catalytic functions of RNA provided a molecule that could combine catalysis and the carrying of information, and bridged the gulf posed by the present partitioned situation where information is carried by one class of molecule (nucleic acids) and proteins are the catalysts. It resolved one of the important problems in how life originated.

The databases of sequence information are now growing at an immense rate and the number and productivity of biological researchers has also vastly increased. There seems to be no limit to the amount of information that we can accumulate, and today, at the end of the millennium, we face the question of what is to be done with all of this information. This problem is now widely debated and there are plans to deal with it electronically, if only to avoid the sheer weight of paper that will be required to document it. Biologists may soon have to spend most of their time in front of their computer screens. It will take a long time—if it can ever be achieved—for computers to become intelligent enough to organize this information into knowledge and to teach it to us. Writing in the last months of this millennium, it is clear that the prime intellectual task of the future lies in constructing an appropriate theoretical framework for biology.

Unfortunately, theoretical biology has a bad name because of its past. Physicists were concerned with questions such as whether biological systems are compatible with the second law of thermodynamics and whether they could be explained by quantum mechanics. Some even expected biology to reveal the presence of new laws of physics. There have also been attempts to seek general mathematical theories of development and of the brain: the application of catastrophe theory is but one example. Even though alternatives have been suggested, such as computational biology, biological systems theory and integrative biology, I have decided to forget and forgive the past and call it theoretical biology.

Now there can be no doubt that parts of biological systems can be treated within the context of physical theories: for example, the passage of ions in membrane channels or the flow of blood in blood vessels. These are physical phenomena, which happen to occur in our bodies and not in artificial membranes or pipes. There is also a considerable body of theory dealing with the chemistry of the molecules in biological systems, and with the physical chemistry of their interactions. But none of this captures the novel feature of biological systems: that, in addition to flows of matter and energy, there is also the flow of information. Biological systems are information-processing machines and this must be an essential part of any theory we may construct. We therefore have to base everything on genes, because they carry the specification of the organism and because they are the entities that record evolutionary changes.

One way of looking at the problem is to ask whether we can compute organisms from their DNA sequences. This computational approach is related to Von Neumann's suggestion that very complex behaviours may be explicable only by providing the algorithm that generates that behaviour, that is, explanation by way of simulation. We need to be very clear that this must not simply be another way of describing the behaviour. For example it is quite easy to write a computer program that will produce a good copy of worms wriggling on a computer screen. But the program, when we examine it, is found to be full of trigonometrical calculations and has nothing in it about neurons or muscles. The program is an imitation; it manipulates the image of a worm rather than the worm object itself. A proper simulation must be couched in the machine language of the object, in genes, proteins, and cells. We notice, in passing, that Turing's test, which is whether an observer could distinguish between a computer and a human being, is a test of an imitation and not of a simulation.

Our analytical tools have become so powerful that complete descriptions of everything can be attained. In fact, obtaining the DNA sequence of an organism can be viewed as the first step, and we could continue by determining the 3D structure of every protein and the quantitative expression of every gene under all conditions. However, not only will this catalogue be indigestible but it will also be incomplete, because we cannot come to the end of different conditions and especially of combinations and permutations of these. Mere description does not allow computation, and novelty cannot be dealt with. On the other hand, a proper simulation would allow us to make predictions, by performing experiments on the model and calculating what it might do. Thus, if this could be carried out successfully an immense amount of information could be derived by calculation from the minimal amount needed. This is essentially the DNA sequence, the shortest description of an organism.

To do this effectively not only must we use the vocabulary of the machine language but we must also pay heed to what may be called the grammar of the biological system. We need to be clear what kind of an information-processing machine it is. It is useful to consider two kinds of such devices. As an example we consider devices that produce the values of mathematical functions. We call one a P-machine because it contains programs. When the value of factorial (5) is requested, a systems procedure invokes the execution of a program that calculates the answer. The other is called a T-machine. It has no programs but tables, and, in response to the same query, a system procedure looks up the fifth entry in the table labelled factorial. Now the T-machine has the advantage that the values in the table can be calculated beforehand by any method whatsoever—by hand, by abacus, by mechanical calculators—and once the answer is known it is stored and the calculation need never be done again. It is clear that at the level we are considering, biological systems are T-machines; evolution has calculated values for the system by the trial and error method of natural selection and the answers are now looked up in the gene tables. There are no imperative commands, and that is why I have avoided using the term genetic program and have called it a description. Of course organisms are P-machines at other levels, for example, in the functioning of our brains. Notice that if memory is a limiting resource, a P-machine will be preferred, as indeed was the case in the evolution of the digital computer. Today, storage is cheaply and abundantly available, and now more and more computer systems employ tables rather than waste valuable processor time in calculations.

There is a second aspect of the grammar that needs comment. Genomes do not contain in any explicit form anything at a level higher than the genes. They do not explicitly define networks, cycles or any other cluster of cell functions. These must be computed by the cell from the properties of the elementary gene products. Biosynthetic pathways exist because individual enzymes carry out defined transformations at specified rates; the pathway drawn in textbooks of biochemistry is an abstraction and does not exist in the same way as the tracks connecting stations in a railway network. We need to be extremely careful in not imposing our constructions on what exists, and it is important to structure information at the atomic gene level to avoid artificial constraints. This becomes evident when we attempt to deal with multiple parallel processes going on in the same space. The coherence of a system, which may be impossible to define at the global level, is assuredly generated by the properties of the elements because the system exists and has survived the test of natural selection. Since it is not possible to start again in evolution, every step must be compatible with what has gone before; biological systems have changed by piecemeal modification and by accretion. Natural selection does not find perfect or elegant or even optimal solutions, all that is required of it is to find satisfactory solutions.

What is the likelihood that we could actually compute a simple organism from its DNA sequence? We can obtain the linear polypeptide chains reliably from the gene sequences. However, the folding problem is unsolved and is very difficult. Indeed, there may be as many different folding problems as there are proteins. However, we can resort to good heuristic solutions in the sense that proteins are composed of smaller substructures called domains, and the sequence signatures of these could be used to compute 3D structures by analogy with other proteins where these structures have been determined. We then have the much more difficult task of computing the interactions of these proteins with other proteins and with their chemical environment. This may well be impossible, but again, we may know enough about related proteins to deduce this. The very detailed properties of proteins, their specific binding constants and, for enzymes, the rates with which they transform substrates may again be beyond computational reach from the gene sequence, since there may be many equivalent solutions to the same problem.

Building theoretical models of cells would be based not on genes but on their protein products and on the molecules produced by these proteins. We do not have to wait to solve all the difficult problems of protein structure and function, but can proceed by measuring the properties that we require. At the level of the organism we would start with cells and, again, measurement could give us what we need. The reader may complain that I have said nothing more than 'carry on with conventional biochemistry and physiology'. I have said precisely that, but I want the new information embedded into biochemistry and physiology in a theoretical framework, where the properties at one level can be produced by computation from the level below.

It may be much easier to compare two genomes. The DNA sequences of any two human genomes differ from each other in one or two of every 1000 bases. If a chimpanzee genome is compared with a human genome the number of differences rises to about ten per 1000 bases. Many of these differences are without significant effect because they occur in regions or in positions where they could be judged to be strictly neutral. It would be fascinating to ask whether we could discover the differences that do count and whether we could reconstruct our common ancestor and thus find out what mutations occurred during the course of evolution to make us different. I believe that this is what we should be trying to do in the next century. It will require theoretical biology.

Steve Jones

from THE LANGUAGE OF THE GENES

■ The Welsh geneticist Steve Jones is a superb scientific raconteur and a wickedly sardonic observer of the scientific scene. He is an amusing lecturer—the phrase 'wry wit' might almost have been invented for him— and his books are as entertaining as they are informative. I have chosen an extract from *The Language of the Genes*, which explains haemophilia, historic scourge of the royal families of Europe. ■

Change or Decay?

By the time you have finished this chapter you will be a different person. I do not mean by this that your views about existence—or even about genes—will alter, although perhaps they may. What I have in mind is simpler. In the next half hour or so your genes, and your life, will be altered by mutation; by errors in your own genetic message. Mutation— change—happens all the time, within ourselves and over the generations. We are constantly corrupted by it; but biology provides an escape from the inevitability of genetic decline.

Evolution is no more than the perpetuation of error. It means that progress can emerge from decay. Mutation is at the heart of human experience, of old age and death but also of sex and of rebirth. All religions share the idea that humanity is a decayed remnant of what was once perfect and that it must be returned to a higher plane by salvation, by starting again from scratch. Mutation embodies what faith demands: each man's decline but mankind's redemption.

The first genes appeared some four thousand million years ago as short strings of molecules which could make rough copies of themselves. At a reckless guess, the original molecule in life's first course, the primeval soup, has passed through four thousand million ancestors before ending up in you or me (or in a chimp or a bacterium). Every one of the untold billions of genes that has existed since then emerged through the process of mutation. A short message has grown to an instruction manual of three thousand million letters. Everyone has a unique edition of the instruction book that differs in millions of ways from that of their fellows. All this comes from the accumulation of errors in an inherited message.

Like random changes to a watch some of these accidents are harmful. But most have no effect and a few may even be useful. Every inherited disease is due to mutation. Now that medicine has, in the Western world at least, almost conquered infection, mutation has become more important. About one child in forty born in Britain has an inborn error of some kind and about a third of all hospital admissions of young children involve a genetic disease. Some damage descends from changes which happened long ago while others are mistakes in the sperm or egg of the parents themselves. Everyone carries single copies of damaged genes which, if two copies were present, would kill. As a result, everyone has at least one mutated skeleton in their genetical cupboard.

Because there are so many different genes the chance of seeing a new genetic accident in one of them is small. Even so, in a few cases, novel errors can be spotted.

Before Queen Victoria, the genetic disease haemophilia (a failure of the blood to clot) had never been seen in the British royal family. Several of her descendants have suffered from it. The biochemical mistake probably took place in the august testicles of her father, Edward, Duke of Kent. The haemophilia gene is on the X chromosome, so that to be a haemophiliac a male needs to inherit just one copy of the gene while a female needs two. The disease is hence much more common among boys. This was known to the Jews three thousand years ago. A mother was allowed not to circumcise her son if his older brother had bled badly at the operation and, more remarkably, if her sister's sons had the same problem.

As well as its obvious effects after a cut, haemophilia does more subtle damage. Affected children often have many bruises and may suffer from internal bleeding which can damage joints and may be fatal. Once, more than half the affected boys died before the age of five. Injection of the clotting factor restores a more or less normal life.

Several of Victoria's grandsons were haemophiliacs, as was one of her sons, Leopold. Two of her daughters—Beatrice and Alice—must have been carriers. The Queen herself said that 'our poor family seems persecuted by this disease, the worst I know'. The most famous sufferer was Alexis, the son of Tsar Nicholas of Russia and Queen Alexandra, Victoria's granddaughter. One reason for Rasputin's malign influence on the Russian court was his ability to calm the unfortunate Alexis. The gene has disappeared from the British royal line, and no haemophiliacs are known among the three hundred descendants of Queen Victoria alive today. In Britain, about one male in five thousand is affected.

Somewhat incidentally, another monarch, George III, may have carried a different mutation. The gene responsible for porphyria can lead to mental illness and might have been responsible for his well-known madness. The retrospective diagnosis was made from the notes of the King's physician, who noticed that the royal urine had the purple 'port-wine' colour characteristic of the disease. A distant descendant also showed signs of the illness. One of the King's less successful appointments was that of his Prime Minister, Lord North, who was largely responsible for the loss of the American Colonies. It is odd to reflect that both the Russian and the American Revolutions may in part have resulted from accidents to royal DNA.

Research on human mutation once involved frustration ameliorated by anecdotes like these. It has been turned on its head by the advance of molecular biology. In the old days, the 1980s, the only way to study it was to find a patient with an inherited disease and to try to work out what had gone wrong in the protein. The change in the DNA was quite unknown. This was as true for haemophilia as for any other gene. In fact, haemophilia seemed a rather simple error. Different patients showed rather different symptoms, but the mode of inheritance was simple and all seemed to share the same disease.

Now whole sections of DNA from normal and haemophiliac families can be compared to show what has happened and, like the genetic map itself, things have got more complicated. Molecular biology has made geneticists' lives much less straightforward. First, uncontrollable bleeding is not one disease, but several. To make a clot is a complicated business that involves several steps. Proteins are arranged in a cascade which responds to the damage, produces and then mobilises the material needed and assembles it into a barrier. A dozen or more different genes scattered all over the DNA take part in the production line.

Two are particularly likely to go wrong. One makes factor VIII in the clotting cascade. Errors in that gene lead to haemophilia A, which accounts for nine tenths of all cases of the disease. The other common type—haemophilia B—involves factor IX. In a rare form of the illness factor VII is at fault.

Factor VIII is a protein of two thousand two hundred and thirty-two amino acids, with a gene larger than most—about 186,000 DNA bases long, which, on the scale from Land's End to John o'Groat's, makes it about a hundred yards long. Just a twentieth of its DNA codes for protein. The gene is divided into dozens of different functional sections separated by segments of uninformative sequence. Much of this extraneous material consists of multiple copies of the same two-letter message, a 'CA repeat'. There is even a 'gene-within-a-gene' (which produces something quite different) in the factor VIII machinery.

The haemophilia A mutation, which once appeared to be a simple change, is in fact complicated. All kinds of mistakes can happen. Nearly a thousand different errors have been found. Their virulence depends on what has gone wrong. Sometimes, just one important letter in the functional part of the structure has changed; usually a different letter in different haemophiliacs. The bits of the machinery which join the working pieces of the product together are very susceptible to accidents of this kind. In more than a third of all patients part, or even the whole, of the factor VIII region has disappeared. A few haemophiliacs have suffered from the insertion of an extra length of DNA into the machinery which has hopped in from elsewhere.

Once, the only way to measure the rate of new mutations to haemophilia (or any other inherited illness) was to count the sufferers, estimate the damage done to their chances of passing on the error and work out from this how often it must happen. Technology has changed everything. Now it is possible to compare the genes of haemophiliac boys with those of parents and grandparents to see when the mutation took place.

If the mother of such a boy already has the haemophilia mutation on one of her two X chromosomes, then she must herself have inherited it and the damage must have occurred at some time in the past. If she has not, then her son's new genetic accident happened when the egg from which he developed was formed within her own body. In a survey of British families with sons with haemophilia B (whose gene, that for factor IX, is 33,000 bases long) many different mutations were found, most unique to one family. Eighty per cent of the mothers of affected boys had themselves inherited a mutation. However, in most cases the damaged gene was not present in their own father (the grandfather of the patient). In other words, the error in the DNA must have taken place when his grandparental sperm was being formed.

A quick calculation of the number of new mutations against the size of the British population gives a rate for the haemophilia B gene of about eight in a million. The difference in the incidence of changes between grandfathers and their daughters suggest that the rate is nine times higher in males than in females. The sex difference is easy to explain. There are many more chances for things to go wrong in men (who—unlike women—produce their sex cells throughout life, rather than making a store of them early on, and hence have many more DNA replications in the germ line than do females). For some genes the rate of mutation among males is fifty times higher than in the opposite sex. Men, it seems, are the source of most of evolution's raw material.

Most people with severe forms of haemophilia have each suffered a different genetic error. Such mistakes happen in a parent's sex cells and disappear at once because the child dies young. Those with milder disease often share the same change in their DNA; an error that took place long ago and has spread to many people. The shared mutation is a clue that these individuals descend from a common ancestor. The non-functional DNA in and around the haemophilia gene is full of changes which appear to have no effect at all and have passed down through hundreds of generations. Near the gene itself is a region with many repeats of the same message. The number of copies often goes up and down, but its high error rate seems to do no damage.

All this hints that mutation is an active process, with plenty of churning round within the DNA. This new fluidity once alarmed geneticists as it violates the idea of gene as particle (admittedly a particle which sometimes makes mistakes) which used to be central to their lives. So powerful is the legacy of Mendel that his followers have sometimes been reluctant to accept results which do not fit.

J. B. S. Haldane

from 'ON BEING THE RIGHT SIZE'

With his legendary pugnacity, J. B. S. Haldane followed in the tradition of Darwin's bulldog, T. H. Huxley. He also believed, like Huxley, in bringing science to working men, and many of his scientific essays (not this one, as it happens, but the style is similar) were first published in his regular column in the *Daily Worker*. Along with R. A. Fisher (also a belligerent character, and they were not best friends) Haldane was one of the giants of the neo-Darwinian synthesis, and he was a famously larger-than-life character. His essay 'On being the right size' exemplifies Haldane's characteristic mixture of biological knowledge, political barbs (not included here as they seem dated—a lesson to us all, which regrettably I have not always heeded), mathematical insight, and literary erudition. His two First Class degrees at Oxford were in Classics and Mathematics, an unusual combination, especially in a man who promptly went on to a brilliant academic career in neither subject. Peter Medawar said of him,

> Haldane could have made a success of any one of half a dozen careers—
> as mathematician, classical scholar, philosopher, scientist, journalist or

imaginative writer. On his life's showing he could not have been a politician, administrator (heavens, no!), jurist or, I think, a critic of any kind. In the outcome he became one of the three or four most influential biologists of his generation. ■

———————

The most obvious differences between different animals are differences of size, but for some reason the zoologists have paid singularly little attention to them. In a large textbook of zoology before me I find no indication that the eagle is larger than the sparrow, or the hippopotamus bigger than the hare, though some grudging admissions are made in the case of the mouse and the whale. But yet it is easy to show that a hare could not be as large as a hippopotamus, or a whale as small as a herring. For every type of animal there is a most convenient size, and a large change in size inevitably carries with it a change of form.

Let us take the most obvious of possible cases, and consider a giant man 60 feet high—about the height of Giant Pope and Giant Pagan in the illustrated *Pilgrim's Progress* of my childhood. These monsters were not only ten times as high as Christian, but ten times as wide and ten times as thick, so that their total weight was a thousand times his, or about eighty to ninety tons. Unfortunately the cross-sections of their bones were only a hundred times those of Christian, so that every square inch of giant bone had to support ten times the weight borne by a square inch of human bone. As the human thigh-bone breaks under about ten times the human weight, Pope and Pagan would have broken their thighs every time they took a step. This was doubtless why they were sitting down in the picture I remember. But it lessens one's respect for Christian and Jack the Giant Killer.

To turn to zoology, suppose that a gazelle, a graceful little creature with long thin legs, is to become large; it will break its bones unless it does one of two things. It may make its legs short and thick, like the rhinoceros, so that every pound of weight has still about the same area of bone to support it. Or it can compress its body and stretch out its legs obliquely to gain stability, like the giraffe. I mention these two beasts because they happen to belong to the same order as the gazelle, and both are quite successful mechanically, being remarkably fast runners.

Gravity, a mere nuisance to Christian, was a terror to Pope, Pagan, and Despair. To the mouse and any smaller animal it presents practically no dangers. You can drop a mouse down a thousand-yard mine shaft; and, on arriving at the bottom, it gets a slight shock and walks away. A rat is killed, a man is broken, a horse splashes. For the resistance presented to movement by the air is proportional to the surface of the moving object. Divide an animal's length, breadth, and height each by ten; its weight is reduced to a thousandth, but its surface only to a hundredth. So the resistance to falling in the case of the small animal is relatively ten times greater than the driving force.

An insect, therefore, is not afraid of gravity; it can fall without danger, and can cling to the ceiling with remarkably little trouble. It can go in for elegant and fantastic forms of support like that of the daddy-long-legs. But there is a force which is as formidable to an insect as gravitation to a mammal. This is surface tension. A man coming out of a bath carries with him a film of water of about one-fiftieth of an inch in thickness. This weighs roughly a pound. A wet mouse has to carry about its own weight of water. A wet fly has to lift many times its own weight and, as everyone knows, a fly once wetted by water or any other liquid is in a very serious position indeed. An insect going for a drink is in as great danger as a man leaning out over a precipice in search of food. If it once falls into the grip of the surface tension of the water—that is to say, gets wet—it is likely to remain so until it drowns. A few insects, such as water-beetles, contrive to be unwettable, the majority keep well away from their drink by means of a long proboscis.

Of course tall land animals have other difficulties. They have to pump their blood to greater heights than a man and, therefore, require a larger blood pressure and tougher blood-vessels. A great many men die from burst arteries, especially in the brain, and this danger is presumably still greater for an elephant or a giraffe. But animals of all kinds find difficulties in size for the following reason. A typical small animal, say a microscopic worm or rotifer, has a smooth skin through which all the oxygen it requires can soak in, a straight gut with sufficient surface to absorb its food, and a simple kidney. Increase its dimensions tenfold in every direction, and its weight is increased a thousand times, so that if it is to use its muscles as efficiently as its miniature counterpart, it will need a thousand times as much food and oxygen per day and will excrete a thousand times as much of waste products.

Now if its shape is unaltered its surface will be increased only a hundredfold, and ten times as much oxygen must enter per minute through each square millimetre of skin, ten times as much food through each square millimetre of intestine. When a limit is reached to their absorptive powers their surface has to be increased by some special device. For example, a part of the skin may be drawn out into tufts to make gills or pushed in to make lungs, thus increasing the oxygen-absorbing surface in proportion to the animal's bulk. A man, for example, has a hundred square yards of lung. Similarly, the gut, instead of being smooth and straight, becomes coiled and develops a velvety surface, and other organs increase in complication. The higher animals are not larger than the lower because they are more complicated. They are more complicated because they are larger. Just the same is true of plants. The simplest plants, such as the green algae growing in stagnant water or on the bark of trees, are mere round cells. The higher plants increase their surface by putting out leaves and roots. Comparative anatomy is largely the story of the struggle to increase surface in proportion to volume.

Some of the methods of increasing the surface are useful up to a point, but not capable of a very wide adaptation. For example, while vertebrates carry the oxygen from the gills or lungs all over the body in the blood, insects take air directly to every part of their body by tiny blind tubes called tracheae which open to the surface at many different points. Now, although by their breathing movements they can renew the air in the outer part of the tracheal system, the oxygen has to penetrate the finer branches by means of diffusion. Gases can diffuse easily through very small distances, not many times larger than the average length travelled by a gas molecule between collisions with other molecules. But when such vast journeys—from the point of view of a molecule—as a quarter of an inch have to be made, the process becomes slow. So the portions of an insect's body more than a quarter of an inch from the air would always be short of oxygen. In consequence hardly any insects are much more than half an inch thick. Land crabs are built on the same general plan as insects, but are much clumsier. Yet like ourselves they carry oxygen around in their blood, and are therefore able to grow far larger than any insects. If the insects had hit on a plan for driving air through their tissues instead of letting it soak in, they might well have become as large as lobsters, though other considerations would have prevented them from becoming as large as man.

Exactly the same difficulties attach to flying. It is an elementary principle of aeronautics that the minimum speed needed to keep an aeroplane of a given shape in the air varies as the square root of its length. If its linear dimensions are increased four times, it must fly twice as fast. Now the power needed for the minimum speed increases more rapidly than the weight of the machine. So the larger aeroplane, which weighs 64 times as much as the smaller, needs 128 times its horsepower to keep up. Applying the same principles to the birds, we find that the limit to their size is soon reached. An angel whose muscles developed no more power weight for weight than those of an eagle or a pigeon would require a breast projecting for about four feet to house the muscles engaged in working its wings, while to economize in weight, its legs would have to be reduced to mere stilts. Actually a large bird such as an eagle or kite does not keep in the air mainly by moving its wings. It is generally to be seen soaring, that is to say balanced on a rising column of air. And even soaring becomes more and more difficult with increasing size. Were this not the case eagles might be as large as tigers and as formidable to man as hostile aeroplanes.

But it is time that we passed to some of the advantages of size. One of the most obvious is that it enables one to keep warm. All warm-blooded animals at rest lose the same amount of heat from a unit area of skin, for which purpose they need a food-supply proportional to their surface and not to their weight. Five thousand mice weigh as much as a man. Their combined surface and food or oxygen consumption are about seventeen times a man's. In fact a mouse eats about one-quarter its own weight of food every day, which is mainly used in keeping it warm. For the same reason small animals cannot live in cold countries. In the arctic regions there are no reptiles or amphibians, and no small mammals. The smallest mammal in Spitzbergen is the fox. The small birds fly away in the winter, while the insects die, though their eggs can survive six months or more of frost. The most successful mammals are bears, seals, and walruses.

Similarly, the eye is a rather inefficient organ until it reaches a large size. The back of the human eye on which an image of the outside world is thrown, and which corresponds to the film of a camera, is composed of a mosaic of 'rods and cones' whose diameter is little more than a length of an average light wave. Each eye has about half a million, and for two objects to be distinguishable their images must fall on separate rods or cones. It is obvious that with fewer but larger rods and cones we should see less distinctly. If they were twice as broad two points would have to be twice as far apart before we could distinguish them at a given distance. But if their size were diminished and their number increased we should see no better. For it is impossible to form a definite image smaller than a wavelength of light. Hence a mouse's eye is not a small-scale model of a human eye. Its rods and cones are not much smaller than ours, and therefore there are far fewer of them. A mouse could not distinguish one human face from another six feet away. In order that they should be of any use at all the eyes of small animals have to be much larger in proportion to their bodies than our own. Large animals on the other hand only require relatively small eyes, and those of the whale and elephant are little larger than our own.

For rather more recondite reasons the same general principle holds true of the brain. If we compare the brain-weights of a set of very similar animals such as the cat, cheetah, leopard, and tiger, we find that as we quadruple the body-weight the brain-weight is only doubled. The larger animal with proportionately larger bones can economize on brain, eyes, and certain other organs.

Such are a very few of the considerations which show that for every type of animal there is an optimum size. Yet although Galileo demonstrated the contrary more than three hundred years ago, people still believe that if a flea were as large as a man it could jump a thousand feet into the air. As a matter of fact the height to which an animal can jump is more nearly independent of its size than proportional to it. A flea can jump about two feet, a man about five. To jump a given height, if we neglect the resistance of the air, requires an expenditure of energy proportional to the jumper's weight. But if the jumping muscles form a constant fraction of the animal's body, the energy developed per ounce of muscle is independent of the size, provided it can be developed quickly enough in the small animal. As a matter of fact an insect's muscles, although they can contract more quickly than our own, appear to be less efficient; as otherwise a flea or grasshopper could rise six feet into the air.

[...]

Mark Ridley

from THE EXPLANATION OF ORGANIC DIVERSITY

■ The title of Mark Ridley's piece is a homage to Haldane, and he shares Haldane's erudition and wit. The chapter from which this extract is taken is a type specimen of the comparative method of research that Ridley could fairly be said to have pioneered: going into the library, locating every published article on some particular phenomenon or its opposite—in this case the tendency of big males to mate with big females, or the opposite—counting the instances of both and analysing them quantitatively to test hypotheses. Such exhaustive research (I've hugely oversimplified it in my summary) does not lend itself to a short extract, and I have reluctantly cut the chapter off at the end of Ridley's entertaining historical overture. He is unrelated to Matt Ridley, by the way, although they are understandably often confused and an enterprising editor once succeeded in getting a review by each, of the other's latest book, in the same issue of his journal. Mark praised Matt's book as a fine addition 'to our joint CV'. ■

On Being the Right Sized Mates

Snapping shrimps owe their name to the pops which they let off by snapping their claws together; they can be heard in shallow waters all around the tropics. They live as adults in monogamous pairs. *Alpheus armatus*, for example, whose mating habits have been watched by Knowlton, lives in pairs on a particular species of anemone in the shallow waters of Discovery Bay, in Jamaica. Knowlton collected up dozens of those pairs, and measured them. She then displayed the sizes of the pairs on a graph, with male size along one axis, and female size up the other. Each pair is a single point. The graph of points for pairs of snapping shrimps always shows a correlation of the sizes of mates: bigger males pair with bigger females, smaller males with smaller females. Such a graph could (in principle) show any of three patterns: homogamy, random mating, and heterogamy. Snapping shrimps are homogamous. Homogamy means 'like mates with like'; it is the prior synonym of assortative mating.[1] Heterogamy is its opposite; it means that unlike forms pair, big males with small females, small males with big females. In fact no examples of heterogamy for size seem to exist.

[...]

Homogamy was only discussed for one species, man, before the twentieth century. Even for man almost no facts were collected before Pearson's research at the turn of the century. The absence of facts had not prevented the development of an almost proverbial belief that in humans 'opposites attract one another'. Darwin was aware of this. In 1837 he wrote in his B notebook (p. 6) 'In man it has been said, there is an instinct for opposites to like each other'. Human heterogamy was later to form a minor part of the opposition to Darwin's theory of evolution. Heterogamy would tend to preserve the type, and prevent evolution. Thus Murray wrote that 'it is a trite to a proverb, that tall men marry little women...a man of genius marries a fool', a habit which Murray explained as 'the effort of nature to preserve the typical medium of the race'. The same thought was expressed by the vast intellect of Jeeves, to explain the otherwise mysterious attractions of Bertie for all those female enthusiasts of Kant and Schopenhauer. The source of this proverbial belief is not certainly known; but one possibility can be ruled out. It did not originate in observation: humans mate homogamously (or perhaps randomly) for both stature and intelligence. As Darwin wrote of Murray's paper to Lyell 'it includes speculations...without a single fact in support'.[2] Darwin again stressed the lack of evidence of selective mating in humans in *The Descent of Man*. It has never been the case, he wrote, that 'certain male and female individuals [have] intentionally been picked out and matched, excepting the well-known case of the Prussian grenadiers' (1894 edn, p. 29). The Prussian grenadiers were renowned for their great stature, which (it was believed) was enhanced by selective breeding. The selection was personally supervised by King Friederich Wilhelm. For the King, Dr Johnson tells us in his biography, 'to review this towering regiment was his daily pleasure; and to perpetuate it was so much his care, that when he met a tall woman he immediately commanded one of his Titanian retinue to marry her, that they might propagate procerity, and produce heirs to the father's habiliments.'[3]

1. 'Assortative mating' is now the more usual term. I prefer homogamy, which has priority, is shorter, is etymologically preferable, and more populist: it is in more dictionaries than is 'assortative mating'. Their antonyms are 'dissassortative mating' and 'heterogamy'. This use of homogamy should not be confused with its botanical meanings.

2. Darwin to Lyell, ?4 January 1860, in F. Darwin (ed.), 1887, Vol. ii, p. 262. W. H. Harvey, a botanist, raised the same objection in reviews of the *Origin* (1860, 1861). There are more remarks on homogamy in Darwin's correspondence. See, for example, F. Darwin and Seward (eds.), 1903, i. 202, 272, 308–9, and 333; ii. 232.

3. In *The Works of Samuel Johnson* (Oxford, 1825) vi. 436. The biography was actually of Frederick the Great.

John Maynard Smith

'THE IMPORTANCE OF THE NERVOUS SYSTEM IN THE EVOLUTION OF ANIMAL FLIGHT'

John Maynard Smith was Haldane's most celebrated pupil, and in my field of evolutionary theory he bestrode the second half of the twentieth century much as Haldane had the first. I never worked at the same institution as Maynard Smith, but that doesn't stop me (and many others) from regarding him as a cherished mentor. He listed his recreation in *Who's Who* as 'talking' and this was a boon at conferences, as I noted when I dedicated my largest book to him:

> The only thing that really matters at a conference is that John Maynard Smith must be in residence and there must be a spacious, convivial bar. If he can't manage the dates you have in mind, you must just reschedule the conference... He will charm and amuse the young research workers, listen to their stories, inspire them, rekindle enthusiasms that might be flagging, and send them back to their laboratories or their muddy fields, enlivened and invigorated, eager to try out the new ideas he has generously shared with them.

I added wistfully, after his death, that 'It isn't only conferences that will never be the same again.' His first career, during the war, was as an aero-engineer, but he then decided that 'aeroplanes were noisy and old-fashioned' and went back to university as a mature student to study biology—under Haldane. This essay fascinatingly brings his two careers together. ■

In order to be able to fly, an animal must not only be able to support its weight but must also be able to control its movements in the air. Since animals do not have to learn to fly, or at most need only to perfect by practice an ability already present, it follows that there has been evolution of the sensory and nervous systems to ensure the correct responses in flight. Although no direct evidence on this point can be obtained from fossils, something can be deduced from the gross morphology of primitive flying animals. This can best be done by comparison with the control of aeroplanes, since the latter problem is well understood.

The Stability of Primitive Flying Animals

If an aeroplane is to be controlled by a pilot, it must be stable. An aeroplane, or a gliding animal, is stable if, when it is disturbed from its course, the forces acting on it tend to restore it to that course without active intervention on the part of the pilot in the case of an aeroplane, or without active muscular contractions in the case of an animal. Although gliding has probably always preceded flapping in the evolution of flight, stability can also be defined for flapping flight. In this case, there is a continuous series of muscular contractions. We may say that such an animal is stable if the forces acting on it tend to restore it to its course without any modification of that cycle of contractions being required. In practice the most important type of stability is that for rotation about the pitching axis; that is, a horizontal axis normal to the flight path. In both gliding and flapping flight, stability in pitch can be ensured by the presence of an adequate horizontal surface behind the centre of gravity.

The stability defined above is referred to as static stability. The response of a stable aeroplane to a disturbance may be a deadbeat subsidence or an oscillation. Such oscillations will normally be damped, but in rather special circumstances long period oscillations may be divergent. Such divergent oscillations can normally be controlled by a pilot, and it seems unlikely that they are of any great importance in animal flight. The response of an unstable aeroplane to a disturbance is a divergence, whose rapidity depends on the degree of instability.

Flight has been perfected by four animal groups, the birds, bats, pterosaurs, and insects. Too little is known of the post-cranial skeleton of primitive bats for them to be discussed with any certainty. However, it is generally assumed that the bats have been evolved from gliding arboreal mammals functionally similar to the modern cobego *Cynocephalus* (syn. *Galeopithecus*), although there is probably no phylogenetic relationship. In this mammal the patagium forms a web connecting the fore and hind limbs, and extending backward as the interfemoral membrane to include the tip of the long tail. There can be little doubt that it is a stable glider. In the bats the length of the tail, and therefore the size of the interfemoral membrane, is reduced, and the forelimbs are greatly elongated. These changes have the effect of shifting forward the horizontal lifting surfaces relative to the centre of gravity and thus reducing stability.

In the other three groups, there are good reasons to suppose that the earliest forms were stable in the sense defined above. The Archaeornithes possessed a long tail bordered on either side by a row of feathers, the whole forming a very effective stabilizing surface. In the case of the pterosaurs, the earliest known forms from the lower Jurassic belong to the suborder Rhamphorhynchoidea. These forms had a long stiff tail which in at least one genus, *Rhamphorhynchus*, is known to have terminated in a stiffened fluke of skin. This tail must have had a stabilizing function. However, the latest worker on these fossils, Gross believes that the fluke of skin was disposed in a vertical plane. If this is the case, it would have acted as a stabilizer for yawing rotations, that is, rotations about a vertical axis. It would, in fact, be analogous to the fin of an aeroplane rather than to the tailplane. This, if confirmed, is a rather surprising fact, since in an aeroplane, although both pitching and yawing stability are necessary, instability in pitch renders an aeroplane more completely uncontrollable than instability in yaw.

There are also several features of fossil insects from the Carboniferous to suggest that they were stable. The oldest and most primitive order of winged insects is the Palaeodictyoptera from the lower and middle strata of the upper Carboniferous. They possessed an elongated abdomen, each segment bearing conspicuous lateral lobes, thus forming an effective stabilizing surface. There was also a pair of slender and often greatly elongated cerci. Although such structures would be ineffective as stabilizers on an aeroplane, they are probably quite effective on an insect, due to the greater importance of air viscosity on a small scale (scale in this sense being measured as the product of length and forward speed).

The Evolution of Instability

It appears, therefore, that primitive flying animals tended to be stable, presumably because in the absence of a highly evolved sensory and nervous system they would have been unable to fly if they were not, just as a pilot cannot control an unstable aeroplane. It is, however, theoretically possible to design an automatic pilot to fly a fundamentally unstable aeroplane. In spite of the obvious practical objections to such a scheme, it would have certain advantages. The first is that an unstable aeroplane could be turned more rapidly. The second advantage lies in the fact that in a stable aeroplane the stabilizing tailplane plays a relatively small part in supporting the weight. In an unstable aeroplane, on the other hand, the elevators would be lowered as the plane flew slower, the tailplane would, therefore, support a larger part of the weight, and thus a lower flying speed could be attained without stalling. (The stalling speed is the minimum speed at which an aeroplane can fly.)

Now it is clear that the practical objections to such a scheme as applied to aeroplanes do not arise in the case of animals. There is, in fact, good evidence that birds do not need to be stable in order to fly. In some birds there is no tail in an aerodynamic sense at all. Other birds, which normally possess a tail, can fly without it; this can often be observed in the case of sparrows which have completely lost their tails. In fact, in most birds the tail does not seem to act as a stabilizer, but as an accessory lifting surface when flying slowly. This can be observed, for example, in the case of gulls. These birds open their tails only when turning sharply or flying slowly. It can then be seen that the slower the bird flies the more the tail is lowered; as mentioned above, this is characteristic of the unstable state.

No such detailed discussion is possible in the case of the pterosaurs, but it is significant that the later upper Jurassic and Cretaceous Pterodactyloidea completely lacked a tail.

In the case of insects, it is impossible to make any generalizations, since there is such a wide adaptive radiation within the group. It is probable that some orders, for example, the Ephemeroptera, are stable. However, in the case of the Diptera the work of Hollick and Pringle has shown the importance of the arista and halteres during flight. Indeed, in the case of the Diptera, so far from being stable, the forces acting on a fly are not even in equilibrium in the absence of sensory input from the arista and muscular response to this input.

To a flying animal there are great advantages to be gained by instability. The greater manoeuvrability is of equal importance to an animal which catches its food in the air and to the animals upon which it preys. A low stalling speed is important in a number of ways, and particularly to larger animals. For a set of geometrically similar animals, the stalling speed increases as the square root of the linear dimensions. Therefore a successful landing may be possible in the case of a large animal only if it can reduce its stalling speed, and instability is one of the ways in which this may be done. The account given above of the way in which gulls use their tails illustrates this point. Thus it is possible that the evolution of a pterosaur as large as *Pteranodon* depended on the prior evolution of instability. In extreme cases a lower stalling speed may make hovering flight possible.

It is also important to realize that we are not concerned with a change from stability to instability which must be made in a single step. Any reduction in the degree of stability will be an advantage provided there is a parallel increase in the efficiency of control. This can be seen by analogy with aeroplanes. Transport aeroplanes are normally designed with a fairly high degree of stability, since safety in steady flight is of greater importance than manoeuvrability. In fighter aircraft, however, manoeuvrability is of first importance, and the stability margin is usually reduced to a minimum. It is, therefore, possible to see how instability may have been evolved gradually.

Palaeontologists will have to solve the question of the relative times taken to evolve stable flight, with the relatively coarse controls needed for it, from walking and climbing; and of unstable from stable flight. Unfortunately we only know Archaeornithes from one horizon; on the other hand the Rhamphorhynchoidea persisted for a time measured in tens of millions of years, as did the stable Paleaodictyoptera. It is possible that the evolutionary changes needed for stable flight could be made rather quickly, while the nervous and sensory adjustments needed for unstable flight were inevitably slower. If, as is also possible, the bats evolved rather quickly to instability, this may be due to the greater adaptability of the mammalian brain.

If the conclusions of this paper are accepted the study of the remains of primitive flying animals, and even experimental studies on full-scale models of them, will acquire a special importance as throwing new light on the functional evolution of nervous systems.

There are, of course, several other cases where similar deductions can be made as to the evolution of systems of which no direct fossil evidence exists. Among the most obvious is the need for a highly developed vasomotor system in large land animals which change their posture. A dinosaur standing on its hind legs without the previous evolution of such a system would have suffered from cerebral anaemia. However, it is doubtful whether the palaeontological evidence for such evolution is as clear in any other case as in that of flight.

Fred Hoyle

from MAN IN THE UNIVERSE

■ Fred Hoyle was a distinguished astrophysicist and cosmologist, whose uncompromisingly blunt Yorkshire character—or so I felt—found expression in the heroes of all his science fiction novels. *The Black Cloud* is such a superb story that I have been waging a one man campaign to have it reissued, only to be rebuffed by publishers on the grounds that the hero is too unpleasant. Apart from being compulsively gripping, the great merit of the novel is its vivid portrayal of the scientific method. Among many other things, it illustrates the serendipitous way in which scientific discoveries are often made simultaneously by two different methods, the importance of testing predictions rather than explaining by hindsight, and the powerful idea of information as a quantitative commodity that is interchangeable from medium to medium. So illuminating is *The Black Cloud* of the way science works that I contemplated including a passage from it in this anthology. Reluctantly, I had to rule that fiction didn't belong here. The passage I have chosen instead is from *Man in the Universe* and is an example of the insight that a physical scientist can bring to biology. It was written before Hoyle began the perverse campaign of his old age, against all aspects of Darwinism including even Darwin's personal honour and the authenticity of the fossil bird *Archaeopteryx*. ▪

———————

Looking back along this chain [of evolution], this incredibly detailed chain of many steps, I am overwhelmingly impressed by the way in which chemistry has gradually given way to electronics. It is not unreasonable to describe the first living creatures as entirely chemical in character. Although electrochemical processes are important in plants, organized electronics, in the sense of data processing, does not enter or operate in the plant world. But primitive electronics begins to assume importance as soon as we have a creature that moves around, instead of being rooted in a particular spot, as a plant is. This is surely what we mean by an animal, a creature that moves around. In order to move in any purposeful way a system capable of analyzing and processing information about the external world, about the lay of the land as one might say, becomes necessary. The first electronic systems possessed by primitive animals were essentially guidance systems, analogous logically to sonar or radar. As we pass to more developed animals we find electronic systems being used not merely for guidance but for directing the animal toward food, particularly toward food in the form of another animal. First we have animals eating plants, then animals eating animals, a second order effect. The situation is analogous to a guided missile, the job of which is to intercept and destroy another missile. Just as in our modern world attack and defense become more and more subtle in their methods, so it was the case with animals. And with increasing subtlety, better and better systems of electronics become necessary. What happened in nature has a close parallel with the development of electronics in modern military applications.

I find it a sobering thought that but for the tooth-and-claw existence of the jungle we should not possess our intellectual capabilities, we should not be able to inquire into the structure of the Universe, or to be able to appreciate a symphony of Beethoven. What happened was that electronic systems gradually outran their original purposes. At first they existed to guide animals with powerful weapons, teeth and claws, toward their victims. The astonishing thing, however, was that at a certain stage of subtlety the teeth and claws became unnecessary. Creatures began to emerge in which the original roles of chemistry and electronics were reversed. Instead of the electronics being servant to the chemistry, the reverse became the case. By the time we reach the human, the body has become the servant of the head, existing very largely to supply the brain with appropriate materials for its operation. In us, the computer in our heads, the computer that we call our brain, has entirely taken control. The same I think is true of most of the higher animals, indeed I think this is how one really defines a higher animal. Viewed in this light, the question that is sometimes asked—can computers think?—is somewhat ironic. Here of course I mean the computers that we ourselves make out of inorganic materials. What on earth do those who ask such a question think they themselves are? Simply computers, but vastly more complicated ones than anything we have yet learned to make. Remember that our man-made computer industry is a mere two or three decades old, whereas we ourselves are the products of an evolution that has operated over hundreds of millions of years.

D'Arcy Thompson

from ON GROWTH AND FORM

■ D'Arcy Thompson was another larger-than-life character who spanned the period of the neo-Darwinian synthesis, but he stood to one side of it, aloof. He is the patron of a minority school of biologists who, while not quite denying natural selection, prefer to emphasize physical forces as direct determinants of living form. Although not agreeing with D'Arcy Thompson's views on evolution, Peter Medawar was of the opinion that his book of 1917, *On Growth and Form*, is 'beyond comparison the finest work of literature in all the annals of science that have been recorded in the English tongue'. Coming from Medawar that is high praise indeed, but I'll say no more here because I am reprinting, later in the book, another part of Medawar's pen portrait of D'Arcy Thompson. The most famous chapter of *On Growth and Form* is the one on the Method of Transformations, but I have here chosen an extract from another chapter, which is a particular favourite of mine (it inspired an entire chapter of *Climbing Mount Improbable*) on spirals. ■

The Equiangular Spiral

SPIRALS IN NATURE

The very numerous examples of spiral conformation which we meet with in our studies of organic form are peculiarly adapted to mathematical methods of investigation. But ere we begin to study them we must take care to define our terms, and we had better also attempt some rough preliminary classification of the objects with which we shall have to deal.

In general terms, a spiral is a curve which, starting from a point of origin, continually diminishes in curvature as it recedes from that point; or, in other words, whose *radius of curvature* continually increases. This definition is wide enough to include a number of different curves, but on the other hand it excludes at least one which in popular speech we are apt to confuse with a true spiral. This latter curve is the simple *screw*, or cylindrical *helix*, which curve neither starts from a definite origin nor changes its curvature as it proceeds. The 'spiral' thickening of a woody plant-cell, the 'spiral' thread within an insect's tracheal tube, or the 'spiral' twist and twine of a climbing stem are not, mathematically speaking, *spirals* at all, but *screws* or *helices*. They belong to a distinct, though not very remote, family of curves.

Of true organic spirals we have no lack.[1] We think at once of horns of ruminants, and of still more exquisitely beautiful molluscan shells—in which (as Pliny says) *magna ludentis Naturae varietas*. Closely related spirals may be traced in the florets of a sunflower; a true spiral, though not, by the way, so easy of investigation, is seen in the outline of a cordiform leaf; and yet again, we can recognise typical though transitory spirals in a lock of hair, in a staple of wool,[2] in the coil of an elephant's trunk, in the 'circling spires' of a snake, in the coils of a cuttle-fish's arm, or of a monkey's or a chameleon's tail.

Among such forms as these, and the many others which we might easily add to them, it is obvious that we have to do with things which, though mathematically similar, are biologically speaking fundamentally different; and not only are they biologically remote, but they are also physically different, in regard to the causes to which they are severally due. For in the first place, the spiral coil of the elephant's trunk or of the chameleon's tail is, as we have said, but a transitory configuration, and is plainly the result of certain muscular forces acting upon a structure of a definite, and normally an essentially different, form. It is rather a position, or an *attitude*, than a *form*, in the sense in which we have been using this latter term; and, unlike most of the forms which we have been studying, it has little or no direct relation to the phenomenon of growth.

Again, there is a difference between such a spiral conformation as is built up by the separate and successive florets in the sunflower, and that which, in the snail or *Nautilus* shell, is apparently a single and indivisible unit. And a similar if not identical difference is apparent between the *Nautilus* shell and the minute shells of the Foraminifera which so closely simulate it: inasmuch as the spiral shells of these latter are composite structures, combined out of successive and separate chambers, while the molluscan shell, though it may (as in *Nautilus*) become secondarily subdivided, has grown as one continuous tube. It follows from all this that there cannot be a physical or dynamical, though there may well be a mathematical *law of growth*, which is common to, and which defines, the spiral form in *Nautilus*, in *Globigerina*, in the ram's horn, and in the inflorescence of the sunflower. Nature at least exhibits in them all '*un reflet des formes rigoureuses qu'étudie la géometrie*'.[3]

Of the spiral forms which we have now mentioned, every one (with the single exception of the cordate outline of the leaf) is an example of the remarkable curve known as the equiangular or logarithmic spiral. But before we enter upon the mathematics of the equiangular spiral, let us carefully observe that the whole of the organic forms in which it is clearly and permanently exhibited, however different they may be from one another in outward appearance, in nature and in origin, nevertheless all belong, in a certain sense, to one particular class of conformations. In the great majority of cases, when we consider an organism in part or whole, when we look (for instance) at our own hand or foot, or contemplate an insect or a worm, we have no reason (or very little) to consider one part of the existing structure as *older* than another; through and through, the newer particles have been merged and commingled among the old; the outline, such as it is, is due to forces which for the most part are still at work to shape it, and which in shaping it have shaped it as a whole. But the horn, or the snail-shell, is curiously different; for in these the presently existing structure is, so to speak, partly old and partly new. It has been conformed by successive and continuous increments; and each successive stage of growth, starting from the origin, remains as an integral and unchanging portion of the growing structure.

We may go further, and see that horn and shell, though they belong to the living, are in no sense alive.[4] They are by-products of the animal; they consist of 'formed material', as it is sometimes called; their growth is not of their own doing, but comes of living cells beneath them or around. The many structures which display the logarithmic spiral increase, or accumulate, rather than grow. The shell of nautilus or snail, the chambered shell of a foraminifer, the elephant's tusk, the beaver's tooth, the cat's claws or the canary-bird's—all these show the same simple and very beautiful spiral curve. And all alike consist of stuff secreted or deposited by living cells; all grow, as an edifice grows, by accretion of accumulated material; and in all alike the parts once formed remain in being, and are thenceforward incapable of change.

In a slightly different, but closely cognate way, the same is true of the spirally arranged florets of the sunflower. For here again we are regarding serially arranged portions of a composite structure, which portions, similar to one another in form, *differ in age*; and differ also in magnitude in the strict ratio of their age. Somehow or other, in the equiangular spiral the *time-element* always enters in; and to this important fact, full of curious biological as well as mathematical significance, we shall afterwards return.

THE SPIRAL OF ARCHIMEDES

In the elementary mathematics of a spiral, we speak of the point of origin as the pole (O); a straight line having its extremity in the pole, and revolving about it, is called the radius vector; and a point (P), travelling along the radius vector under definite conditions of velocity, will then describe our spiral curve.

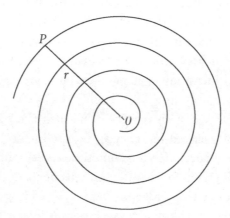

Figure 3. The spiral of Archimedes.

Of several mathematical curves whose form and development may be so conceived, the two most important (and the only two with which we need deal) are those which are known as (1) the equable spiral, or spiral of Archimedes, and (2) the equiangular or logarithmic spiral.

The former may be roughly illustrated by the way a sailor coils a rope upon the deck; as the rope is of uniform thickness, so in the whole spiral coil is each whorl of the same breadth as that which precedes and as that which follows it. Using its ancient definition, we may define it by saying, that 'If a straight line revolve uniformly about its extremity, a point which likewise travels uniformly along it will describe the equable spiral.'[5] Or, putting the same thing into our more modern words, 'If, while the radius vector revolve uniformly about the pole, a point (P) travel with uniform velocity along it, the curve described will be that called the equable spiral, or spiral of Archimedes.' It is plain that the spiral of Archimedes may be compared, but again roughly, to a *cylinder* coiled up. It is plain also that a radius ($r = OP$), made up of the successive and equal whorls, will increase in *arithmetical* progression: and will equal a certain constant quantity (*a*) multiplied by the whole number of whorls or (more strictly speaking) multiplied by the whole angle (θ) through which it has revolved: so that $r = a\theta$. And it is also plain that the radius meets the curve (or its tangent) at an angle which changes slowly but continuously, and which tends towards a right angle as the whorls increase in number and become more and more nearly circular.

THE EQUIANGULAR SPIRAL

But, in contrast to this, in the equiangular spiral of the *Nautilus* or the snail-shell or *Globigerina*, the whorls continually increase in breadth, and do so in a steady and unchanging ratio. Our definition is as follows: 'If, instead of travelling with a *uniform* velocity, our point moves along the radius vector with a velocity *increasing as its distance from the pole*, then the path described is called an equiangular spiral.' Each whorl which the radius vector intersects will be broader than its predecessor in a definite ratio; the radius vector will increase in length in *geometrical* progression, as it sweeps through successive equal angles; and the equation to the spiral will be $r = a^\theta$. As the spiral of Archimedes, in our example of the coiled rope, might be looked upon as a coiled cylinder,

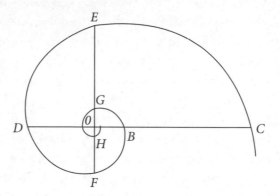

Figure 4. The equiangular spiral.

so (but equally roughly) may the equiangular spiral, in the case of the shell, be pictured as a *cone* coiled upon itself; and it is the conical shape of the elephant's trunk or the chameleon's tail which makes them coil into a rough simulacrum of an equiangular spiral.

While the one spiral was known in ancient times, and was investigated if not discovered by Archimedes, the other was first recognised by Descartes, and discussed in the year 1638 in his letters to Mersenne.[6] Starting with the conception of a growing curve which should cut each radius vector at a constant angle—just as a circle does—Descartes showed how it would necessarily follow that radii at equal angles to one another at the pole would be in continued proportion; that the same is therefore true of the parts cut off from a common radius vector by successive whorls or convolutions of the spire; and furthermore, that distances measured along the curve from its origin, and intercepted by any radii, as at *B, C*, are proportional to the lengths of these radii, *OB, OC*. It follows that the sectors cut off by successive radii, at equal vectorial angles, are similar to one another in every respect; and it further follows that the figure may be conceived as growing continuously without ever changing its shape the while.

The many specific properties of the equiangular spiral are so interrelated to one another that we may choose pretty well any one of them

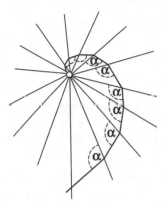

Figure 5. Spiral path of an insect, as it draws towards a light. From Wigglesworth (after van Buddenbroek).

as the basis of our definition, and deduce the others from it either by analytical methods or by elementary geometry. In algebra, when $m^x = n$, x is called the logarithm of n to the base m. Hence, in this instance, the equation $r = a^\theta$ may be written in the form $\log r = \theta \log a$, or $\theta = \log r / \log a$, or (since a is a constant) $\theta = k \log r$.[7] Which is as much as to say that (as Descartes discovered) the vector angles about the pole are proportional to the logarithms of the successive radii; from which circumstance the alternative name of the 'logarithmic spiral' is derived.

Moreover, for as many properties as the curve exhibits, so many names may it more or less appropriately receive. James Bernoulli called it the logarithmic spiral, as we still often do; P. Nicolas called it the geometrical spiral, because radii at equal polar angles are in geometrical progression; Halley, the proportional spiral, because the parts of a radius cut off by successive whorls are in continued proportion; and lastly, Roger Cotes, going back to Descartes' first description or first definition of all, called it the equiangular spiral.[8] We may also recall Newton's remarkable demonstration that, had the force of gravity varied inversely as the *cube* instead of the *square* of the distance, the planets, instead of being bound to their ellipses, would have been shot off in spiral orbits from the sun, the equiangular spiral being one case thereof.[9]

A singular instance of the same spiral is given by the route which certain insects follow towards a candle. Owing to the structure of their compound eyes, these insects do not look straight ahead but make for a light which they see abeam, at a certain angle. As they continually adjust their path to this constant angle, a spiral pathway brings them to their destination at last.[10]

In mechanical structures, *curvature* is essentially a mechanical phenomenon. It is found in flexible structures as the result of *bending*, or it may be introduced into the construction for the purpose of resisting such a bending-moment. But neither shell nor tooth nor claw are flexible structures; they have not been *bent* into their peculiar curvature, they have *grown* into it.

In the growth of a shell, we can conceive no simpler law than this, namely, that it shall widen and lengthen in the same unvarying proportions: and this simplest of laws is that which Nature tends to follow. The shell, like the creature within it, grows in size *but does not change its shape*; and the existence of this constant relativity of growth, or constant similarity of form, is of the essence, and may be made the basis of a definition, of the equiangular spiral.[11]

Such a definition, though not commonly used by mathematicians, has been occasionally employed; and it is one from which the other properties of the curve can be deduced with great ease and simplicity. In mathematical language it would run as follows: 'Any [plane] curve proceeding from a fixed point (which is called the pole), and such that the arc intercepted between any two radii at a given angle to one another is always similar to itself, is called an equiangular, or logarithmic, spiral.'

In this definition, we have the most fundamental and 'intrinsic' property of the curve, namely the property of continual similarity, and the very property by reason of which it is associated with organic growth in such structures as the horn or the shell. For it is peculiarly characteristic of the spiral shell, for instance, that it does not alter as it grows; each increment is similar to its predecessor, and the whole, after every spurt of growth, is just like what it was before. We feel no surprise when the animal which secretes the shell, or any other animal whatsoever, grows by such symmetrical expansion as to preserve its form unchanged; though even there, as we have already seen, the unchanging form denotes a nice balance between the rates of growth in various directions, which is but seldom accurately maintained for long. But the shell retains its unchanging form in spite of its *asymmetrical* growth; it grows at one end only, and so does the horn. And this remarkable property of increasing by *terminal* growth, but nevertheless retaining unchanged the form of the entire figure, is characteristic of the equiangular spiral, and of no other mathematical curve. It well deserves the name, by which James Bernoulli was wont to call it, of *spira mirabilis*.

1. A great number of spiral forms, both organic and artificial, are described and beautifully illustrated in Sir T. A. Cook's *Spirals in Nature and Art* (1903) and *Curves of Life* (1914).
2. On this interesting case see, for example, J. E. Duerden, in *Science* (25 May 1934).
3. Haton de la Goupillière, in the introduction to his important study of the *Surfaces Nautiloides, Annaes sci. da Acad. Polytechnica do Porto*, Coimbra, iii, 1908.
4. For Oken and Goodsir the logarithmic spiral had a profound significance, for they saw in it a manifestation of life itself. For a like reason Sir Theodore Cook spoke of the *Curves of Life*; and Alfred Lartigues says (in his *Biodynamique générale*, 1930, p. 60): 'Nous verrons la Conchyliologie apporter une magnifique contribution à la Stéréo-dynamique du tourbillon vital.' The fact that the spiral is always formed of non-living matter helps to contradict these mystical conceptions.
5. Leslie's *Geometry of Curved Lines* (1821), p. 417. This is practically identical with Archimedes' own definition (ed. Torelli, p. 219); cf. Cantor, *Geschichte der Mathematik* (1880), 1, 262.
6. *Œuvres*, ed. Adam et Tannery (Paris, 1898), p. 360.
7. Instead of $r = a^\theta$, we might write $r = r_0 a^\theta$; in which case r_0 is the value of r for zero value of θ.
8. James Bernoulli, in *Acta Eruditorum* (1691), p. 282; P. Nicolas, *De novis spiralibus* (Tolosae, 1693), 27; E. Halley, *Phil. Trans.* 19 (1696), 58; Roger Cotes, ibid. (1714), and *Harmonia Mensurarum* (1722), 19. For the further history of the curve see (e.g.) Gomes de Teixeira, *Traité des courbes remarquables* (Coimbra, 1909), 76–86; Gino Loria, *Spezielle algebräische Kurven* (1911), 11, 60 seq.; R. C. Archibald (to whom I am much indebted) in *Amer. Math. Mon.* 25 (1918), 189–93, and in Jay Hambidge's *Dynamic Symmetry* (1920), 146–57.
9. *Principia*, 1, 9, 11, 15. On these 'Cotes's spirals' see Tait and Steele, 147.
10. Cf. W. Buddenbroek, *Sitzungsber. Heidelb. Akad.* (1917); V. B. Wigglesworth, *The Principles of Insect Physiology* (1933), 167.
11. See an interesting paper by W. A. Whitworth, 'The Equiangular Spiral, its Chief Properties Proved Geometrically', *Messenger of Mathematics* (1), 1 (1862), 5. The celebrated Christian Wiener gave an explanation on these lines of the logarithmic spiral of the shell, in his highly original *Grundzüge der Weltordnung* (1863).

G. G. Simpson

from THE MEANING OF EVOLUTION

■ George Gaylord Simpson was the palaeontologist among the founding fathers of the neo-Darwinian synthesis. Simpson's fossils may have been dry and dusty, but his approach to them was anything but. Like my own palaeontology teacher at Oxford, Harold Pusey, Simpson always thought of the living flesh and skin that clothed the bones, and of the ecological circumstances in which his old fossils lived out their lives. This particular extract pursues the functional correlations between different parts of the body, as evolution marches in parallel in many different lineages across a broad front of ancestral mammals. ■

Outlines of the History Of Mammals

The rise of the mammals involved the development of numerous interrelated anatomical and physiological characters that proved in the long run to be more effective in many (not all) of the spheres of life occupied by the reptiles. These also were, in the course of time, a basis for the development of new ways of life never achieved by reptiles or by other forms arising within earlier adaptive radiations. The evolution of these new and, as the outcome proved, potent features began among certain of the reptiles, and very early in reptilian history. In a sense the mammals, and the birds too, are simply glorified reptiles. But in a similar sense the reptiles are glorified amphibians, the amphibians glorified fishes, and so on back until all forms of life might be called glorified amebas,[1] and the very amebas could be considered glorified protogenes or protoviruses. The point is that a particular sort of reptilian development turned out to have such unusual possibilities for diversification and for the rise of novel and successful types of organization that its outcome came to overshadow that of all other reptiles put together. The zoologists therefore label that outcome as a distinct Class Mammalia,

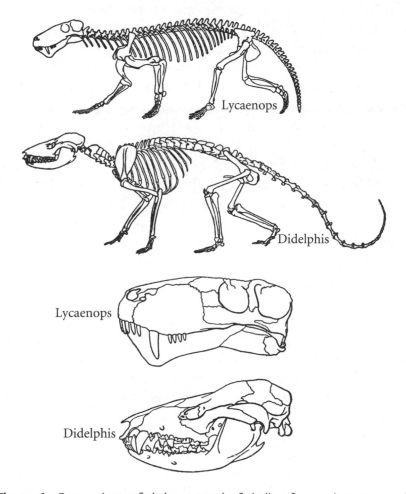

Figure 6. Comparison of skeletons and of skulls of an extinct mammal-like reptile and of a primitive living mammal. (*Lycaenops* is from the Permian of South Africa; its skeleton was reconstructed under the direction of Dr. E. H. Colbert. *Didelphis* is the Virginia opossum. The drawings are to different scales: *Lycaenops* was considerably larger than *Didelphis*.)

on a level with the Class Reptilia of which it is a particularly flowery branch.

Among the many developments within this potent reptile-mammal line, care of the young must be given high place. Eggs were no longer deposited and left at the mercy of an egg-hungry world, nor even given such lesser care as external (as in birds) or internal (as in some reptiles) incubation. The embryo developing from the egg was continuously nourished, by intricate and marvelous means, within the body of the mother.[2] After being introduced to the world the young still receive care from their mother and are nourished for a time by milk from her. These animals also came to be adapted for a higher or more sustained level of activity and for a more constant level of metabolism. Most of them maintain a body temperature relatively independent of temporary activity or of external heat and cold. (This is what is meant by being 'warm blooded'; a 'cold-blooded' animal may have warmer blood than a mammal if it has been exercising violently or is exposed to the hot sun, but its blood cools down again when it stops muscular action or when it moves out of the sun.) The bones of the skeleton grow in a way that maintains firm, bony joints even while they are growing. Growth ceases and the bones knit firmly at an approximate size characteristic for each kind of mammal. These arrangements are mechanically stronger than in animals, like the typical reptiles, in which the joints are more cartilaginous and continue to grow at decreasing rates through most or all of the animal's life. In connection with these features, the basic type of mammal, which was quadrupedal, came to have the legs drawn in directly under the body and to hold the body well up off the ground. This stance led to characteristic modifications of almost every muscle and bone in the body in comparison with those of the typically sprawling basic reptiles.

More sustained activity and more constant metabolism require considerable regularity of food intake and efficient utilization of the food. Evidently connected with this was the development in the reptile-mammal line of teeth specialized by regions: nipping incisors in front, then larger, pointed, piercing or tearing canines, and then a row of cheek teeth (premolars and molars), diversely fashioned for seizing, cutting, pounding, or grinding the food before it is swallowed. Early in the definitely mammalian part of the history a particularly important basic cheek tooth type was developed: the tribosphenic type, with several points or cusps, crests, and valleys on each tooth, so that seizing, cutting, and pounding can all be performed at once. Evolutionary modification of such a tooth, with emphasis of one part and function or another and duplication or extension of the pertinent parts, can and has led to the divergent development of teeth particularly suited for almost any conceivable diet. These dental developments were accompanied by direct jointing of the tooth-bearing jawbone to the skull and by increased strength of jaw action and versatility in directions of jaw movement. This change was, in turn, correlated with an extraordinary change in the ear. The single vibration-transmitting bone in the middle part of the reptilian ear was replaced by a chain of three small bones in the mammals and the two extra bones are probably parts of the old reptilian complex jaw joint. Other changes in the mouth region included development of a hard secondary palate between the mouth and the nose passage with the result that mammals can easily chew and breathe at the same time.

Many other changes were involved in the reptile-mammal transition, but enough have been noted to exemplify, in a general way, the sort of thing involved in the rise of a new grade of animal organization. Many of these changes were already under way among the *therapsid reptiles* of the Triassic. As far as can be shown by fossils, all had been *essentially established in the Jurassic*, in which *four* quite distinct sorts (orders) of mammals are known from unfortunately scanty remains.

1. To use 'ameba' in a merely figurative sense for the hypothetical, perhaps not specifically amebalike, archaic protozoan level of life where supramolecular, multigene organization had been reached.

2. Everyone knows that this is another of the vast majority of generalizations that are open to exception. The platypus and the echidna lay eggs and are called mammals, but viviparity as opposed to oviparity is nevertheless a mammalian characteristic. For all we know it may even have arisen already in mammalian ancestors that we call reptiles, and the egg-laying mammals may derive from some other, allied, line of (nominal) reptiles that did not happen to develop this particular mammalian character. It is well known, too, that in one group of mammals, the marsupials, protection and nourishment of the developing young are less perfected than in the great majority of mammals, the many placental groups — but the fact that an evolutionary development may occur in greater or less degree or under different forms does not make it less characteristic of a group as a whole or lessen the importance of its degree and form in the majority of the members of the group. Incidentally, the platypus and echidna do have milk and are not exceptions to the next statement.

Richard Fortey

from TRILOBITE!

■ Richard Fortey has the same virtues as a palaeontologist, and he loves his trilobites as Simpson loved his mammals and reptiles. This extract is about the wonderful crystalline eyes of the trilobites—and it is already wonderful enough to think that trilobites had eyes at all, so long ago, and to imagine the Palaeozoic coral gardens they gazed upon and the long forgotten fights and flights that their eyes initiated and mediated. Fortey reappears towards the end of this book. ■

Trilobite eyes are made of calcite. This makes them unique in the animal kingdom.

Calcite is one of the most abundant minerals. The white cliffs of Dover are calcite; the bluffs along the Mississippi river are largely calcite; the mountains stacked like giant termite mounds in Guilin Province, China, are composed of calcite that has resisted millennia of weathering. Limestones (which are calcite) have been used to build many of the most monumental and enduring buildings: the sublime crescents of Bath, the pyramids of Gizeh, the amphitheatres and Corinthian columns of classical times. Polished slabs composed of calcite deck the floors of Renaissance churches in Italy, still grace the interiors of Hyatt–Regency hotels, or conference halls, or wherever architects wish to suggest the dignity that only real rock seems to confer. Rubbly limestone builds our rockeries; its finer, whiter counterpart provides the raw material from which great sculpture grows. Only silica sand seems as ubiquitous. Surely one could expect no surprises from a substance so common and so familiar. Yet it was calcite transformed that allowed the trilobite to see.

The purest forms of calcite are transparent. In building stones and decorative slabs it is the impurities and fine crystal masses that provide the colour and design: the yellows and greys and fine mottling.

The dark red of the *scaglio rosso* so typical of Italian church floors is a deep stain of ferric iron. Purge calcite of all these impurities and it is colourless. But it may not be transparent even then. Chalk is almost pure calcite, but it is a mass of tiny grains—fossil fragments most of them—which scatter and reflect the light: hence its almost indecent whiteness. When the Seven Sisters on the southern English coast emerge from a sea mist it is like observing a line of undulating starched sheets, so frigid is their purity. But when a calcite crystal grows more slowly in nature, then it may acquire its perfect crystal form, and be glassy clear. The chemical composition, calcium carbonate ($CaCO_3$), is simple as minerals go. As the crystal grows the constituent atoms stack together in a lopsided way, and do not allow other stray atoms to intrude to cloud its mineral exactitude. Layer builds on layer to reveal the crystalline form, the macrocosm of the gem reflecting exactly the microcosm of atomic sructure. As with the handiwork of a master mason, there is no mistake permissible in the atomic brickwork. Large, fine crystals often grow in mineral veins. These are often rejected by miners in search of rarer booty, for precious metals sometimes hide in grey and opaque minerals that seem dull by comparison with calcite's perfectly formed spar. Some of these crystals are sharply pointed and then are described as dog's tooth spar, looking much like the zig-zag ornament favoured by Norman craftsmen over church doors; others, blunt-tipped, are termed nail head spar. But the clearest crystal, transparent as a toddler's motives, is Iceland spar.

Look into a crystal of Iceland spar and you can see the secret of the trilobite's vision. For trilobites used clear calcite crystals to make lenses in their eyes; in this they were unique. Other arthropods have mostly developed 'soft' eyes, the lenses made of cuticle similar to that constructing the rest of the body. Within this limitation there is enormous variety: many-lensed eyes like those of the fly; large complex eyes such as those of most spiders; eyes that can see in the dark; eyes that function best in brilliant sunshine. The octopus among the molluscs has an eye that is famously like that of backbone-bearing animals, and provides one of the best examples of convergent evolution in the animal kingdom. Most of us will have contemplated the sorry eyes of a dead fish, and remarked the comparison with our own, large, focusing eyes. Trilobites alone have used the transparency of calcite as a means of transmitting light. The trilobite eye is in continuity with the rest of its shelly armour. It sits on top of the cheek of the animal, an *en suite* eyeglass, tough as clamshell.

The science of the eye demands a little explanation. It all depends on the optical properties of calcite, and this depends in turn on its crystallography. If you break a large piece of crystalline calcite it will fracture in a fashion related to its fine atomic structure: such cleavage of the mineral does the bidding of the invisible arrangement of matter itself. You are left with a regular, six-sided chunk of the mineral in your hand, termed a rhomb. Neither foursquare like a cube, nor rectangular like a chunk of chocolate, the sides of a rhomb lean away from the perpendicular. The geometry of mineral shape can be described quite simply by the orientation of a few main axes passing through the centre of the crystal: the simplest case is the cube, in which axes passing through the centre of the faces and meeting at the middle are all at right angles and all the same length. These axes are termed a, b, and c, respectively, a case of science for once taking the simplest route to make a name. In the structure of calcite, one major axis has three axes perpendicular to it set at 120 degrees from one another, hence the configuration of the rhomb. The clear calcite of this not-quite-a-cube treats light in a peculiar way. If a beam of light is shone at the sides of the rhomb it splits in two; this is known as double refraction. The rays of light so produced are the 'ordinary' and the 'extraordinary' rays: their course is determined, just like the shape of the rhomb, by the stacking of the individual atoms. There is a huge specimen of Iceland spar on the first floor of London's Natural History Museum through which you may peer to see two images of a Maltese cross, one generated by the extraordinary, and the other by the ordinary rays. But there is one direction, and one direction only, in which light is not subjected to this optical splitting. This is where a ray of light approaches along the c crystallographic axis; from this direction it does not split into two rays at all but passes straight through.

The way that calcite treats light might have remained no more than an odd fact to be trotted out as an esoteric answer in tests of general knowledge. But what the selectivity of the c axis guarantees is that light approaching from the angle at which it points is specially favoured. If a crystal is elongated in parallel to the c axis into the shape of a prism light will still pass unrefracted through the crystal along the long axis of the prism. But light approaching the same prism from other angles will be split into ordinary and extraordinary rays, which will in turn be deflected to reach the edge of the prism, where they might be partly internally reflected, or refracted yet again. When the prism is long enough the *only* light to pass clearly through to the far side of the prism is that which approaches from the direction of the c crystallographic axis. To put it the other way round, the light that such a crystal 'sees' approaches from one particular direction. It is an astonishing fact that trilobites have hijacked the special properties of calcite for their own ends. They have crystal eyes.

The eyes of trilobites are composed of elongate prisms of clear calcite. Most eyes have many such prisms stacked side by side. By comparison with dozens of other kinds of arthropods the prisms obviously functioned as individual lenses, in just the same way as a fly's eye is a honeycomb of hexagons, each one a lens—or the dragonfly's, or the lobster's. The trilobite carries on its head another example of such an arthropod compound eye—an eye composed of numerous small ocular units, which had to collaborate to paint a portrait of the world. Each component unit is a lens. The unique difference is that the trilobite's lenses are composed of a rock-forming mineral. It would be no less than the truth to say that the trilobite could give you a stony stare. One is reminded of the strange lines from that strangest of Shakespeare's plays, *The Tempest*:

> Full fathom five thy father lies:
> Of his bones are coral made:
> Those are pearls that were his eyes:
> Nothing of him that doth fade,
> But doth suffer a sea-change
> Into something rich and strange.

If to travel back to the time of the trilobite is a historical sea-change then there can be nothing stranger than the calcareous eyes of the trilobite. And pearls are chemically the same as the trilobite's unblinking lenses, being yet another manifestation of calcium carbonate, although pearls

are exquisite reflectors of light rather than transmitters of it. The weirdness of Shakespeare's line results from his suggestions of pearly opacity,
the hints of a corpse transformed; dead, yet seeing. The trilobite saw the
submarine world with eyes tessellated into a mosaic of calcified lenses;
unlike the dead seafarer, his stony eyes read the world through the
medium of the living rock.

Colin Blakemore

from THE MIND MACHINE

■ The theme of eyes continues with the next extract by Colin Blakemore, British physiologist and head of the Medical Research Council,
who always seems to me to be an incongruous (in the nicest possible
way) combination of glamorous whiz kid and eminent elder statesman of
science. The use of 'Grandma' as the object perceived in this hypothetical
story of a photon is not as arbitrary as it sounds, by the way, but is an
in-joke. Ever since the American neurophysiologist Jerry Lettvin epitomized
highly specific pattern-detecting neurones as 'grandmother detectors',
scientists have used 'Lettvin's grandmother' as an affectionate shorthand
for the idea. It is Blakemore's witty tribute to Lettvin to use this vivid
image without actually spelling out its history. ■

Sight Unseen

A single atom of gas, baking in the unimaginable heat at the surface
of the sun, suddenly shifts from one energy state to another, and spits
out the surplus energy as a photon—the smallest, indivisible unit of
light. This tiny pellet of energy is thrown into space at the highest speed
that Einstein could conceive of. Eight minutes or so after its birth, our

chosen photon slows down a little as it hits the atmosphere of the Earth and a fraction of a second later it reaches the surface. It strikes the wrinkled skin of an old woman but, as chance has it, the wavelength of our charmed photon of light is such that it is not captured by the pigments of her skin. It is reflected, and 10 microseconds later it shoots into a tiny black hole, just 3 millimetres across. This hole is the pupil of a man's eye.

The photon slips past the transparent window that covers the front of his eye, through the lens within it and on, between the particles of the gelatinous mass behind the lens, even across the membranes and cytoplasm of the nerve cells of the retina in the back of the eye. But time is running out. It penetrates a strange, thin cell at the back of the retina and its existence ends as it strikes a single molecule of pigment inside that cell, which captures the photon, destroying it by stealing its energy.

'Hello, Grandma.' The man, whose retina has caught our hero, the photon, has recognised his grandmother. He sees her wrinkled face and her blue gingham dress.

She smiles and opens her mouth. As she exhales, the folds of her larynx vibrate as the air rushes past them. Her breath rushes around her moving tongue as it darts skilfully back and forth within her mouth, occasionally touching her lips or her teeth. She is speaking. The rich mixture of tones and noises pulses through the air towards her grandson's head. Some of the vibrating particles in the air are caught by the crevices of his outer ear and funnelled into the narrow tube that leads to his eardrum. They beat on it, setting up a rhythm in a chain of minute bones, which rattle at another membrane, setting up waves in the liquid inside a tiny coiled tube. And these vibrations, in turn, tickle hairs on tiny specialised nerve cells that stand like a regiment of sharp-eared soldiers along the length of the tube. 'Hello dear.' The man hears his grandmother speaking.

This everyday scene sets the stage for a detective story. The detective is the human brain; the story is our perception of the world around us.

KNOWLEDGE FROM MOLECULES AND WAVES

To the inner eye and ear of the conscious mind, our senses give us windows through which we see, hear, touch, taste, and smell the physical world.

The job of the sense organs is to convert light, sound, heat, pressure, and molecules into the tiny electrical impulses that scurry along nerve fibres—their currency of communication. We are blissfully unaware of the machinery of nerves within our sense organs and our brains; all that we know directly is the impression of reality. Perception is an invention of nerve cells inside our heads.

For more than 2000 years, philosophers and scientists have been deceived by the apparent simplicity of perception. The great Greek geometer, Euclid, who lived about 300 BC, thought that we see the world because light flows *out* of the eyes, like an invisible hand, feeling the reality of the physical world. But Plato, who lived 100 years earlier, realised that knowledge—even knowledge of the outside world—comes from *within*. He described a fable, told by Socrates, about people living in a strange underground world:

> Behold! Human beings living in an underground den, which has a mouth open towards the light…Here they have been from their child-hood…Above and behind them a fire is blazing at a distance, and between the fire and the prisoners there is a raised path; and you will see, if you look, a low wall built along the path, like the screen which marionette players have in front of them, over which they show the puppets.

Socrates described the way in which the human beings trapped in the cave could see the people and objects that were out of sight behind the wall only by virtue of the flickering shadows of them thrown on the opposite wall of the cave by the light of the fire. In such a frightful world, Socrates said that 'the truth would be literally nothing but the shadows of the images'. He went on to explain his allegory: 'The prison-house is the world of sight, the light of the fire is the sun.' And so it is; our understanding of the world around us comes from the merest echoes of reality—the photons of light that bombard the eye, the vibrations in the air that strike the ear, the floating molecules that rush into the nostrils.

Aristotle, the first great biologist, wrote that each sense organ 'receives the form of the object without its matter'. The immaterial qualities (such as colour, shape and smell) of physical things in the world strike the sense organs and evoke the perception of the world. Galileo's eyes, peering through his telescope at the heavens, changed our entire understanding of the universe, yet he wrote, in 1623, fourteen years before he lost his sight, that sensations 'are nothing more than names when separated from living beings, just as tickling and titillation are nothing but names in the absence of such things as noses and armpits'.

PICTURES IN THE HEAD

How, then, can nerve cells create our knowledge of the world? To start to answer that question, we can peer backwards in time and downwards through the animal world to simple creatures with no more than a few thousand neurons to help them find their food, avoid their enemies and manage their lives. Any animal that moves must understand something of the world around it. Many simple animals *detect* light but learn little from it: they merely move towards or away from the light, depending on their particular style of life. The light-sensitive nerve cells in the human retina, on which those photons that enter the pupil fall, can also do nothing more than signal the intensity of light. All the richness of our visual perceptions, all the information needed to recognise a grandmother's face, comes from those tiny cells in the retina that know nothing but the number of photons hitting them.

Richard Gregory

from MIRRORS IN MIND

For years as a college tutor at Oxford, I would try the intelligence and reasoning powers of entrance candidates by asking them at interview to muse aloud on the conundrum of why mirror images appear left-right reversed but not upside down. It is a provocative puzzle, which is hard to situate among academic disciplines. Is it a question in psychology, in physics, in philosophy, in geometry, or just commonsense? I wasn't necessarily expecting my candidates to 'know the right answer'. I wanted to hear them think aloud, wanted to see if the question piqued their interest and their curiosity. If it did, they would probably be fun to teach. Richard Gregory, psychologist, engineer, philosopher, historian, and enthusiastic overgrown schoolboy has an unrivalled capacity to think aloud, and to pique interest and curiosity, his own and that of others. He explains and solves this mirror riddle, among many other reflections, in his book *Mirrors in Mind*. ▪

Why are Looking-glass Images Right-Left Reversed?

This most famous mirror puzzle has confused bright people for centuries. So, why is everything in a looking-glass right-left reversed yet not reversed up-down? For example, why does *writing* appear as horizontally reversed though not upside down—as 'mirror writing'? The reader may find this simply obvious. Most people, however, go through their lives without ever considering it. Once considered, it can remain a puzzle for life.

How can a mere *mirror* distinguish right-left from up-down, even though many *people* don't know their right from their left? One's first thought is likely to be that *there is no problem*—because the top of the object is reflected from the top of the mirror, the bottom from bottom of the mirror, the left from the left and the right from the right. But this is just where the puzzle (if it is a puzzle for the reader) starts: the mirror is optically symmetrical—up-down and right-left—yet the reflected image is only reversed one way: right-left.

[...]

To make sure that a plane mirror is indeed optically symmetrical (up-down and right-left), we may rotate it around its centre. There is no change of the image: one's face remains upright, and right-left reversed. Words continue to be upright in 'mirror writing', like this:

words in a mirror

But they are not not-reversed:

words in a mirror

nor upside down only:

words in a mirror

nor right-left reversed and upside down:

words in a mirror

Why should this be so? One's first thought is likely to be that the answer lies in a *ray diagram*. But this can't be—for a diagram can be held any way round. So it can not show what is vertical or horizontal. This is the same for a map. A map can't show where north (or west or east, or whatever) is without a compass bearing, because a map can be held any way round. This is just the same for an optical ray diagram: a ray diagram can not explain why a mirror reverses right-left but not up-down because it can be held any way up.

Now let's ask: do plane mirrors *always* switch right and left? Let's try a little experiment—with a match and its box. Hold a match horizontal and parallel to a vertical mirror. What happens? When the head of the match is to the right, its image is also to the right. It is *not* right-left reversed. Now take the match-box and view it in the mirror. Its writing is right-left reversed. So the match and its box behave differently! Why is the match *not* reversed though the match-box writing *is* reversed?

Here is a related puzzle: hold a mug with writing on it to a mirror. What do you see in the mirror? The reflection of the *handle* is unchanged— but the *writing* is right-left reversed. Can a mirror *read*?!

[...]

Let's list what seems both *true* and *relevant*:

- Writing on a *transparent* sheet (such as an overhead transparency) does not show reversal when held before a mirror.
- When writing on a transparent sheet is turned around (so its front and back surfaces are switched over) the image *is* reversed.
- An *opaque* sheet of writing (or a book) must be turned around to face the mirror—then the image is reversed.
- When the transparent sheet is rotated around its *horizontal* axis, it is vertically and not horizontally reversed.

A mirror allows us to see the back of an opaque object though we are in front of it. But to see its front, in a mirror behind it, the object must be rotated. When, say, a book is rotated around its *vertical* axis to face the mirror, its *left and right* switch over. It is this that produces mirror reversal. It is really object reversal.

What happens when a book is turned around its *horizontal* axis to face the mirror? It then appears upside down. Because it *is* upside down and not right-left reversed.

Object-rotation does not produce these effects without a mirror because the front of the object gets hidden, to be replaced by its back as the object is rotated. So the mirror is *necessary*, though it doesn't *cause* the reversal. It is necessary because without it we can't see the front of the rotated object.

When looking behind while driving a car, one looks in the rear-view mirror to avoid turning one's head. When an ambulance is behind the car, by rotating one's head one sees AMBULANCE. In the mirror this appears as:

ƎƆИАⅬUꓭMA

So it is often printed or stencilled reversed—to appear non-reversed in the mirror.

A mirror shows us *ourselves* right-left reversed (reversed from how others see us without the mirror) because we have to turn around to face it. Normally we turn round vertically, keeping our feet on the ground. But we can face the mirror by standing on our head—then, we are upside down in the mirror and *not* left-right reversed. So again the reversal is rotation of the *object*—oneself.

[...]

Alice

The author of *Alice in Wonderland* (1865) and *Alice Through the Looking-glass* (1872)—Charles Lutwidge Dodgson (1832–98)—was a logician, a creator of mathematical puzzles, an expert photographer, and a conjuror with simple apparatus, including mirrors. He spent his days protected from normal reality, as a Fellow of Christ Church College in the University of Oxford. Lewis Carroll was, of course, his pseudonym. A particular friend was John Henry Pepper, who invented 'Pepper's Ghost'—a large part-reflecting mirror, that allowed actors on stage to appear and disappear and become transparent. This very likely inspired the disappearing grin of the Cheshire Cat:

> This time it vanished quite slowly, beginning with the end of the tail, and ending with the grin, which remained some time after the rest of it had gone.

There were two Alices. The Alice in *Wonderland* was the second daughter of a distinguished Oxford don, Henry George Liddell. The Alice who ventured through the looking-glass was a distant cousin, Alice Raikes. Six years after *Wonderland*, it was a chance encounter with Alice Raikes that suggested *Through the Looking-glass*. Staying in London, Lewis Carroll happened to overhear children playing in one of the Kensington squares and heard them call a little girl, 'Alice!' He called her over and told her he liked Alices. He invited her indoors, then put an orange into her hand and asked which hand she was holding it in:

> 'In my right hand', she told him.
> 'Now go and look at the little girl in the glass over there,' he said, 'and tell me which hand she is holding the orange in.'
> Alice went to the mirror and stood before it thoughtfully.
> 'She is holding it in her left hand.'

Carroll asked if she could explain that to him, and she hesitated before replying:

> 'Supposing I was on the other side of the glass, wouldn't the orange still be in my right hand?'

He was delighted with her answer, and that decided him. The make-believe world for his new book should be that on the other side of the looking-glass (see Figure 7). At the start of *Through the Looking-glass*, this was the result:

> Now if you'll only attend, Kitty, and not talk so much, I'll tell you all my ideas about Looking-glass House. First, there's the room you can see through the glass—that's just the same as our drawing room, only the things go the other way. I can see all of it when I get upon a chair—all but the bit just behind the fireplace. Oh! I do so wish I could see *that* bit! I want so much to know whether they've got a fire in the winter: you never *can* tell, you know, unless our fire smokes, and the smoke comes into that room too—but that may be pretence, just to make it look as if they had a fire. Well then, the books are something like our books, only the words go the wrong way. I know that because I've held up one of our books to the glass and then they hold up one in the other room.

When Alice jumped lightly through the mirror above the fireplace, to the Looking-glass room, she was: 'quite pleased to find that there was a real [fire], blazing away as brightly as the one she had left behind'. Then

Figure 7. Alice through the mirror.
From *Alice Through the Looking-glass.*

she noticed that the clock on the chimney piece had got the face of a little old man, which grinned at her. This was surprising, for as Alice said: 'You can only see the back of it in the Looking-glass.' When Alice, with her kitten, returned (by *re-turning*, and so *reversed*!) from Looking-glass Land, she mused:

'It is a very inconvenient habit of kittens...that whatever you say to them, they *always* purr. If they would only purr "yes", and mew for "no", or any rule of that sort...one could keep up a conversation!
But how *can* you talk with a person if they always say the same thing?'

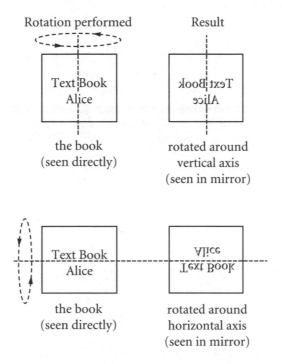

Figure 8. What happens for reversal, diagrammatically.

On this occasion the kitten only purred: it was impossible to guess whether it meant 'yes' or 'no'.

Then Alice considers confronting the normal with the mirror world, by trying to get the kitten to look at the Red Queen:

> So Alice…put the kitten and the Queen to look at each other. 'Now, Kitty!' she cried, clapping her hands triumphantly. 'Confess that was what you turned into!'
>
> 'But it wouldn't look at it,' she said, when she was explaining the thing afterwards to her sister: 'it turned away its head and pretended not to see it, but it looked a *little* ashamed of itself, so I think it *must* have been the Red Queen.'

So just looking wasn't quite adequate. How did Alice, at the end of the dream, turn the Red Queen into the white kitten? By *shaking* her:

'I'll shake you into a kitten, that I will!' Lewis Carroll describes what happened in vivid detail:

> She took her off the table as she spoke and shook her backwards and for-wards with all her might.
> The Red Queen made no resistance whatever: only her face grew very small, and her eyes got large and green: and still, as Alice went on shaking her, she kept growing shorter—and fatter—and softer—and rounder—and—and it really *was* a kitten after all.

If we could pick up and shake *images* as we can *objects* we would not be bemused by mirrors.

Nicholas Humphrey

'ONE SELF: A MEDITATION ON THE UNITY OF CONSCIOUSNESS'

■ The psychologist and evolutionist Nicholas Humphrey is one of the more graceful and literate of modern scientist writers. Here we have a probing meditation on the hard problem of consciousness. It may be too hard for anybody to solve, but Humphrey makes a promising approach through speculating that a baby might be a kind of federation of separate subjectivities, which only gradually come together during the first years. Humphrey's question—why does each of us feel like a single subjective unit?—is one of those questions that seems too obvious to ask until you take the trouble to think about it, and then the more you think, the more profound and tantalizing it becomes. He may not have solved the riddle, but it is an achievement to show that there is a question here to worry us at all. Science doesn't have all the answers, but it is good at spotting the important questions when they are camouflaged against a background of common sense. ■

I am looking at my baby son as he thrashes around in his crib, two arms flailing, hands grasping randomly, legs kicking the air, head and eyes turning this way and that, a smile followed by a grimace crossing his face…And I'm wondering: what is it like to be him? What is he feeling now? What kind of experience is he having of *himself*?

Then a strong image comes to me. I am standing now, not at the rail of a crib, but in a concert hall at the rail of the gallery, watching as the orchestra assembles. The players are arriving, section by section—strings, percussion, woodwind—taking their separate places on the stage. They pay little if any attention to each other. Each adjusts his chair, smoothes his clothes, arranges the score on the rack in front of him. One by one they start to tune their instruments. The cellist draws his bow darkly across the strings, cocks his head as if savouring the resonance, and slightly twists the screw. The harpist leans into the body of her harp, runs her fingers trippingly along a scale, relaxes and looks satisfied. The oboist pipes a few liquid notes, stops, fiddles with the reed and tries again. The tympanist beats a brief rally on his drum. Each is, for the moment, entirely in his own world, playing only to and for himself, oblivious to anything but his own action and his own sound. The noise from the stage is a medley of single notes and snatches of melody, out of time, out of harmony. Who would believe that all these independent voices will soon be working in concert under one conductor to create a single symphony.

Now, back in the nursery, I seem to be seeing another kind of orchestra assembling. It is as if, with this baby, all the separate agencies of which he is composed still have to settle into place and do *their* tuning up: nerves need tightening and balancing, sense organs calibrating, pipes clearing, airways opening, a whole range of tricks and minor routines have to be practised and made right. The subsystems that will one day be a system have as yet hardly begun to acknowledge one another, let alone to work together for one common purpose. And as for the conductor who one day will be leading all these parts in concert into life's *Magnificat*: he is still nowhere to be seen.

I return to my question: what kind of experience is this baby having of himself? But, as I ask it, I realize I do not like the answer that suggests itself. If there is no conductor inside him yet, perhaps there is in fact no self yet, and if no self perhaps no experience either—perhaps nothing at all.

If I close my eyes and try to think like a hard-headed philosophical sceptic, I can almost persuade myself it could be so. I must agree that, in theory,
there could be no kind of consciousness within this little body, no inner
life, nobody at home to have an inner life. But then, as I open my eyes and
look at him again, any such scepticism melts. *Someone* in there is surely
looking back at me, someone is smiling, someone seems to know my face,
someone is reaching out his tiny hand…Philosophers think one way, but
fathers think another. I can hardly doubt sensations are registering inside
this boy, willed actions initiating, memories coming to the surface. However disorganized his life may be, he is surely not totally unconscious.

Yet I realize I cannot leave it there. If these experiences are occurring
in the baby boy, they presumably have to belong to an *experiencer*. Every
experience has to have a corresponding subject whose experience it is.
The point was well made by the philosopher Gottlob Frege, a hundred
years ago: it would be absurd, he wrote, to suppose 'that a pain, a mood,
a wish should rove about the world without a bearer, independently.
An experience is impossible without an experient. The inner world
presupposes the person whose inner world it is.'

But if that is the case, I wonder what to make of it. For it seems to
imply that those 'someones' that I recognize inside this boy—the someone who is looking, the someone who is acting, the someone who is
remembering—must all be genuine subjects of experience (subjects;
note the plural). If indeed he does not yet possess a single Self—that Self
with a capital S which will later mould the whole system into one—then
perhaps he must in fact possess a set of relatively independent sub-selves,
each of which must be counted a separate centre of subjectivity, a separate experiencer. Not yet being one person, perhaps he is in fact *many*.

But, isn't this idea bizarre? A lot of independent experiencers? Or—
to be clear about what this has to mean—a lot of independent con-
sciousnesses? And all within one body? I confess I find it hard to see
how it would work. I try to imagine what it would be like for me to be
fractionated in this way and I simply cannot make sense of the idea.

Now, I agree that I myself have many kinds of 'lesser self' inside
me: I can, if I try, distinguish a part of me that is seeing, a part that
is smelling, a part raising my arm, a part recalling what day it is, and
so on. These are certainly different types of mental activity, involving

different categories of subjective experience, and I am sure they can properly be said to involve different dimensions of my Self.

I can even agree that these parts of me are a relatively loose confederation that do not all have to be present at one time. Parts of my mind can and do sometimes wander, get lost, and return. When I have come round from a deep sleep, for example, I think it is even true that I have found myself having to gather myself together—which is to say *my selves* together—piecemeal.

Marcel Proust, in *À la recherche du temps perdu*, provides a nice description of just this peculiar experience: 'When I used to wake up in the middle of the night,' he writes,

> not knowing where I was, I could not even be sure at first who I was; I had only the most rudimentary sense of existence, such as may lurk and flicker in the depths of an animal's consciousness...But then...out of a blurred glimpse of oil-lamps, of shirts with turned-down collars, [I] would gradually piece together the original components of my ego.

So it is true, if I think about this further, that the idea of someone's consciousness being dispersed in different places is not completely unfamiliar to me. And yet I can see that this kind of example will hardly do to help me understand the baby. For what distinguishes my case from the baby's is precisely that these 'parts of me' that separate and recombine do not, while separate, exist as distinct and self-sufficient subjects of experience. When I come together on waking, it is surely not a matter of my bringing together various sub-selves that are already separately conscious. Rather, these sub-selves only come back into existence as and when I plug them back, as it were, into the main me.

As I stand at the crib watching my baby boy, trying to find the right way in, I now realize I am up against an imaginative barrier. I will not say that, merely because I can't imagine it, it could make no sense at all to suppose that this baby has got all those separate conscious selves within him. But I will say I do not know what to say next.

Yet, I am beginning to think there is the germ of some real insight here. Perhaps the reason why I cannot imagine the baby's case is tied into that very phrase, '*I* can't imagine...'. Indeed, as soon as I try to imagine the baby as split into several different selves, I make him back into one again by virtue of imagining it. I imagine each set of experiences as *my* experiences—but, just to the extent that they are all *mine*, they are no longer separate.

And doesn't this throw direct light on what may be the essential difference between my case and the baby's? For doesn't it suggest that it is all a matter of how a person's experiences are owned—to whom they belong?

With *me* it seems quite clear that every experience that any of my sub-selves has is *mine*. And, to paraphrase Frege, in my case it would certainly make no sense to suppose that a pain, a mood, a wish should rove about my inner world without the bearer in every case being *me*! But maybe with the baby every experience that any of his sub-selves has is not yet *his*. And maybe in his case it does make perfect sense to suppose that a pain, a mood, a wish should rove about inside his inner world without the bearer in every case being *him*.

How so? What kind of concept of 'belonging' can this be, such that I can seriously suggest that, while my experiences belong to me, the baby's do not belong to him? I think I know the answer intuitively; yet I need to work it through.

Let me return to the image of the orchestra. In their case, I certainly want to say that the players who arrive on stage as isolated individuals come to belong to a single orchestra. As an example of 'belonging', this seems as clear as any. But, if there is indeed something that binds the players to belong together, what kind of something is this?

The obvious answer would seem to be the one I have hinted at already: that there is a 'conductor'. After each player settles in and has his period of free play, a dominant authority mounts the stage, lifts his baton, and proceeds to take overall control. Yet, now I am beginning to realize that this image of the conductor as 'chief self' is not the one I want—nor, in fact, was it a good or helpful image to begin with.

Ask any orchestral player, and he'll tell you: although it may perhaps look to an outsider as if the conductor is totally in charge, in reality he often has a quite minor—even a purely decorative—role. Sure, he can provide a common reference point to assist the players with the timing and punctuation of their playing. And he can certainly influence the overall style and interpretation of a work. But that is not what gets the players to belong together. What truly binds them into one organic unit and creates the flow between them is something much deeper and more magical, namely, the very act of making music: that they are together creating a single work of art.

Doesn't this suggest a criterion for 'belonging' that should be much more widely applicable: that parts come to belong to a whole just in so far as they are *participants in a common project*?

Try the definition where you like: what makes the parts of an oak tree belong together—the branches, roots, leaves, acorns? They share a common interest in the tree's survival. What makes the parts of a complex machine like an aeroplane belong to the aeroplane—the wings, the jet engines, the radar? They participate in the common enterprise of flying.

Then, here's the question: what makes the parts of a person belong together—if and when they do? The clear answer has to be that the parts will and do belong together *just in so far as they are involved in the common project of creating that person's life.*

This, then, is the definition I was looking for. And, as I try it, I immediately see how it works in my own case. I may indeed be made up of many separate sub-selves, but these selves have come to belong together as the one Self that I am because they are engaged in one and the same enterprise: the enterprise of steering me—body and soul—through the physical and social world. Within this larger enterprise each of my selves may indeed be doing its own thing: providing me with sensory information, with intelligence, with past knowledge, goals, judgements, initiatives, and so on. But the point—the wonderful point—is that each self doing its own thing shares a final common path with all the other selves doing their own things. And it is for this reason that these selves are all *mine*, and for this reason that their experiences are all *my experiences*. In short, my selves have become co-conscious through collaboration.

But the baby? Look at him again. There he is, thrashing about. The difference between him and me is precisely that he has as yet no common project to unite the selves within him. Look at him. See how he has hardly started to do anything for himself as a whole: how he is still completely helpless, needy, dependent—reliant on the projects of other people for his survival. Of course, his selves are beginning to get into shape and function on their own. But they do not yet share a final common path. And it is for that reason his selves are not yet all of them *his*, and for that reason their experiences are not yet *his* experiences. His selves are not co-conscious because there is as yet no co-laboration.

Even as I watch, however, I can see things changing. I realize the baby boy is beginning to come together. Already there are hints of small collaborative projects getting under way: his eyes and his hands working together, his face and his voice, his mouth and his tummy. As time goes by, some of these mini-projects will succeed; others will be abandoned. But inexorably over days and weeks and months he will become one coordinated, centrally conscious human being. And, as I anticipate this happening, I begin to understand how in fact he may be going to achieve this miracle of unification. It will not be, as I might have thought earlier, through the power of a supervisory Self who emerges from nowhere and takes control, but through the power inherent in all his sub-selves for, literally, their own *self-organization*.

Then, stand with me again at the rail of the orchestra, watching those instrumental players tuning up. The conductor has not come yet, and maybe he is not ever going to come. But it hardly matters: for the truth is, *it is of the nature of these players to play*. See, one or two of them are already beginning to strike up, to experiment with half-formed melodies, to hear how they sound for themselves, and—remarkably—to find and recreate *their* sound in the *group sound* that is beginning to arise around them. See how several little alliances are forming, the strings are coming into register, and the same is happening with the oboes and the clarinets. See, now, how they are joining together across different sections, how larger structures are emerging.

Perhaps I can offer a better picture still. Imagine, at the back of the stage, above the orchestra, a lone dancer. He is the image of Nijinsky in *The Rite of Spring*. His movements are being shaped by the sounds of the instruments, his body absorbing and translating everything he hears. At first his dance seems graceless and chaotic. His body cannot make one dance of thirty different tunes. Yet, something is changing. See how each of the instrumental players is watching the dancer—looking to find how, within the chaos of those body movements, the dancer is dancing to his tune. And each player, it seems, now wants the dancer to be *his*, to have the dancer give form to *his* sound. But see how, in order to achieve this, each must take account of all the other influences to which the dancer is responding—how each must accommodate to and join in harmony with the entire group. See, then, how, at last, this group of players is becoming *one orchestra* reflected in *the one body of the dancer*—and how the music they are making and the dance that he is dancing have indeed become a *single work of art*.

And my boy, Samuel? His body has already begun to dance to the sounds of his own selves. Soon enough, as these selves come together in creating him, he too will become a single, self-made human being.

Steven Pinker

■ Steven Pinker, linguist and evolutionary psychologist, is one of science's most compelling writers today. His stylistic mastery lies not in cadences of lyrical prose poetry such as we find in Carl Sagan, Peter Atkins or Loren Eiseley. I think Pinker's power as a writer comes from his vivacious choice of words and images, sometimes strung together in arrestingly surprising lists. It is these lists that make Pinker an interesting challenge to read aloud, as my wife, the actress Lalla Ward, discovered when she recorded *The Language Instinct* for the audio book publication. Also characteristic of Pinker's style is a racy familiarity with popular culture and jokes, which leaven the literary allusions. The two passages extracted here are the opening paragraphs of his classic *The Language Instinct*, and a part of *How the Mind Works*, which exemplifies the best of the unjustly maligned discipline of Evolutionary Psychology. ■

---

from THE LANGUAGE INSTINCT

An Instinct to Acquire an Art

As you are reading these words, you are taking part in one of the wonders of the natural world. For you and I belong to a species with a remarkable ability: we can shape events in each other's brains with exquisite precision. I am not referring to telepathy or mind control or the other obsessions of fringe science; even in the depictions of believers these are blunt instruments compared to an ability that is uncontroversially present in every one of us. That ability is language. Simply by making noises with our mouths, we can reliably cause precise new combinations of ideas to arise in each other's minds. The ability comes so naturally that we are apt to forget what a miracle it is. So let me remind you with some simple demonstrations. Asking you only to surrender your imagination to my words for a few moments, I can cause you to think some very specific thoughts:

> When a male octopus spots a female, his normally grayish body suddenly becomes striped. He swims above the female and begins caressing her with seven of his arms. If she allows this, he will quickly reach toward her and slip his eighth arm into her breathing tube. A series of sperm packets moves slowly through a groove in his arm, finally to slip into the mantle cavity of the female.

> Cherries jubilee on a white suit? Wine on an altar cloth? Apply club soda immediately. It works beautifully to remove the stains from fabrics.

> When Dixie opens the door to Tad, she is stunned, because she thought he was dead. She slams it in his face and then tries to escape. However, when Tad says, 'I love you', she lets him in. Tad comforts her, and they become passionate. When Brian interrupts, Dixie tells a stunned Tad that she and Brian were married earlier that day. With much difficulty, Dixie informs Brian that things are nowhere near finished between her and Tad. Then she spills the news that Jamie is Tad's son. 'My what?' says a shocked Tad.

Think about what these words have done. I did not simply remind you of octopuses; in the unlikely event that you ever see one develop stripes, you now know what will happen next. Perhaps the next time you are in a supermarket you will look for club soda, one out of the tens of thousands of items available, and then not touch it until months later when a particular substance and a particular object accidentally come together. You now share with millions of other people the secrets of protagonists in a world that is the product of some stranger's imagination, the daytime drama *All My Children*. True, my demonstrations depended on our ability to read and write, and this makes our communication even more impressive by bridging gaps of time, space, and acquaintanceship. But writing is clearly an optional accessory; the real engine of verbal communication is the spoken language we acquired as children.

In any natural history of the human species, language would stand out as the pre-eminent trait. To be sure, a solitary human is an impressive problem-solver and engineer. But a race of Robinson Crusoes would not give an extraterrestrial observer all that much to remark on. What is truly arresting about our kind is better captured in the story of the Tower of Babel, in which humanity, speaking a single language, came so close to reaching heaven that God himself felt threatened. A common language connects the members of a community into an information-sharing network with formidable collective powers. Anyone can benefit from the strokes of genius, lucky accidents, and trial-and-error wisdom accumulated by anyone else, present or past. And people can work in teams, their efforts coordinated by negotiated agreements. As a result, *Homo sapiens* is a species, like blue-green algae and earthworms, that has wrought far-reaching changes on the planet. Archeologists have discovered the bones of ten thousand wild horses at the bottom of a cliff in France, the remains of herds stampeded over the clifftop by groups of paleolithic hunters seventeen thousand years ago. These fossils of ancient cooperation and shared ingenuity may shed light on why saber-tooth tigers, mastodons, giant woolly rhinoceroses, and dozens of other large mammals went extinct around the time that modern humans arrived in their habitats. Our ancestors, apparently, killed them off.

Language is so tightly woven into human experience that it is scarcely possible to imagine life without it. Chances are that if you find two or more people together anywhere on earth, they will soon be exchanging words. When there is no one to talk with, people talk to themselves, to their dogs, even to their plants. In our social relations, the race is not to the swift but to the verbal—the spellbinding orator, the silver-tongued seducer, the persuasive child who wins the battle of wills against a brawnier parent. Aphasia, the loss of language following brain injury, is devastating, and in severe cases family members may feel that the whole person is lost forever.

* * *

from HOW THE MIND WORKS

The Smell of Fear

Language-lovers know that there is a word for every fear. Are you afraid of wine? Then you have *oenophobia*. Tremulous about train travel? You suffer from *siderodromophobia*. Having misgivings about your mother-in-law is *pentheraphobia*, and being petrified of peanut butter sticking to the roof of your mouth is *arachibutyrophobia*. And then there's Franklin Delano Roosevelt's affliction, the fear of fear itself, or *phobophobia*.

But just as not having a word for an emotion doesn't mean that it doesn't exist, having a word for an emotion doesn't mean that it does exist. Word-watchers, verbivores, and sesquipedalians love a challenge. Their idea of a good time is to find the shortest word that contains all the vowels in alphabetical order or to write a novel without the letter *e*. Yet another joy of lex is finding names for hypothetical fears. That is where these improbable phobias come from. Real people do not tremble at the referent of every euphonious Greek or Latin root. Fears and phobias fall into a short and universal list.

Snakes and spiders are always scary. They are the most common objects of fear and loathing in studies of college students' phobias, and have been so for a long time in our evolutionary history. D. O. Hebb found that chimpanzees born in captivity scream in terror when they first see a snake, and the primatologist Marc Hauser found that his laboratory-bred cotton-top tamarins (a South American monkey) screamed out alarm calls when they saw a piece of plastic tubing on the floor. The reaction of foraging peoples is succinctly put by Irven DeVore: 'Hunter-gatherers will not suffer a snake to live.' In cultures that revere snakes, people still treat them with great wariness. Even Indiana Jones was afraid of them!

The other common fears are of heights, storms, large carnivores, darkness, blood, strangers, confinement, deep water, social scrutiny, and leaving home alone. The common thread is obvious. These are the situations that put our evolutionary ancestors in danger. Spiders and snakes are often venomous, especially in Africa, and most of the others are obvious hazards to a forager's health, or, in the case of social scrutiny, status. Fear is the emotion that motivated our ancestors to cope with the dangers they were likely to face.

Fear is probably several emotions. Phobias of physical things, of social scrutiny, and of leaving home respond to different kinds of drugs, suggesting that they are computed by different brain circuits. The psychiatrist Isaac Marks has shown that people react in different ways to different frightening things, each reaction appropriate to the hazard. An animal triggers an urge to flee, but a precipice causes one to freeze. Social threats lead to shyness and gestures of appeasement. People really do faint at the sight of blood, because their blood pressure drops, presumably a response that would minimize the further loss of one's own blood. The best evidence that fears are adaptations and not just bugs in the nervous system is that animals that have evolved on islands without predators lose their fear and are sitting ducks for any invader—hence the expression 'dead as a dodo'.

Fears in modern city-dwellers protect us from dangers that no longer exist, and fail to protect us from dangers in the world around us. We ought to be afraid of guns, driving fast, driving without a seatbelt, lighter fluid, and hair dryers near bathtubs, not of snakes and spiders. Public safety officials try to strike fear in the hearts of citizens using everything from statistics to shocking photographs, usually to no avail. Parents scream and punish to deter their children from playing with matches or chasing a ball into the street, but when Chicago schoolchildren were asked what they were most afraid of, they cited lions, tigers, and snakes, unlikely hazards in the Windy City.

Of course, fears do change with experience. For decades psychologists thought that animals learn new fears the way Pavlov's dogs learned to salivate to a bell. In a famous experiment, John B. Watson, the founder of behaviorism, came up behind an eleven-month-old boy playing with a tame white rat and suddenly clanged two steel bars together. After a few more clangs, the boy became afraid of the rat and other white furry things, including rabbits, dogs, a sealskin coat, and Santa Claus. The rat, too, can learn to associate danger with a previously neutral stimulus. A rat shocked in a white room will flee it for a black room every time it is dumped there, long after the shocker has been unplugged.

But in fact creatures cannot be conditioned to fear just any old thing. Children are nervous about rats, and rats are nervous about bright rooms, before any conditioning begins, and they easily associate them with danger. Change the white rat to some arbitrary object, like opera glasses, and the child never learns to fear it. Shock the rat in a black room instead of a white one, and that nocturnal creature learns the association more slowly and unlearns it more quickly. The psychologist Martin Seligman suggests that fears can be easily conditioned only when the animal is evolutionarily prepared to make the association.

Few, if any, human phobias are about neutral objects that were once paired with some trauma. People dread snakes without ever having seen one. After a frightening or painful event, people are more prudent around the cause, but they do not fear it; there are no phobias for electrical outlets, hammers, cars, or air-raid shelters. Television clichés notwithstanding, most survivors of a traumatic event do not get the screaming meemies every time they face a reminder of it. Vietnam veterans resent the stereotype in which they hit the dirt whenever someone drops a glass.

A better way to understand the learning of fears is to think through the evolutionary demands. The world is a dangerous place, but our ancestors could not have spent their lives cowering in caves; there was food to gather and mates to win. They had to calibrate their fears of typical dangers against the actual dangers in the local environment (after all, not *all* spiders are poisonous) and against their own ability to neutralize the danger: their know-how, defensive technology, and safety in numbers.

Marks and the psychiatrist Randolph Nesse argue that phobias are innate fears that have never been unlearned. Fears develop spontaneously in children. In their first year, babies fear strangers and separation, as well they should, for infanticide and predation are serious threats to the tiniest hunter-gatherers. (The film *A Cry in the Dark* shows how easily a predator can snatch an unattended baby. It is an excellent answer to every parent's question of why the infant left alone in a dark bedroom is screaming bloody murder.) Between the ages of three and five, children become fearful of all the standard phobic objects—spiders, the dark, deep water, and so on—and then master them one by one. Most adult phobias are childhood fears that never went away. That is why it is city-dwellers who most fear snakes.

As with the learning of safe foods, the best guides to the local dangers are the people who have survived them. Children fear what they see their parents fear, and often unlearn their fears when they see other children coping. Adults are just as impressionable. In wartime, courage and panic are both contagious, and in some therapies, the phobic watches as an aide plays with a boa constrictor or lets a spider crawl up her arm. Even monkeys watch one another to calibrate their fear. Laboratory-raised rhesus macaques are not afraid of snakes when they first see them, but if they watch a film of another monkey being frightened by a snake, they fear it, too. The monkey in the movie does not instil the fear so much as awaken it, for if the film shows the monkey recoiling from a flower or a bunny instead of a snake, the viewer develops no fear.

The ability to conquer fear selectively is an important component of the instinct. People in grave danger, such as pilots in combat or Londoners during the blitz, can be remarkably composed. No one knows why some people can keep their heads when all about them are losing theirs, but the main calming agents are predictability, allies within shouting distance, and a sense of competence and control, which the writer Tom Wolfe called The Right Stuff. In his book by that name about the test pilots who became Mercury astronauts, Wolfe defined the right stuff as 'the ability [of a pilot] to go up in a hurtling piece of machinery and put his hide on the line and then have the moxie, the reflexes, the experience, the coolness, to pull it back in the last yawning moment'. That sense of control comes from 'pushing the outside of the envelope': testing, in small steps, how high, how fast, how far one can go without bringing on disaster. Pushing the envelope is a powerful motive. Recreation, and the emotion called 'exhilaration', come from enduring relatively safe events that look and feel like ancestral dangers. These include most non-competitive sports (diving, climbing, spelunking, and so on) and the genres of books and movies called 'thrillers'. Winston Churchill once said, 'Nothing in life is so exhilarating as to be shot at without result.'

Jared Diamond

from THE RISE AND FALL OF THE
THIRD CHIMPANZEE

■ Jared Diamond is another distinguished scientist who is also a highly successful writer on science for a general audience. Physiologist, ornithologist, anthropologist, geographer and explorer, he brings wide reading and a generous measure of what can only be called deep wisdom to every topic he investigates. His books constitute major publishing events, and have won many prizes. The theme of the passage I have chosen was more fully developed in *Guns, Germs and Steel,* but this extract is from Diamond's earlier book, *The Rise and Fall of the Third Chimpanzee.* ■

By around 4000 BC western Eurasia already had its 'Big Five' domestic livestock that continue to dominate today: sheep, goats, pigs, cows, and horses. Eastern Asians domesticated four other cattle species that locally replace cows: yaks, water buffalo, gaur, and banteng. As already mentioned, these animals provided food, power, and clothing, while the horse was also of incalculable military value. (It was both the tank, the truck, and the jeep of warfare until the nineteenth century.) Why did American Indians not reap similar benefits by domesticating the corresponding native American mammal species, such as mountain sheep, mountain goats, peccaries, bison, and tapirs? Why did Indians mounted on tapirs, and native Australians mounted on kangaroos, not invade and terrorize Eurasia?

The answer is that, even today, it has proved possible to domesticate only a tiny fraction of the world's wild mammal species. This becomes clear when one considers all the attempts that failed. Innumerable species reached the necessary first step of being kept captive as tame pets. In New Guinea villages I routinely find tamed possums and kangaroos, while I saw tamed monkeys and weasels in Amazonian Indian villages. Ancient Egyptians had tamed gazelles, antelopes, cranes, and even hyenas and possibly giraffes. Romans were terrorized by the tamed African elephants with which Hannibal crossed the Alps (*not* Asian elephants, the tame elephant species in circuses today).

But all these incipient efforts at domestication failed. Since the domestication of horses around 4000 BC and reindeer a few thousand years later, no large European mammal has been added to our repertoire of successful domesticates. Thus, our few modern species of domestic mammals were quickly winnowed from hundreds of others that had been tried and abandoned.

Why have efforts at domesticating most animal species failed? It turns out that a wild animal must possess a whole suite of unusual characteristics for domestication to succeed. Firstly, in most cases it must be a social species living in herds. A herd's subordinate individuals have instinctive submissive behaviours that they display towards dominant individuals, and that they can transfer towards humans. Asian mouflon sheep (the ancestors of domestic sheep) have such behaviour but North American bighorn sheep do not—a crucial difference that prevented Indians from domesticating the latter. Except for cats and ferrets, solitary territorial species have not been domesticated.

Secondly, species such as gazelles and many deer and antelopes, which instantly take flight at signs of danger instead of standing their ground, prove too nervous to manage. Our failure to domesticate deer is especially striking, since there are few other wild animals with which humans have been so closely associated for tens of thousands of years. Although deer have always been intensively hunted and often tamed, reindeer alone among the world's forty-one deer species were successfully domesticated. Territorial behaviour, flight reflexes, or both eliminated the other forty species as candidates. Only reindeer had the necessary tolerance of intruders and gregarious, non-territorial behaviour.

Finally, domestication requires being able to breed an animal in captivity. As zoos often discover to their dismay, captive animals that are docile and healthy may nevertheless refuse to breed in cages. You yourself would not want to carry out a lengthy courtship and copulate under the watchful eyes of others; many animals do not want to either.

This problem has derailed persistent attempts to domesticate some potentially very valuable animals. For example, the finest wool in the world comes from the vicuña, a small camel species native to the Andes. But neither the Incas nor modern ranchers have ever been able to domesticate it, and wool must still be obtained by capturing wild vicuñas. Many potentates, from ancient Assyrian kings to nineteenth-century Indian maharajahs, have tamed cheetahs, the world's swiftest land mammal, for hunting. However, every prince's cheetah had to be captured from the wild, and not even zoos were able to breed them until 1960.

Collectively, these reasons help explain why Eurasians succeeded in domesticating the Big Five but not other closely related species, and why American Indians did not domesticate bison, peccaries, tapirs, and mountain sheep or goats. The military value of the horse is especially interesting in illustrating what seemingly slight differences make one species uniquely prized, another useless. Horses belong to the group of mammals termed Perissodactyla, which consists of the hoofed mammals with an odd number of toes: horses, tapirs, and rhinoceroses. Of the seventeen living species of Perissodactyla, all four tapirs and all five rhinos, plus five of the eight wild horse species, have never been domesticated. Africans or Indians mounted on rhinos or tapirs would have trampled any European invaders, but it never happened.

A sixth wild horse relative, the wild ass of Africa, gave rise to domestic donkeys, which proved splendid as pack animals but useless as military chargers. The seventh wild horse relative, the onager of western Asia, may have been used to pull wagons for some centuries after 3000 BC. But all accounts of the onager blast its vile disposition with adjectives like 'bad-tempered', 'irascible', 'unapproachable', 'unchangeable', and 'inherently intractable'. The vicious beasts had to be kept muzzled to prevent them from biting their attendants. When domesticated horses reached the Middle East around 2300 BC, onagers were finally kicked onto the scrapheap of failed domesticates.

Horses revolutionized warfare in a way that no other animal, not even elephants or camels, ever rivalled. Soon after their domestication, they may have enabled herdsmen speaking the first Indo-European languages to begin the expansion that would eventually stamp their languages on much of the world. A few millenia later, hitched to battle chariots, horses became the unstoppable Sherman tanks of ancient war. After the invention of saddles and stirrups, they enabled Attila the Hun to devastate the Roman Empire,

Genghis Khan to conquer an empire from Russia to China, and military kingdoms to arise in West Africa. A few dozen horses helped Cortes and Pizarro, leading only a few hundred Spaniards each, to overthrow the two most populous and advanced New World states, the Aztec and Inca empires. With futile Polish cavalry charges against Hitler's invading armies in September 1939, the military importance of this most universally prized of all domestic animals finally came to an end after 6,000 years.

Ironically, relatives of the horses that Cortes and Pizarro rode had formerly been native to the New World. Had those horses survived, Montezuma and Atahuallpa might have shattered the conquistadores with cavalry charges of their own. But, in a cruel twist of fate, America's horses had become extinct long before that, along with eighty or ninety per cent of the other large animal species of the Americas and Australia. It happened around the time that the first human settlers—ancestors of modern Indians and native Australians—reached those continents. The Americas lost not only their horses but also other potentially domesticatable species like large camels, ground sloths, and elephants. Australia lost all its giant kangaroos, giant wombats, and rhinoceros-like diprotodonts. Australia and North America ended up with no domesticatable mammal species at all, unless Indian dogs were derived from North American wolves. South America was left with only the guineapig (used for food), alpaca (used for wool), and llama (used as a pack animal, but too small to carry a rider).

As a result, domestic mammals made no contribution to the protein needs of native Australians and Americans except in the Andes, where their contribution was still much slighter than in the Old World. No native American or Australian mammal ever pulled a plough, cart, or war chariot, gave milk, or bore a rider. The civilizations of the New World limped forward on human muscle power alone, while those of the Old World ran on the power of animal muscle, wind, and water.

Scientists still debate whether the prehistoric extinctions of most large American and Australian mammals were due to climatic factors or were caused by the first human settlers themselves. Whichever was the case, the extinctions may have virtually ensured that the descendants of those first settlers would be conquered over 10,000 years later by people from Eurasia and Africa, the continents that retained most of their large mammal species.

David Lack

from THE LIFE OF THE ROBIN

■ David Lack was a schoolmaster and gentleman ornithologist, who later became Director of the Edward Grey Institute of Field Ornithology at Oxford and one of the leading population ecologists of his generation, principal architect of our understanding of the natural regulation of animal numbers (the title of his major scholarly book). His earlier title *Darwin's Finches* has entered the language as the name for the finches of Galapagos. That is not the only feat of ornithological renaming he can claim. When he submitted his paper on 'Polygamy in a Bishop', the killjoy editor changed it to 'Bishop bird'. I remember him as a somewhat forbidding figure in the Oxford zoology department, seldom called by his Christian name and known in his Institute as 'Doctor Lack' or 'The Boss'. Yet he was capable of great kindness, as when, though not a rich man, he personally financed a Japanese graduate student whose grant ran out before he could complete his research. The passage I have chosen is from *The Life of the Robin* and it exemplifies, like *Darwin's Finches* and *Swifts in a Tower*, David Lack's habit (perceptively noted by John Maynard Smith) of writing simultaneously for a technical and a popular audience. It describes Lack's experiments with a stuffed Robin, which both cock and hen birds vigorously attacked. ■

––––––––

It was before breakfast on a cold October morning that the strangest of all the results with a stuffed robin was achieved. The stuffed bird had been erected in the territory of a hen robin previously known to be exceptionally fierce, and for the record time of forty minutes this bird continued to posture, strike, and sing at the specimen. She was still continuing to do so when the sound of the distant breakfast gong caused me to interrupt proceedings by removing the specimen from its perch and walking off. By chance I looked back, to see the hen robin return, hover in the air, and deliver a series of violent pecks at the empty air. I was able to get to the exact place where I had previously stood, so could see that the bird was attacking the identical spot formerly occupied by the specimen. Three more attacks were delivered in rapid succession, but on the last two the bird was about a foot out in position. She then sang hard but returned for a final attack, now three feet out of position, while her violent singing continued for some time longer. As Pliny noted, 'Verily, for mine own part, the more I look into Nature's workes, the sooner am I induced to believe of her, even those things that seem incredible.'

Niko Tinbergen

from CURIOUS NATURALISTS

■ Eventually, David Lack's stuffed robin came to a bad end, but a productive one. A particularly aggressive hen bird beheaded it and then continued to belabour the body. This gave Lack the idea of dissecting the stuffed bird to see precisely which bits of it were needed to elicit the attacks. He finally narrowed it down to a tuft of red feathers: the rest of the bird wasn't necessary. This was the kind of experiment that my old maestro Niko Tinbergen made characteristically his own. Both before and after Lack's experiment with the disembodied red feathers, Tinbergen did the same kind of thing with solitary wasps, baby thrushes, sticklebacks, and especially with various species of gull. The kindly, smiling, avuncular Tinbergen was the master of naturalist experiments, proper, controlled experiments but done in the wild rather than in the laboratory. He pioneered the technique as a young man during his famous series of experiments on digger wasps in the sand dunes of his native Holland. This is the subject of the following extract from Tinbergen's scientific autobiography, *Curious Naturalists*. Like Ernst Mayr and Theodosius Dobzhansky, Tinbergen wrote English better than most anglophone scientists. ■

Settling down to work, I started spending the wasps' working days (which lasted from about 8 a.m. till 6 p.m. and so did not put too much of a strain on me) on the 'Philanthus plains', as we called this part of the sands as soon as we had found out that *Philanthus triangulum Fabr.* was the official name of this bee-killing digger wasp. Its vernacular name was 'Bee-Wolf'.

An old chair, field glasses, note-books, and food and water for the day were my equipment. The local climate of the open sands was quite amazing, considering that ours is a temperate climate. Surface temperatures of 110° F were not rare, and judging from the response of my skin, which developed a dark tan, I got my share of ultraviolet radiation.

My first job was to find out whether each wasp was really limited to one burrow, as I suspected from the unhesitating way in which the home-coming wasps alighted on the sand patches in front of the burrows. I installed myself in a densely populated quarter of the colony, five yards or so from a group of about twenty-five nests. Each burrow was marked and mapped. Whenever I saw a wasp at work at a burrow, I caught it and, after a short unequal struggle, adorned its back with one or two colour dots (using quickly drying enamel paint) and released it. Such wasps soon returned to work, and after a few hours I had ten wasps, each marked with a different combination of colours, working right in front of me. It was remarkable how this simple trick of marking my wasps changed my whole attitude to them. From members of the species *Philanthus triangulum* they were transformed into personal acquaintances, whose lives from that very moment became affairs of the most personal interest and concern to me.

While waiting for events to develop, I spent my time having a close look at the wasps. A pair of lenses mounted on a frame that could be worn as spectacles enabled me, by crawling up slowly to a working wasp, to observe it, much enlarged, from a few inches away. When seen under such circumstances most insects reveal a marvellous beauty, totally unexpected as long as you observe them with the unaided eye. Through my lenses I could look at my *Philanthus* right into their huge compound eyes; I saw their enormous, claw-like jaws which they used for crumbling up the sandy crust; I saw their agile black antennae in continuous, restless movement; I watched their yellow, bristled legs rake away the loose sand with such vigour that it flew through the air in rhythmic Ruffs, landing several inches behind them.

Soon several of my marked wasps stopped working at their burrows, raked loose sand back over the entrance, and flew off. The take
off was often spectacular. Before leaving they circled a little while over
the burrow, at first low above the ground, soon higher, describing
ever widening loops; then flew away, but returned to cruise more low
over the nest. Finally, they would set out in a bee-line, fifteen to thirty
feet above the ground, a rapidly vanishing speck against the blue sky.
All the wasps disappeared towards the south-east. Half a mile away in
that direction the bare sands bordered upon an extensive heath area,
buzzing with bees. This, as I was to see later, was the wasps' hunting
area.

The curious loops my wasps described in the air before leaving
their home area had been described by other observers of many other
digger wasps. Philip Rau had given them the name of 'locality studies'.
Yet so far nobody had proved that they deserved that name; that the
wasps actually took in the features of the burrow's surroundings while
circling above them. To check this if possible was one of my aims—I
thought that it was most probable that the wasps would use landmarks, and that this locality study was what the name implied. First,
however, I had to make sure that my marked wasps would return to
their own holes.

When my wasps had left, there began one of those periods of patient
waiting which are usual in this kind of work. It was, of course, necessary to be continually on the look-out for returning wasps; at the same
time it was tempting to look round and watch the multitude of other
creatures that were busy on the hot plains. For I soon discovered that
I had many neighbours. First of all, there were other diggers about.
Among the *Philanthus* burrows there were some that looked different—
the sand patches were a little larger and less regular. These belonged
to the largest of our digger wasps, the fly-killing *Bembex*, that is almost
the size of a Hornet. Buzzing loudly, flying with terrific speed low
over the ground, these formidable wasps dashed to and fro; and it took
me a long time before I saw more of them than a momentary sulphur-
yellow flash. A Leafcutter Bee was coming home carrying its 'wall
paper', a neat circular disc cut out of a rose leaf. Its burrow, scarcely
visible, was in the carpet of dry moss just beyond the *Philanthus*

settlement. Robber Flies (*Asilus crabroniformis*) whizzed past, catching flies and other insects in the air. Sometimes they would make a mistake and attack *Philanthus*. A brief but furious buzzing struggle, and the two fell apart, *Asilus* darting off to find a less petulant prey, *Philanthus* returning to its burrow. Small grasshoppers came walking by, greedy and single-minded, devouring one grass shoot after the other, or spending hours chirping their song, or courting, with pathetic perseverance, seemingly unconcerned females.

On some days we saw endless processions of migrating Cabbage White butterflies crossing the plains in a continuous stream of scattered formations, usually going north-west. Hobbies came and levied their toll on them, as well as on the many dragonflies and the clumsy Dung Beetles. Bumblebees often zoomed past on their long, mysterious trips and, like the Cabbage Whites, would show their interest in nectar-filled blue flowers by alighting on anything blue in our equipment.

In August the monotony of the deep blue sky might be broken by a lonely Osprey, flying in from his fishing grounds north of us—the coastal waters of the Zuiderzee—to settle in the crown of an old Pine, there to dream away the hours of digestion. Or a group of lovely, black-and-white Storks, on migration to Africa, might come sailing by, stopping in the rising air over the hot sands to soar up in wide circles, higher and higher, until they resumed their glide south in search of the next 'thermal'. Thus sitting and waiting for the wasps was never dull, if only one kept one's eyes open.

To return to my marked wasps: before the first day was over, each of them had returned with a bee; some had returned twice or even three times. At the end of that day it was clear that each of them had its own nest, to which it returned regularly.

On subsequent days I extended these observations and found out some more facts about the wasps' daily life. As in other species, the digging of the large burrows and the capturing of prey that served as food for the larvae was exclusively the task of the females. And a formidable task it was. The wasps spent hours digging the long shafts, and throwing the sand out. Often they stayed down for a long time and, waiting for them to reappear, my patience was often put to a hard test. Eventually, however, there would be some almost imperceptible movement in the sand, and a small mound of damp soil was gradually lifted up, little by little, as if a miniature Mole were at work. Soon the wasp emerged, tail first, and all covered with sand. One quick shake, accompanied by a sharp staccato buzz, and the wasp was clean. Then it began to mop up, working as if possessed, shovelling the sand several inches away from the entrance.

I often tried to dig up the burrows to see their inner structure. Usually the sand crumbled and I lost track of the passage before I was ten inches down, but sometimes, by gently probing with a grass shoot first, and then digging down along it, I succeeded in getting down to the cells. These were found opening into the far end of the shaft, which itself was a narrow tube, often more than 2 ft. long. Each cell contained an egg or a larva with a couple of Honey Bees, its food store. A burrow contained from one to five cells. Each larva had its own living room-cum-larder in the house, provided by the hard-working female. From the varying number of cells I found in the nests, and the varying ages of the larvae in one burrow, I concluded that the female usually filled each cell with bees before she started to dig a new cell, and I assumed that it was the tunnelling out of a new cell that made her stay down for such long spells.

I did not spend much time digging up the burrows, for I wanted to observe the wasps while they were undisturbed. Now that I was certain that each wasp returned regularly to her own burrow, I was faced with the problem of her orientation. The entire valley was littered with the yellow sand patches; how could a wasp, after a hunting trip of about a mile in all, find exactly her own burrow?

Having seen the wasps make their 'locality studies', I naturally believed that each female actually did what this term implied: take her bearings. A simple test suggested that this was correct. While a wasp was away I brushed over the ground surrounding the nest entrance, moving all possible landmarks such as pebbles, twigs, tufts of grass, Pine cones, etc, so that over an area of 3-4 square metres none of them remained in exactly the same place as before. The burrow itself, however, I left intact. Then I awaited the wasp's return. When she came, slowly descending from the skies, carrying her bee, her behaviour was striking. All went well until she was about 4 ft. above the ground. There she suddenly stopped, dashed back and forth as if in panic, hung motionless in the air for a while, then flew back and up in a wide loop, came slowly down again in the same way, and again shied at the same distance from the nest.

Obviously she was severely disturbed. Since I had left the nest itself, its entrance, and the sand patch in front of it untouched, this showed that the wasp was affected by the change in the surroundings.

Gradually she calmed down, and began to search low over the disturbed area. But she seemed to be unable to find the nest. She alighted now here, now there, and began to dig tentatively at a variety of places at the approximate site of the nest entrance.

After a while she dropped her bee and started a thorough trial-and-error search. After twenty-five minutes or so she stumbled on the nest entrance as if by accident, and only then did she take up her bee and drag it in. A few minutes later she came out again, closed the entrance, and set off. And now she had a nice surprise in store for me: upon leaving she made an excessively long 'locality study': for fully two minutes she circled and circled, coming back again and again to fly over the disturbed area before she finally zoomed off.

I waited for another hour and a half, and had the satisfaction of seeing her return once more. And what I had hoped for actually happened: there was scarcely a trace of hesitation this time. Not only had the wasp lost her shyness of the disturbed soil, but she now knew her way home perfectly well.

I repeated this test with a number of wasps, and their reactions to my interference were roughly the same each time. It seemed probable, therefore, that the wasps found their way home by using something like landmarks in the environment, and not by responding to some stimulus (visual or otherwise) sent out by the nest itself. I had now to test more critically whether this was actually the case.

The test I did next was again quite simple. If a wasp used landmarks it should be possible to do more than merely disturb her by throwing her beacons all over the place; I ought to be able to mislead her, to make her go to the wrong place, by moving the whole constellation of her landmarks over a certain distance. I did this at a few nests that were situated on bare sandy soil and that had only a few, but conspicuous, objects nearby, such as twigs, or tufts of grass. After the owner of such a nest was gone, I moved these two or three objects a foot to the south-west, roughly at right angles to the expected line of approach. The result was as I had hoped for and expected, and yet I could not help being surprised as well as delighted: each wasp missed her own nest, and alighted at exactly the spot where the nest 'ought' to be according to the landmarks' new positions! I could vary my tests by very cautiously shooing the wasp away, then moving the beacons a foot in another direction, and allowing the wasp to alight again. In whatever position I put the beacons, the wasp would follow them. At the end of such a series of tests I replaced the landmarks in their original position, and this finally enabled the wasp to return to her home. Thus the tests always had a happy ending—for both of us. This was no pure altruism on my part—I could now use the wasp for another test if I wished.

When engaged in such work, it is always worth observing oneself as well as the animals, and to do it as critically and as detachedly as possible—which, of course, is a tall order. I have often wondered why the outcome of such a test delighted me so much. A rationalist would probably like to assume that it was the increased predictability resulting from the test. This was a factor of considerable importance, I am sure. But a more important factor still (not only to me, but to many other people I have watched in this situation) is of a less dignified type: people enjoy, they relish the satisfaction of their desire for power. The truth of this was obvious, for instance, in people who enjoyed seeing the wasps being misled without caring much for the intellectual question whether they used landmarks or not. I am further convinced that even the joy of gaining insight was not often very pure either; it was mixed with pride at having had success with the tests.

To return to the wasps: next I tried to make the wasps use landmarks which I provided. This was not only for the purpose of satisfying my lust for power, but also for nobler purposes, as I hope to show later. Since changing the environment while the wasp was away disturbed her upon her return and even might prevent her from finding her nest altogether, I waited until a wasp had gone down into her nest, and then put my own landmarks round the entrance—sixteen Pine cones arranged in a circle of about eight inches diameter.

The first wasp to emerge was a little upset, and made a rather long locality study. On her return home, she hesitated for some time, but eventually alighted at the nest. When next she went out she made a really thorough locality study, and from then on everything went smoothly.

Figure 9. A homing test. Philanthus returns to the displaced Pine cones and fails to find her burrow.

Other wasps behaved in much the same way, and next day regular work was going on at five burrows so treated. I now subjected all five wasps, one by one, to a displacement test similar to those already described. The results, however, were not clear-cut. Some wasps, upon returning, followed the cones; but others were not fooled, and went straight home, completely ignoring my beacons. Others again seemed to be unable to make up their minds, and oscillated between the real nest and the ring of cones. This half-hearted behaviour did not disturb me, however, for if my idea was correct—that the wasps use landmarks—one would rather expect that my tests put the wasps in a kind of conflict situation: the natural landmarks which they must have been using before I gave them the Pine cones were still in their original position; only the cones had been moved. And while the cones were very conspicuous landmarks, they had been there for no more than one day. I therefore put all the cone-rings back and waited for two more days before testing the wasps again. And sure enough, this time the tests gave a hundred per cent preference for the Pine cones; I had made the wasps train themselves to my landmarks.

Robert Trivers

from SOCIAL EVOLUTION

■ If Mayr and Dobzhansky, Fisher and Haldane gave us (what Julian Huxley called) the Modern Synthesis of neo-Darwinism in the 1930s and 40s, was there a Postmodern Synthesis in the period that followed? Well yes, there was, but unfortunately we can't use the word because it has been debased in literary and social science circles by a pretentious school of *haute* francophonyism that has become grotesquely influential. The postscript to the Modern Synthesis could be said to have begun in the 1960s with George C. Williams and W. D. Hamilton, but it really hit its stride in the 1970s. The bright shooting star of that period was the young Robert Trivers. He introduced (what E. O. Wilson called) the 'unsentimental calculus' of parental investment theory, and nowhere was it more unsentimental than in Trivers's theory of parent offspring conflict.

Trivers thinks about evolutionary strategy in economic terms. His concept of parental investment is defined as an 'opportunity cost': the amount invested by a parent in any one offspring (food, time, risk, etc.) is measured as the lost opportunity to invest in other offspring. One remarkable feature of the definition is that superficially very different kinds of investment (food, time, risk) can all be measured in the same currency, the currency of lost alternative offspring. The definition immediately predisposes us to think of sibling rivalry, but Trivers went on to point out that it also leads to conflict between parents and offspring: because parents tend to value all their offspring equally, while each offspring values himself more than his siblings. I remember when I first read Trivers's definition of parental investment, how the idea instantly hit me with its power: Yes! I think I grasped, for the first time, some of the fascination of economic thinking; and the idea of parental investment—foreshadowed to some extent by R. A. Fisher—is immensely influential in evolutionary theory. I wanted to include Trivers's own exposition of parent offspring conflict in this anthology, but I couldn't find a brief extract that did the subject justice. Instead, I have chosen a different passage of Trivers, where he explains the alluring and pernicious fallacy of 'group selection'. ■

The Group Selection Fallacy

After 1859 there were three main ways in which people tried to blunt the force of Darwin's argument for evolution and natural selection. The first was to raise doubts about whether evolution had occurred, incidentally distracting attention from the more important concept, natural selection, the force directing evolution. The second was to acknowledge natural selection but to minimize its significance by making it appear to be very weak in its effect. In a long-lived species such as humans, death and reproduction are fairly rare events, so it is easy to imagine that in day-to-day life, most of what we do has very little effect on our survival or reproduction. But...even apparently trivial traits such as birth order in piglets may be subject to strong selection.

The third approach was to replace the idea that natural selection acts on the individual with the notion that natural selection acts for the benefit of the group or the species. This fallacy has been so widespread and so powerful in its repercussions that it deserves special treatment. Indeed, I remember well the grip it had on my own mind—and the confusion it generated—while I was still an undergraduate. So many disciplines conceptualized the human condition in terms of individual versus society. Sociology and anthropology seemed to claim that the larger unit was the key to understanding the smaller one. Societies, groups, species— all evolved mechanisms by which individuals are merely unconscious tools in their larger designs. In the extreme position, the larger groups were imagined to have the cohesiveness and interconnectedness usually associated with individual organisms. We call this the 'group selection fallacy' or 'species benefit fallacy', which claims that selection has operated at a higher level than the individual, that is, at the level of the group or species, favoring traits that allow these larger units to survive. In this chapter we review some examples of species-benefit reasoning in biology, and describe their flaws.

SPECIES-ADVANTAGE REASONING WITHIN BIOLOGY

Darwin was very clear on the idea that natural selection favors traits that benefit the individual possessing them but that are not necessarily beneficial for larger groups, such as the species itself. Indeed, on several occasions he explicitly rejected species-advantage reasoning. For example, Darwin gathered evidence showing that in many species the two sexes are produced in a roughly 50:50 ratio, yet he could not see how natural selection might affect the sex ratio. He concluded:

> I formerly thought that when a tendency to produce the two sexes in equal numbers was advantageous for the species, it would follow from natural selection, but I now see that the whole problem is so intricate that it is safer to leave its solution for the future.

That is, when Darwin could not solve the problem in terms of natural selection, he held his peace and did not invent a higher-level explanation. Fisher solved the problem in 1930.

Darwin regarded his work as a clear break with past biology, which believed in an instantaneous creation of an unchanging world whose various parts functioned together like so many parts of a clock. In this view it was easy to imagine that individuals could be created to subserve the interest of the species or, indeed, the interest of other species. Darwin's concept required that life be created over long periods of time by a natural process based on *individual* differences in reproductive success.

We might imagine that Darwin's successors clearly grasped his concept and began to apply it systematically to all biological phenomena, but quite the opposite seems to have happened. After a brief period, biologists returned en masse to the species-advantage view, only to cite Darwin as their support! For example, since it was then clear that numerous species had become extinct, it was easy to imagine that natural selection refers not to differential individual success but to differential success of species; that is, natural selection favors traits that permit species to survive. It certainly seems true that extinction is a selection process; that is, species that become extinct are not a random set of species available at the time. But this selection does not explain the traits of the species that do survive. To explain the traits of the species that do survive, we must understand how natural selection acts *within* each species.

Thus, for over 100 years after Darwin, most biologists believed that natural selection favors traits that are good for the species. In retrospect, there seem to be at least three reasons for this curious development. First, the existence of altruistic traits in nature seemed to require the concept of species advantage. As we have already seen, this is an illusion. Most examples of altruism can easily be explained by some benefit to kin or return benefit to the altruist, but these explanations were not well developed until the 1960's and '70's.

The second reason for species-advantage reasoning is that biologists have mostly studied non-social traits and for these traits it matters little whether we imagine they evolve for the benefit of the species or for the benefit of those possessing the traits. For example, the human kneecap locks in place when we are standing upright, thereby saving us energy. Chimpanzees and gorillas lack such a locking device. The device evolved because it benefitted those possessing it, not because it helped the species itself survive, but the latter notion leads to no misunderstanding of how the kneecap operates, since we assume the kneecap benefits the species by benefitting those possessing it. By contrast, social traits immediately pose a problem, because a benefit may be conferred on one individual at a cost to another. Considering the benefit of the species, we may imagine that social traits will evolve as long as the net effect on everyone is positive. But—in the absence of kinship or return effects—selection on the actor favors selfish traits no matter how large the cost that is inflicted and never favors altruistic acts no matter how great the benefit conferred.

Finally, the early application of Darwinian thinking to human social problems frightened people back to thinking on the level of the species. Shortly after Darwin's work, 'Social Darwinists' argued in favor of 19th-century capitalism in the following sort of way: The poor are less fit than the rich because they have already lost out in competition for resources (that is, because they are poor). Therefore, the rich should do nothing to ameliorate the condition of the poor since this would interfere with natural selection and, therefore, with nature's plan. Indeed, we can improve on nature's plan by actively selecting against the interests of the poor. Thus, for the good of the species, the poor should suffer their poverty, augmented by a biological prejudice against them!

But, in fact, fitness refers to reproductive success, or the production of surviving offspring. It can only be demonstrated after the fact. In principle, we cannot look at two people and say which is more fit until both are dead and their surviving offspring have been counted. If the poor leave more surviving offspring than do the rich, as is sometimes true, then they are, by evolutionary definition, more fit, and the whole argument can be stood on its head. In reaction to this kind of thinking, I believe, people returned to species-advantage reasoning, partly because it was incapable of saying much about social interactions *within* species, especially social conflict.

Alister Hardy

from THE OPEN SEA

■ The oceans cover more than 70 per cent of our planet's surface. The majority of the Sun's photons that are available for photosynthesis fall in the sea, where, in the green cells of the phytoplankton, they drive chemical reactions 'uphill' (thermodynamically speaking) and synthesize carbon compounds that later fuel the ecosystems. Nobody had a better feel for the great rolling pastures, sunlit green meadows and waving prairies of *The Open Sea* than Alister Hardy, my first professor. His paintings for that book still adorn the corridors of the Oxford Zoology Department, and the images seem to dance with enthusiasm, just as the old man himself danced boyishly around the lecture hall, a strabismically beaming cross between Peter Pan and the Ancient Mariner. Yea, slimy things did crawl with legs upon the slimy sea—and across the blackboard in coloured chalk with Sir Alister bobbing and weaving in pursuit. In this extract, he lights up the page with his description of the remarkable phenomenon of marine phosphorescence. ■

Some of the general phosphorescence of the sea may possibly be caused at times by bacteria, but it is usually due to vast numbers of little flagellates. The Dinoflagellates, such as *Ceratium* and *Peridinium* and the aberrant globular form *Noctiluca*, give rise to the most brilliant displays of this general lighting up of the sea. If you like fireworks it is always an entertaining experience to take a rowboat out on a dark night when some of these little flagellates are really abundant, as they often are in August and September. Every time the oar touches the sea there is a splash of flame, and as it is drawn through the water it leaves a trail of fire behind it—as does the boat itself. Let Charles Darwin give his account of such a night when on his famous voyage of the *Beagle*; it is an entry in his journal under the date of 6 December 1833:

> While sailing a little south of the Plata on one very dark night, the sea presented a wonderful and most beautiful spectacle. There was a fresh breeze, and every part of the surface, which during the day is seen as foam, now glowed with a pale light. The vessel drove before her bows two billows of liquid phosphorus, and in her wake she was followed by a milky train. As far as the eye reached, the crest of every wave was bright, and the sky above the horizon, from the reflected glare of these livid flames, was not so utterly obscure as over the vault of the heavens.

There is, of course, no need to voyage across the world to see such displays—sometimes they may be equally brilliant in our own seas. I have already described how I once saw every fish in a small shoal outlined in 'fire' and it is not at all rare to see, especially in late summer, every wave breaking on the beach with a flash of pale greenish light. The little flagellates flash with light whenever they are violently agitated. Mr George Atkinson of Lowestoft recently told me of an interesting occurrence during the first world war. A zeppelin dropped some bombs which exploded in the sea a mile or two from land; after each explosion there was a flash of phosphorescence through the sea along the shore on which he was standing.

Among the coelenterates many of the small hydroid medusae are said to be luminous, and among the larger jellyfish there is a very striking example in *Pelagia noctiluca* already referred to.

It is the comb-jellies—the Ctenophora—which give us some of the most spectacular displays of brilliant flashing light in our waters.... They are nearly all capable of emitting sudden vivid flashes. The sea is often full of very small young specimens, each of which may give off quite a bright flash. They are excellent animals to use for demonstrations of spontaneous luminescence. A plankton sample containing these animals can nearly always be relied upon to give a good show—but we must remember that they do not perform at all until they have been in the dark for almost twenty minutes. If you intend to show your friends a good display you must keep your sample of plankton completely covered with light-proof cloth, or in a light-proof cupboard, for this length of time before bringing it out for exhibition in the darkened room.

As a young student I once had an amusing demonstration of this inhibitory effect of light. I had gone over to Brightlingsea to hunt at low tide in the thick Essex mud for the rare and curious worm-like animal *Priapulus*. It was nearly dark before I had found any and it was too late to return to Oxford that night, so I put up at a very old inn where I slept in a four-poster bed in an oak-panelled room. After a strenuous day digging in the mud I retired early and soon dropped to sleep after blowing out my candle. Later in the night I was awakened by some reveller coming noisily to bed in the room next door. I opened my eyes and blinked them with astonishment, for a number of little blue lights were bobbing about in the darkness just over the end of my bed. It was as if there were a lot of little goblins dancing up and down in the air. Before coming to bed I had of course celebrated the finding of *Priapulus*—but only with a pint of bitter; clearly there must be some more objective explanation! I struck a match and lit the candle. I now saw that, level with the end of my bed, was the top of the chimney-piece on which I had placed a row of large glass jars filled with sea-water, with a little mud at the bottom of each containing my precious animals. Getting up and switching on the electric light I examined them closely and then saw that the water was full of very young ctenophores—*Pleurobrachia*, I think—actively swimming up and down. They had certainly not been flashing when I first turned out the light and got into bed; nor were there any flashes when I settled into bed for the second time—or rather not at once. I was now well awake, and it was some time before I could get off to sleep again; before I did so, after about twenty minutes in the dark, the little 'blue devils' began their dance again.

Rachel Carson

from THE SEA AROUND US

■ The poetry of the sea continues with Rachel Carson, the first great prophet of environmental doom. Jim Watson, in his memoir *Avoid Boring People*, another part of which is anthologized here, gives a fascinating reminiscence about the hostility that Rachel Carson aroused when *Silent Spring* was first published. She was invited to testify to a government advisory panel of which Watson was a member. She turned out to be

> ...perfectly even-tempered...giving no indication of the nutty hysterical naturalist that agricultural and chemical lobbyists had portrayed her to be. The chemical giant Monsanto had distributed five thousand copies of a brochure parodying *Silent Spring* entitled 'The Desolate Years,' describing a pesticide-free world devastated by famine, disease, and insects. The attack was mirrored in *Time Magazine*'s review of *Silent Spring* deploring Carson's oversimplification and downright inaccuracy.

Watson, belying his reputation as a scientistic zealot, strongly supported Carson along with the rest of the committee, and he now reaffirms his relief that President Kennedy accepted their report, with its strong final sentence paying tribute to Carson for alerting the public to the problem of pesticides.

Rachel Carson's concern for the future of the living planet was at least partly driven by a deeply felt love of it. Her own specialism, like Alister Hardy's, was marine biology, and before *Silent Spring* she devoted her considerable literary talent to books about the sea. This extract is from her bestseller, *The Sea Around Us*. ■

The Changing Year

For the sea as a whole, the alternation of day and night, the passage of the seasons, the procession of the years, are lost in its vastness, obliterated in its own changeless eternity. But the surface waters are different. The face of the sea is always changing. Crossed by colors, lights, and moving shadows, sparkling in the sun, mysterious in the twilight, its aspects and its moods vary hour by hour. The surface waters move with the tides, stir to the breath of the winds, and rise and fall to the endless, hurrying forms of the waves. Most of all, they change with the advance of the seasons. Spring moves over the temperate lands of our Northern Hemisphere in a tide of new life, of pushing green shoots and unfolding buds, all its mysteries and meanings symbolized in the northward migration of the birds, the awakening of sluggish amphibian life as the chorus of frogs rises again from the wet lands, the different sound of the wind stirs the young leaves where a month ago it rattled the bare branches. These things we associate with the land, and it is easy to suppose that at sea there could be no such feeling of advancing spring. But the signs are there, and seen with understanding eye, they bring the same magical sense of awakening.

In the sea, as on land, spring is a time for the renewal of life. During the long months of winter in the temperate zones the surface waters have been absorbing the cold. Now the heavy water begins to sink, slipping down and displacing the warmer layers below. Rich stores of minerals have been accumulating on the floor of the continental shelf—some freighted down the rivers from the lands; some derived from sea creatures that have died and whose remains have drifted down to the bottom; some from the shells that once encased a diatom, the streaming protoplasm of a radiolarian, or the transparent tissues of a pteropod. Nothing is wasted in the sea; every particle of material is used over and over again, first by one creature, then by another. And when in spring the waters are deeply stirred, the warm bottom water brings to the surface a rich supply of minerals, ready for use by new forms of life.

Just as land plants depend on minerals in the soil for their growth, every marine plant, even the smallest, is dependent upon the nutrient salts or minerals in the sea water. Diatoms must have silica, the element of which their fragile shells are fashioned. For these and all other micro-plants, phosphorus is an indispensable mineral. Some of these elements are in short supply and in winter may be reduced below the minimum necessary for growth. The diatom population must tide itself over this

season as best it can. It faces a stark problem of survival, with no opportunity to increase, a problem of keeping alive the spark of life by forming
tough protective spores against the stringency of winter, a matter of
existing in a dormant state in which no demands shall be made on an
environment that already withholds all but the most meagre necessities
of life. So the diatoms hold their place in the winter sea, like seeds of
wheat in a field under snow and ice, the seeds from which the spring
growth will come.

These, then, are the elements of the vernal blooming of the sea: the
'seeds' of the dormant plants, the fertilizing chemicals, the warmth of
the spring sun.

In a sudden awakening, incredible in its swiftness, the simplest plants
of the sea begin to multiply. Their increase is of astronomical proportions. The spring sea belongs at first to the diatoms and to all the other
microscopic plant life of the plankton. In the fierce intensity of their
growth they cover vast areas of ocean with a living blanket of their
cells. Mile after mile of water may appear red or brown or green, the
whole surface taking on the color of the infinitesimal grains of pigment
contained in each of the plant cells.

The plants have undisputed sway in the sea for only a short time.
Almost at once their own burst of multiplication is matched by a similar increase in the small animals of the plankton. It is the spawning
time of the copepod and the glassworm, the pelagic shrimp and the
winged snail. Hungry swarms of these little beasts of the plankton roam
through the waters, feeding on the abundant plants and themselves
falling prey to larger creatures. Now in the spring the surface waters
become a vast nursery. From the hills and valleys of the continent's
edge lying far below, and from the scattered shoals and banks, the eggs
or young of many of the bottom animals rise to the surface of the sea.
Even those which, in their maturity, will sink down to a sedentary life
on the bottom, spend the first weeks of life as freely swimming hunters
of the plankton. So as spring progresses new batches of larvae rise
into the surface each day, the young of fishes and crabs and mussels
and tube worms, mingling for a time with the regular members of the
plankton.

Under the steady and voracious grazing, the grasslands of the surface are soon depleted. The diatoms become more and more scarce, and with them the other simple plants. Still there are brief explosions of one or another form, when in a sudden orgy of cell division it comes to claim whole areas of the sea for its own. So, for a time each spring, the waters may become blotched with brown, jelly-like masses, and the fishermen's nets come up dripping a brown slime and containing no fish, for the herring have turned away from these waters as though in loathing of the viscid, foul-smelling algae. But in less time than passes between the full moon and the new, the spring flowering of Phaeocystis is past and the waters have cleared again.

In the spring the sea is filled with migrating fishes, some of them bound for the mouths of great rivers, which they will ascend to deposit their spawn. Such are the spring-run chinooks coming in from the deep Pacific feeding grounds to breast the rolling flood of the Columbia, the shad moving into the Chesapeake and the Hudson and the Connecticut, the alewives seeking a hundred coastal streams of New England, the salmon feeling their way to the Penobscot and the Kennebec. For months or years these fish have known only the vast spaces of the ocean. Now the spring sea and the maturing of their own bodies lead them back to the rivers of their birth.

Other mysterious comings and goings are linked with the advance of the year. Capelin gather in the deep, cold water of the Barents Sea, their shoals followed and preyed upon by flocks of auks, fulmars, and kittiwakes. Cod approach the banks of Lofoten, and gather off the shores of Ireland. Birds whose winter feeding territory may have encompassed the whole Atlantic or the whole Pacific converge upon some small island, the entire breeding population arriving within the space of a few days. Whales suddenly appear off the slopes of the coastal banks where the swarms of shrimplike krill are spawning, the whales having come from no one knows where, by no one knows what route.

With the subsiding of the diatoms and the completed spawning of many of the plankton animals and most of the fish, life in the surface waters slackens to the slower pace of midsummer. Along the meeting places of the currents the pale moon jelly Aurelia gathers in thousands, forming sinuous lines or windrows across miles of sea, and the birds see their pale forms shimmering deep down in the green water. By mid-summer the large red jellyfish Cyanea may have grown from the size of a thimble to that of an umbrella. The great jellyfish moves through the sea with rhythmic pulsations, trailing long tentacles and as likely as not shepherding a little group of young cod or haddock, which find shelter under its bell and travel with it.

A hard, brilliant, coruscating phosphorescence often illuminates the summer sea. In waters where the protozoa Noctiluca is abundant it is the chief source of this summer luminescence, causing fishes, squids, or dolphins to fill the water with racing flames and to clothe themselves in a ghostly radiance. Or again the summer sea may glitter with a thousand thousand moving pinpricks of light, like an immense swarm of fireflies moving through a dark wood. Such an effect is produced by a shoal of the brilliantly phosphorescent shrimp Meganyctiphanes, a creature of cold and darkness and of the places where icy water rolls upward from the depths and bubbles with white ripplings at the surface.

Out over the plankton meadows of the North Atlantic the dry twitter of the phalaropes, small brown birds, wheeling and turning, dipping and rising, is heard for the first time since early spring. The phalaropes have nested on the arctic tundras, reared their young, and now the first of them are returning to the sea. Most of them will continue south over the open water far from land, crossing the equator into the South Atlantic. Here they will follow where the great whales lead, for where the whales are, there also are the swarms of plankton on which these strange little birds grow fat.

As the fall advances, there are other movements, some in the surface, some hidden in the green depths, that betoken the end of summer. In the fog-covered waters of Bering Sea, down through the treacherous passes between the islands of the Aleutian chain and southward into the open Pacific, the herds of fur seals are moving. Left behind are two small islands, treeless bits of volcanic soil thrust up into the waters of Bering Sea. The islands are silent now, but for the several months of summer they resounded with the roar of millions of seals come ashore to bear and rear their young—all the fur seals of the eastern Pacific crowded into a few square miles of bare rock and crumbling soil. Now once more the seals turn south, to roam down along the sheer underwater cliffs of the continent's edge, where the rocky foundations fall away steeply into the deep sea. Here, in a blackness more absolute than that of arctic winter, the seals will find rich feeding as they swim down to prey on the fishes of this region of darkness.

Autumn comes to the sea with a fresh blaze of phosphorescence, when every wave crest is aflame. Here and there the whole surface may glow with sheets of cold fire, while below schools of fish pour through the water like molten metal. Often the autumnal phosphorescence is caused by a fall flowering of the dinoflagellates, multiplying furiously in a short-lived repetition of their vernal blooming.

Sometimes the meaning of the glowing water is ominous. Off the Pacific coast of North America, it may mean that the sea is filled with the dinoflagellate Gonyaulax, a minute plant that contains a poison of strange and terrible virulence. About four days after Gonyaulax comes to dominate the coastal plankton, some of the fishes and shellfish in the vicinity become toxic. This is because, in their normal feeding, they have strained the poisonous plankton out of the water. Mussels accumulate the Gonyaulax toxins in their livers, and the toxins react on the human nervous system with an effect similar to that of strychnine. Because of these facts, it is generally understood along the Pacific coast that it is unwise to eat shellfish taken from coasts exposed to the open sea when Gonyaulax may be abundant, in summer or early fall. For generations before the white men came, the Indians knew this. As soon as the red streaks appeared in the sea and the waves began to flicker at night with the mysterious blue-green fires, the tribal leaders forbade the taking of mussels until these warning signals should have passed. They even set guards at intervals along the beaches to warn inlanders who might come down for shellfish and be unable to read the language of the sea.

But usually the blaze and glitter of the sea, whatever its meaning for those who produce it, implies no menace to man. Seen from the deck of a vessel in open ocean, a tiny, man-made observation point in the vast world of sea and sky, it has an eerie and unearthly quality. Man, in his vanity, subconsciously attributes a human origin to any light not of moon or stars or sun. Lights on the shore, lights moving over the water, mean lights kindled and controlled by other men, serving purposes understandable to the human mind. Yet here are lights that flash and fade away, lights that come and go for reasons meaningless to man, lights that have been doing this very thing over the eons of time in which there were no men to stir in vague disquiet.

[...]

Like the blazing colors of the autumn leaves before they wither and fall, the autumnal phosphorescence betokens the approach of winter. After their brief renewal of life the flagellates and the other minute algae dwindle away to a scattered few; so do the shrimps and the cope-pods, the glassworms and the comb jellies. The larvae of the bottom fauna have long since completed their development and drifted away to take up whatever existence is their lot. Even the roving fish schools have deserted the surface waters and have migrated into warmer latitudes or have found equivalent warmth in the deep, quiet waters along the edge of the continental shelf. There the torpor of semi-hibernation descends upon them and will possess them during the months of winter.

The surface waters now become the plaything of the winter gales. As the winds build up the giant storm waves and roar along their crests, lashing the water into foam and flying spray, it seems that life must forever have deserted this place.

For the mood of the winter sea, read Joseph Conrad's description:

> The greyness of the whole immense surface, the wind furrows upon the faces of the waves, the great masses of foam, tossed about and waving, like matted white locks, give to the sea in a gale an appearance of hoary age, lustreless, dull, without gleams, as though it had been created before light itself.[1]

But the symbols of hope are not lacking even in the grayness and bleakness of the winter sea. On land we know that the apparent lifelessness of winter is an illusion. Look closely at the bare branches of a tree, on which not the palest gleam of green can be discerned. Yet, spaced along each branch are the leaf buds, all the spring's magic of swelling green concealed and safely preserved under the insulating, overlapping layers. Pick off a piece of the rough bark of the trunk; there you will find hibernating insects. Dig down through the snow into the earth. There are the eggs of next summer's grasshoppers; there are the dormant seeds from which will come the grass, the herb, the oak tree.

So, too, the lifelessness, the hopelessness, the despair of the winter sea are an illusion. Everywhere are the assurances that the cycle has come to the full, containing the means of its own renewal. There is the promise of a new spring in the very iciness of the winter sea, in the chilling of the water, which must, before many weeks, become so heavy that it will plunge downward, precipitating the overturn that is the first act in the drama of spring. There is the promise of new life in the small plantlike things that cling to the rocks of the underlying bottom, the almost formless polyps from which, in spring, a new generation of jellyfish will bud off and rise into the surface waters. There is unconscious purpose in the sluggish forms of the copepods hibernating on the bottom, safe from the surface storms, life sustained in their tiny bodies by the extra store of fat with which they went into this winter sleep.

Already, from the gray shapes of cod that have moved, unseen by man, through the cold sea to their spawning places, the glassy globules of eggs are rising into the surface waters. Even in the harsh world of the winter sea, these eggs will begin the swift divisions by which a granule of protoplasm becomes a living fish-let.

Most of all, perhaps, there is assurance in the fine dust of life that remains in the surface waters, the invisible spores of the diatoms, needing only the touch of warming sun and fertilizing chemicals to repeat the magic of spring.

1. From *The Mirror of the Sea*, Kent edition, 1925, Doubleday-Page, p. 71.

Loren Eiseley

from 'HOW FLOWERS CHANGED THE WORLD'

■ Rachel Carson's exact contemporary Loren Eiseley was another American scientist with a flair for lyrical writing. His style derives its poetry from the science itself, and from the author's scientifically informed imagination. This passage from an essay in *The Immense Journey* encourages the reader's poetic response to a watershed event in the history of the world, the rise of the flowering plants. All poets know that flowers are beautiful, but not all poets get the important point—also articulated by the physicist Richard Feynman:

> The beauty that is there for you is also available for me, too. But I see a deeper beauty that isn't so readily available to others. I can see the complicated interactions of the flower. The color of the flower is red. Does the fact that the plant has color mean that it evolved to attract insects? This adds a further question. Can insects see color? Do they have an aesthetic sense? And so on. I don't see how studying a flower ever detracts from its beauty. It only adds.

Feynman here spoke for all scientists (though, alas, most insect eyes don't see red), but Loren Eiseley perhaps expressed it better. ■

When the first simple flower bloomed on some raw upland late in the Dinosaur Age, it was wind pollinated, just like its early pine-cone relatives. It was a very inconspicuous flower because it had not yet evolved the idea of using the surer attraction of birds and insects to achieve the transportation of pollen. It sowed its own pollen and received the pollen of other flowers by the simple vagaries of the wind. Many plants in regions where insect life is scant still follow this principle today. Nevertheless, the true flower—and the seed that it produced—was a profound innovation in the world of life.

In a way, this event parallels, in the plant world, what happened among animals. Consider the relative chance for survival of the exteriorly deposited egg of a fish in contrast with the fertilized egg of a mammal, carefully retained for months in the mother's body until the young animal (or human being) is developed to a point where it may survive. The biological wastage is less—and so it is with the flowering plants. The primitive spore, a single cell fertilized in the beginning by a swimming sperm, did not promote rapid distribution, and the young plant, moreover, had to struggle up from nothing. No one had left it any food except what it could get by its own unaided efforts.

By contrast, the true flowering plants (angiosperm itself means 'encased seed') grew a seed in the heart of a flower, a seed whose development was initiated by a fertilizing pollen grain independent of outside moisture. But the seed, unlike the developing spore, is already a fully equipped *embryonic plant* packed in a little enclosed box stuffed full of nutritious food. Moreover, by featherdown attachments, as in dandelion or milkweed seed, it can be wafted upward on gusts and ride the wind for miles; or with hooks it can cling to a bear's or a rabbit's hide; or like some of the berries, it can be covered with a juicy, attractive fruit to lure birds, pass undigested through their intestinal tracts and be voided miles away.

The ramifications of this biological invention were endless. Plants traveled as they had never traveled before. They got into strange environments heretofore never entered by the old spore plants or stiff pine-cone-seed plants. The well-fed, carefully cherished little embryos raised their heads everywhere. Many of the older plants with more primitive reproductive mechanisms began to fade away under this unequal contest. They contracted their range into secluded environments. Some, like the giant redwoods, lingered on as relics; many vanished entirely.

The world of the giants was a dying world. These fantastic little seeds skipping and hopping and flying about the woods and valleys brought with them an amazing adaptability. If our whole lives had not been spent in the midst of it, it would astound us. The old, stiff, sky-reaching wooden world had changed into something that glowed here and there with strange colors, put out queer, unheard-of fruits and little intricately carved seed cases, and, most important of all, produced concentrated foods in a way that the land had never seen before, or dreamed of back in the fish-eating, leaf-crunching days of the dinosaurs.

That food came from three sources, all produced by the reproductive system of the flowering plants. There were the tantalizing nectars and pollens intended to draw insects for pollenizing purposes, and which are responsible also for that wonderful jeweled creation, the humming-bird. There were the juicy and enticing fruits to attract larger animals, and in which tough-coated seeds were concealed, as in the tomato, for example. Then, as if this were not enough, there was the food in the actual seed itself, the food intended to nourish the embryo. All over the world, like hot corn in a popper, these incredible elaborations of the flowering plants kept exploding. In a movement that was almost instantaneous, geologically speaking, the angiosperms had taken over the world. Grass was beginning to cover the bare earth until, today, there are over six thousand species. All kinds of vines and bushes squirmed and writhed under new trees with flying seeds.

The explosion was having its effect on animal life also. Special-ized groups of insects were arising to feed on the new sources of food and, incidentally and unknowingly, to pollinate the plant. The flowers bloomed and bloomed in ever larger and more spectacular varieties. Some were pale unearthly night flowers intended to lure moths in the evening twilight, some among the orchids even took the shape of female spiders in order to attract wandering males, some flamed redly in the light of noon or twinkled modestly in the meadow grasses. Intricate mechanisms splashed pollen on the breasts of humming-birds, or stamped it on the bellies of black, grumbling bees droning assiduously from blossom to blossom. Honey ran, insects multiplied, and even the descendants of that toothed and ancient lizard-bird had become strangely altered. Equipped with prodding beaks instead of biting teeth they pecked the seeds and gobbled the insects that were really converted nectar.

Across the planet grasslands were now spreading. A slow continental upthrust which had been a part of the early Age of Flowers had cooled the world's climates. The stalking reptiles and the leather-winged black imps of the seashore cliffs had vanished. Only birds roamed the air now, hot-blooded and high-speed metabolic machines.

The mammals, too, had survived and were venturing into new do-mains, staring about perhaps a bit bewildered at their sudden eminence now that the thunder lizards were gone, Many of them, beginning as small browsers upon leaves in the forest, began to venture out upon this new sunlit world of the grass. Grass has a high silica content and demands a new type of very tough and resistant tooth enamel, but the seeds taken incidentally in the cropping of the grass are highly nutritious. A new world had opened out for the warm-blooded mammals. Great herbivores like the mammoths, horses and bisons appeared. Skulking about them had arisen savage flesh-feeding carnivores like the now extinct dire wolves and the saber-toothed tiger.

Flesh eaters though these creatures were, they were being sustained on nutritious grasses one step removed. Their fierce energy was being maintained on a high, effective level, through hot days and frosty nights, by the concentrated energy of the angiosperms. That energy, thirty per cent or more of the weight of the entire plant among some of the cereal grasses, was being accumulated and concentrated in the rich proteins and fats of the enormous game herds of the grasslands.

On the edge of the forest, a strange, old-fashioned animal still hesitated. His body was the body of a tree dweller, and though tough and knotty by human standards, he was, in terms of that world into which he gazed, a weakling. His teeth, though strong for chewing on the tough fruits of the forest, or for crunching an occasional unwary bird caught with his prehensile hands, were not the tearing sabers of the great cats. He had a passion for lifting himself up to see about, in his restless, roving curiosity. He would run a little stiffly and uncertainly, perhaps, on his hind legs, but only in those rare moments when he ventured out upon the ground. All this was the legacy of his climbing days; he had a hand with flexible fingers and no fine specialized hoofs upon which to gallop like the wind.

If he had any idea of competing in that new world, he had better forget it; teeth or hooves, he was much too late for either. He was a ne'er-do-well, an in-betweener. Nature had not done well by him. It was as if she had hesitated and never quite made up her mind. Perhaps as a consequence he had a malicious gleam in his eye, the gleam of an outcast who has been left nothing and knows he is going to have to take what he gets. One day a little band of these odd apes—for apes they were—shambled out upon the grass; the human story had begun.

Apes were to become men, in the inscrutable wisdom of nature, because flowers had produced seeds and fruits in such tremendous quantities that a new and totally different store of energy had become available in concentrated form. Impressive as the slow-moving, dim-brained dinosaurs had been, it is doubtful if their age had supported anything like the diversity of life that now rioted across the planet or flashed in and out among the trees. Down on the grass by a streamside, one of those apes with inquisitive fingers turned over a stone and hefted it vaguely. The group clucked together in a throaty tongue and moved off through the tall grass foraging for seeds and insects. The one still held, sniffed, and hefted the stone he had found. He liked the feel of it in his fingers. The attack on the animal world was about to begin.

If one could run the story of that first human group like a speeded-up motion picture through a million years of time, one might see the stone in the hand change to the flint ax and the torch. All that swarming grass-land world with its giant bison and trumpeting mammoths would go down in ruin to feed the insatiable and growing numbers of a carnivore who, like the great cats before him, was taking his energy indirectly from the grass. Later he found fire and it altered the tough meats and drained their energy even faster into a stomach ill adapted for the ferocious turn man's habits had taken.

His limbs grew longer, he strode more purposefully over the grass. The stolen energy that would take man across the continents would fail him at last. The great Ice Age herds were destined to vanish. When they did so, another hand like the hand that grasped the stone by the river long ago would pluck a handful of grass seed and hold it contemplatively.

In that moment, the golden towers of man, his swarming millions, his turning wheels, the vast learning of his packed libraries, would glimmer dimly there in the ancestor of wheat, a few seeds held in a muddy hand. Without the gift of flowers and the infinite diversity of their fruits, man and bird, if they had continued to exist at all, would be today unrecognizable. Archaeopteryx, the lizard-bird, might still be snapping at beetles on a sequoia limb; man might still be a nocturnal insectivore gnawing a roach in the dark. The weight of a petal has changed the face of the world and made it ours.

Edward O. Wilson

from THE DIVERSITY OF LIFE

■ Edward O. Wilson's prolific hard work led to his achieving the standing of America's most distinguished living naturalist, while still too young to be a grand old man. He is the leading authority on one of the natural economy's most important groups, the ants. He has made major contributions to entomology, physiology, ecology, taxonomy and biogeography. Withal, he is learned in anthropology, a deep and wise thinker on human affairs, and also, as it happens, a very fine writer, as these opening pages from *The Diversity of Life* show. ■

Storm Over the Amazon

In the amazon basin the greatest violence sometimes begins as a flicker of light beyond the horizon. There in the perfect bowl of the night sky, untouched by light from any human source, a thunderstorm sends its premonitory signal and begins a slow journey to the observer, who thinks: the world is about to change. And so it was one night at the edge of rain forest north of Manaus, where I sat in the dark, working my mind through the labyrinths of field biology and ambition, tired, bored, and ready for any chance distraction.

Each evening after dinner I carried a chair to a nearby clearing to escape the noise and stink of the camp I shared with Brazilian forest workers, a place called Fazenda Dimona. To the south most of the forest had been cut and burned to create pastures. In the daytime cattle browsed in remorseless heat bouncing off the yellow clay and at night animals and spirits edged out onto the ruined land. To the north the virgin rain forest began, one of the great surviving wildernesses of the world, stretching 500 kilometers before it broke apart and dwindled into gallery woodland among the savannas of Roraima.

Enclosed in darkness so complete I could not see beyond my out-stretched hand, I was forced to think of the rain forest as though I were seated in my library at home, with the lights turned low. The forest at night is an experience in sensory deprivation most of the time, black and silent as the midnight zone of a cave. Life is out there in expected abundance. The jungle teems, but in a manner mostly beyond the reach of the human senses. Ninety-nine percent of the animals find their way by chemical trails laid over the surface, puffs of odor released into the air or water, and scents diffused out of little hidden glands and into the air downwind. Animals are masters of this chemical channel, where we are idiots. But we are geniuses of the audiovisual channel, equaled in this modality only by a few odd groups (whales, monkeys, birds). So we wait for the dawn, while they wait for the fall of darkness; and because sight and sound are the evolutionary prerequisites of intelligence, we alone have come to reflect on such matters as Amazon nights and sensory modalities.

I swept the ground with the beam from my headlamp for signs of life, and found—diamonds! At regular intervals of several meters, intense pinpoints of white light winked on and off with each turn-ing of the lamp. They were reflections from the eyes of wolf spiders, members of the family Lycosidae, on the prowl for insect prey. When spotlighted the spiders froze, allowing me to approach on hands and knees and study them almost at their own level. I could distinguish a wide variety of species by size, color, and hairiness. It struck me how little is known about these creatures of the rain forest, and how deeply satisfying it would be to spend months, years, the rest of my life in this place until I knew all the species by name and every detail of their lives. From specimens beautifully frozen in amber we know that the Lycosidae have survived at least since the beginning of the Oligocene epoch, forty million years ago, and probably much longer. Today a riot of diverse forms occupy the whole world, of which this was only the minutest sample, yet even these species turning about now to watch me from the bare yellow clay could give meaning to the lifetimes of many naturalists.

The moon was down, and only starlight etched the tops of the trees. It was August in the dry season. The air had cooled enough to make the humidity pleasant, in the tropical manner, as much a state of mind as a physical sensation. The storm I guessed was about an hour away. I thought of walking back into the forest with my headlamp to hunt for new treasures, but was too tired from the day's work. Anchored again to my chair, forced into myself, I welcomed a meteor's streak and the occasional courtship flash of luminescent click beetles among the nearby but unseen shrubs. Even the passage of a jetliner 10,000 meters up, a regular event each night around ten o'clock, I awaited with pleasure. A week in the rain forest had transformed its distant rumble from an urban irritant into a comforting sign of the continuance of my own species.

But I was glad to be alone. The discipline of the dark envelope summoned fresh images from the forest of how real organisms look and act. I needed to concentrate for only a second and they came alive as eidetic images, behind closed eyelids, moving across fallen leaves and decaying humus. I sorted the memories this way and that in hope of stumbling on some pattern not obedient to abstract theory of textbooks. I would have been happy with *any* pattern. The best of science doesn't consist of mathematical models and experiments, as textbooks make it seem. Those come later. It springs fresh from a more primitive mode of thought, wherein the hunter's mind weaves ideas from old facts and fresh metaphors and the scrambled crazy images of things recently seen. To move forward is to concoct new patterns of thought, which in turn dictate the design of the models and experiments. Easy to say, difficult to achieve.

The subject fitfully engaged that night, the reason for this research trip to the Brazilian Amazon, had in fact become an obsession and, like all obsessions, very likely a dead end. It was the kind of favorite puzzle that keeps forcing its way back because its very intractability makes it perversely pleasant, like an overly familiar melody intruding into the relaxed mind because it loves you and will not leave you. I hoped that some new image might propel me past the jaded puzzle to the other side, to ideas strange and compelling.

Bear with me for a moment while I explain this bit of personal esoterica; I am approaching the subject of central interest. Some kinds of plants and animals are dominant, proliferating new species and spreading over large parts of the world. Others are driven back until they become rare and threatened by extinction. Is there a single formula for this biogeographic difference, for all kinds of organisms? The process, if articulated, would be a law or at least a principle of dynastic succession in evolution. I was intrigued by the circumstance that social insects, the group on which I have spent most of my life, are among the most abundant of all organisms. And among the social insects, the dominant subgroup is the ants. They range 20,000 or more species strong from the Arctic Circle to the tip of South America. In the Amazon rain forest they compose more than 10 per cent of the biomass of all animals. This means that if you were to collect, dry out, and weigh every animal in a piece of forest, from monkeys and birds down to mites and roundworms, at least 10 per cent would consist of these insects alone. Ants make up almost half of the insect biomass overall and 70 percent of the individual insects found in the treetops. They are only slightly less abundant in grasslands, deserts, and temperate forests throughout the rest of the world.

It seemed to me that night, as it has to others in varying degrees of persuasion many times before, that the prevalence of ants must have something to do with their advanced colonial organization. A colony is a superorganism; an assembly of workers so tightly knit around the mother queen as to act like a single, well-coordinated entity. A wasp or other solitary insect encountering a worker ant on its nest faces more than just another insect. It faces the worker and all her sisters, united by instinct to protect the queen, seize control of territory, and further the growth of the colony. Workers are little kamikazes, prepared—eager—to die in order to defend the nest or gain control of a food source. Their deaths matter no more to the colony than the loss of hair or a claw tip might to a solitary animal.

There is another way to look at an ant colony. Workers foraging around their nest are not merely insects searching for food. They are a living web cast out by the superorganism, ready to congeal over rich food finds or shrink back from the most formidable enemies. Superorganisms can control and dominate the ground and treetops in competition with ordinary, solitary organisms, and that is surely why ants live everywhere in such great numbers.

I heard around me the Greek chorus of training and caution: *How can you prove that is the reason for their dominance? Isn't the connection just another shaky conclusion that because two events occur together, one causes the other? Something else entirely different might have caused both. Think about it—greater individual fighting ability? Sharper senses? What?*

Such is the dilemma of evolutionary biology. We have problems to solve, we have clear answers—too many clear answers. The difficult part is picking out the right answer. The isolated mind moves in slow circles and breakouts are rare. Solitude is better for weeding out ideas than for creating them. Genius is the summed production of the many with the names of the few attached for easy recall, unfairly so to other scientists. My mind drifted into the hourless night, no port of call yet chosen.

The storm grew until sheet lightning spread across the western sky. The thunderhead reared up like a top-heavy monster in slow motion, tilted forward, blotting out the stars. The forest erupted in a simulation of violent life. Lightning bolts broke to the front and then closer, to the right and left, 10,000 volts dropping along an ionizing path at 800 kilometers an hour, kicking a countersurge skyward ten times faster, back and forth in a split second, the whole perceived as a single flash and crack of sound. The wind freshened, and rain came stalking through the forest.

In the midst of chaos something to the side caught my attention. The lightning bolts were acting like strobe flashes to illuminate the wall of the rain forest. At intervals I glimpsed the storied structure: top canopy 30 meters off the ground, middle trees spread raggedly below that, and a lowermost scattering of shrubs and small trees. The forest was framed for a few moments in this theatrical setting. Its image turned surreal, projected into the unbounded wildness of the human imagination, thrown back in time 10,000 years. Somewhere close I knew spear-nosed bats flew through the tree crowns in search of fruit, palm vipers coiled in ambush in the roots of orchids, jaguars walked the river's edge; around them eight hundred species of trees stood, more than are native to all of North America; and a thousand species of butterflies, 6 percent of the entire world fauna, waited for the dawn.

About the orchids of that place we knew very little. About flies and beetles almost nothing, fungi nothing, most kinds of organisms nothing. Five thousand kinds of bacteria might be found in a pinch of soil, and about them we knew absolutely nothing. This was wilderness in the sixteenth-century sense, as it must have formed in the minds of the Portuguese explorers, its interior still largely unexplored and filled with strange, myth-engendering plants and animals. From such a place the pious naturalist would send long respectful letters to royal patrons about the wonders of the new world as testament to the glory of God. And I thought: there is still time to see this land in such a manner.

The unsolved mysteries of the rain forest are formless and seductive. They are like unnamed islands hidden in the blank spaces of old maps, like dark shapes glimpsed descending the far wall of a reef into the abyss. They draw us forward and stir strange apprehensions. The unknown and prodigious are drugs to the scientific imagination, stirring insatiable hunger with a single taste. In our hearts we hope we will never discover everything. We pray there will always be a world like this one at whose edge I sat in darkness. The rain forest in its richness is one of the last repositories on earth of that timeless dream.

PART II

WHO
SCIENTISTS
ARE

Arthur Eddington

from THE EXPANDING UNIVERSE

▓ The first section of this book was on science's subject matter, and we began with the astronomer Sir James Jeans. In the second section we turn to scientists themselves and the nature of science, beginning with Jeans's near contemporary Sir Arthur Eddington. Their names are often bracketed together, as eminent astronomers of the early twentieth century who went out of their way to communicate the romance of their subject to the interested public. Eddington is also famous for his expedition to the island of Principe in 1919 to exploit the total eclipse of the sun and make observations of a distant star to test Einstein's general theory of relativity (see also Paul Davies, below). The prediction was confirmed, and Eddington was able to announce to the world, in Banesh Hoffman's phrase, that Germany was host to the greatest scientist of the age. Einstein himself is reported to have been indifferent to Eddington's dramatic vindication. Any other result, and he would have been '…sorry for the dear Lord. The theory is correct.' Perhaps Einstein should have been more ready, in the words of this extract from Eddington's *The Expanding Universe*, 'to see what is going on in the workshops' of science, rather than to rely on his aesthetic intuition, amazingly gifted though it was. ▓

———

Now I have told you 'everything right as it fell out'.

How much of the story are we to believe?

Science has its showrooms and its workshops. The public today, I think rightly, is not content to wander round the showrooms where the tested products are exhibited; the demand is to see what is going on in the workshops. You are welcome to enter; but do not judge what you see by the standards of the showroom.

We have been going round a workshop in the basement of the building of science. The light is dim, and we stumble sometimes. About us is confusion and mess which there has not been time to sweep away. The workers and their machines are enveloped in murkiness. But I think that something is being shaped here—perhaps something rather big. I do not quite know what it will be when it is completed and polished for the showroom. But we can look at the present designs and the novel tools that are being used in its manufacture; we can contemplate too the little successes which make us hopeful.

A slight reddening of the light of distant galaxies, an adventure of the mathematical imagination in spherical space, reflections on the underlying principles implied in all measurement, nature's curious choice of certain numbers such as 137 in her scheme—these and many other scraps have come together and formed a vision. As when the voyager sights a distant shore, we strain our eyes to catch the vision. Later we may more fully resolve its meaning. It changes in the mist; sometimes we seem to focus the substance of it, sometimes it is rather a vista leading on and on till we wonder whether aught can be final.

Once more I have recourse to Bottom the weaver—

> I have had a most rare vision. I have had a dream, past the wit of man to say what dream it was: man is but an ass, if he go about to expound this dream....Methought I was,—and methought I had,—but man is but a patched fool, if he will offer to say what methought I had....
>
> It shall be called Bottom's Dream, because it hath no bottom.

C. P. Snow

from the Foreword to G. H. Hardy's
A MATHEMATICIAN'S APOLOGY

■ C. P. Snow was better known as a novelist than as a scientist, one of rather few novelists who included sympathetically familiar portraits of scientists. The following extract is not fiction, but is from Snow's biographical Foreword to *A Mathematician's Apology*, the mathematician being the eccentric, cricket-loving G. H. Hardy, from whom we shall hear later. Snow tells the tale of how Hardy discovered the Indian mathematical genius Ramanujan and brought him to Cambridge. Hardy had earlier been a Fellow of my own college at Oxford, where he seems to have been a party to most of the low stake wagers that can still be seen in the Betting Book of the Senior Common Room. The following is typical: 'The subwarden bets Professor Hardy his fortune till death to one halfpenny that the sun will rise tomorrow (7th Feb 1923).' A couple of days later Hardy took the same bet again, but the odds had shortened significantly—for reasons at which we can only guess. This time it was only half his fortune till death, against one whole penny. ■

———————

About his discovery of Ramanujan, Hardy showed no secrecy at all. It was, he wrote, the one romantic incident in his life: anyway, it is an admirable story, and one which showers credit on nearly everyone (with two exceptions) in it. One morning early in 1913, he found, among the letters on his breakfast table, a large untidy envelope decorated with Indian stamps. When he opened it, he found sheets of paper by no means fresh, on which, in a non-English holograph, were line after line of symbols. Hardy glanced at them without enthusiasm. He was by this time, at the age of thirty-six, a world famous mathematician: and world famous mathematicians, he had already discovered, are unusually exposed to cranks. He was accustomed to receiving manuscripts from strangers, proving the prophetic wisdom of the Great Pyramid, the revelations of the Elders of Zion, or the cryptograms that Bacon had inserted in the plays of the so-called Shakespeare.

So Hardy felt, more than anything, bored. He glanced at the letter, written in halting English, signed by an unknown Indian, asking him to give an opinion of these mathematical discoveries. The script appeared to consist of theorems, most of them wild or fantastic looking, one or two already well-known, laid out as though they were original. There were no proofs of any kind. Hardy was not only bored, but irritated. It seemed like a curious kind of fraud. He put the manuscript aside, and went on with his day's routine. Since that routine did not vary throughout his life, it is possible to reconstruct it. First he read *The Times* over his breakfast. This happened in January, and if there were any Australian cricket scores, he would start with them, studied with clarity and intense attention.

Maynard Keynes, who began his career as a mathematician and who was a friend of Hardy's, once scolded him: if he had read the stock exchange quotations half an hour each day with the same concentration he brought to the cricket scores, he could not have helped becoming a rich man.

Then, from about nine to one, unless he was giving a lecture, he worked at his own mathematics. Four hours creative work a day is about the limit for a mathematician, he used to say. Lunch, a light meal, in hall. After lunch he loped off for a game of real tennis in the university court. (If it had been summer, he would have walked down to Fenner's to watch cricket.) In the late afternoon, a stroll back to his rooms. That particular day, though, while the timetable wasn't altered, internally things were not going according to plan. At the back of his mind, getting in the way of his complete pleasure in his game, the Indian manuscript nagged away. Wild theorems. Theorems such as he had never seen before, nor imagined. A fraud of genius? A question was forming itself in his mind. As it was Hardy's mind, the question was forming itself with epigrammatic clarity: is a fraud of genius more probable than an unknown mathematician of genius? Clearly the answer was no. Back in his rooms in Trinity, he had another look at the script. He sent word to Littlewood (probably by messenger, certainly not by telephone, for which, like all mechanical contrivances including fountain pens, he had a deep distrust) that they must have a discussion after hall.

When the meal was over, there may have been a slight delay. Hardy liked a glass of wine, but, despite the glorious vistas of 'Alan St. Aubyn' which had fired his youthful imagination, he found he did not really enjoy lingering in the combination-room over port and walnuts. Littlewood, a good deal more *homme moyen sensuel*, did. So there may have been a delay. Anyway, by nine o'clock or so they were in one of Hardy's rooms, with the manuscript stretched out in front of them.

That is an occasion at which one would have liked to be present. Hardy, with his combination of remorseless clarity and intellectual panache (he was very English, but in argument he showed the characteristics that Latin minds have often assumed to be their own): and Littlewood, imaginative, powerful, humorous. Apparently it did not take them long. Before midnight they knew, and knew for certain. The writer of these manuscripts was a man of genius. That was as much as they could judge, that night. It was only later that Hardy decided that Ramanujan was, in terms of *natural* mathematical genius, in the class of Gauss and Euler: but that he could not expect, because of the defects of his education, and because he had come on the scene too late in the line of mathematical history, to make a contribution on the same scale.

It all sounds easy, the kind of judgment great mathematicians should have been able to make. But I mentioned that there were two persons who do not come out of the story with credit. Out of chivalry Hardy concealed this in all that he said or wrote about Ramanujan. The two people concerned have now been dead, however, for many years, and it is time to tell the truth. It is simple. Hardy was not the first eminent mathematician to be sent the Ramanujan manuscripts. There had been two before him, both English, both of the highest professional standard. They had each returned the manuscripts without comment. I don't think history relates what they said, if anything, when Ramanujan became famous. Anyone who has been sent unsolicited material will have a sneaking sympathy with them.

Anyway, the following day Hardy went into action. Ramanujan must be brought to England, he decided. Money was not a major problem. Trinity has usually been good at supporting unorthodox talent (the college did the same for Kapitsa a few years later). Once Hardy was determined, no human agency could have stopped Ramanujan, but they needed a certain amount of help from a superhuman one.

Ramanujan turned out to be a poor clerk in Madras, living with his wife on twenty pounds a year. But he was also a Brahmin, unusually strict about his religious observances, with a mother who was even stricter. It seemed impossible that he could break the proscriptions and cross the water. Fortunately his mother had the highest respect for the goddess of Namakkal. One morning Ramanujan's mother made a startling announcement. She had had a dream on the previous night, in which she saw her son seated in a big hall among a group of Europeans, and the goddess of Namakkal had commanded her not to stand in the way of her son fulfilling his life's purpose. This, say Ramanujan's Indian biographers, was a very agreeable surprise to all concerned.

In 1914 Ramanujan arrived in England. So far as Hardy could detect (though in this respect I should not trust his insight far) Ramanujan, despite the difficulties of breaking the caste proscriptions, did not believe much in theological doctrine, except for a vague pantheistic benevolence, any more than Hardy did himself. But he did certainly believe in ritual. When Trinity put him up in college—within four years he became a Fellow—there was no 'Alan St. Aubyn' apolausticity for him at all. Hardy used to find him ritually changed into his pyjamas, cooking vegetables rather miserably in a frying pan in his own room.

Their association was a strangely touching one. Hardy did not forget that he was in the presence of genius: but genius that was, even in mathematics, almost untrained. Ramanujan had not been able to enter Madras University because he could not matriculate in English. According to Hardy's report, he was always amiable and good-natured, but no doubt he sometimes found Hardy's conversation outside mathematics more than a little baffling. He seems to have listened with a patient smile on his good, friendly, homely face. Even inside mathematics they had to come to terms with the difference in their education. Ramanujan was self-taught: he knew nothing of the modern rigour: in a sense he didn't know what a proof was. In an uncharacteristically sloppy moment, Hardy once wrote that if he had been better educated, he would have been less Ramanujan. Coming back to his ironic senses, Hardy later corrected himself and said that the statement was nonsense. If Ramanujan had been better educated, he would have been even more wonderful than he was. In fact, Hardy was obliged to teach him some formal mathematics as though Ramanujan had been a scholarship candidate at Winchester. Hardy said that this was the most singular experience of his life: what did modern mathematics look like to someone who had the deepest insight, but who had literally never heard of most of it?

Anyway, they produced together five papers of the highest class, in which Hardy showed supreme originality of his own (more is known of the details of this collaboration than of the Hardy–Littlewood one). Generosity and imagination were, for once, rewarded in full.

This is a story of human virtue. Once people had started behaving well, they went on behaving better. It is good to remember that England gave Ramanujan such honours as were possible. The Royal Society elected him a Fellow at the age of 30 (which, even for a mathematician, is very young). Trinity also elected him a Fellow in the same year. He was the first Indian to be given either of these distinctions. He was amiably grateful. But he soon became ill. It was difficult, in war-time, to move him to a kinder climate.

Hardy used to visit him, as he lay dying in hospital at Putney. It was on one of those visits that there happened the incident of the taxi-cab number. Hardy had gone out to Putney by taxi, as usual his chosen method of conveyance. He went into the room where Ramanujan was lying. Hardy, always inept about introducing a conversation, said, probably without a greeting, and certainly as his first remark: 'I thought the number of my taxicab was 1729. It seemed to me rather a dull number.' To which Ramanujan replied: 'No, Hardy! No, Hardy! It is a very interesting number. It is the smallest number expressible as the sum of two cubes in two different ways.'

That is the exchange as Hardy recorded it. It must be substantially accurate. He was the most honest of men; and further, no one could possibly have invented it.

Freeman Dyson

from DISTURBING THE UNIVERSE

■ The distinguished mathematical physicist Freeman Dyson is one of the most adventurous thinkers in all of science, not afraid to throw his mind far into the distant future, somewhat in the manner of science fiction but exceptionally well-informed science fiction. In this extract, however, he turns to his own past, with an engaging story of his days on the threshold of a scientific career, in a time of promise at the end of the Second World War. His near fatal mishap on the only occasion he was let loose in a laboratory dramatically illustrates the deep divide—natural to physicists but mysterious to biologists—between theorist and experimenter. Dyson, we learn, is a type specimen of the subspecies *Homo sapiens theoreticus*. The experiment that Dyson tried to repeat was originally made famous by the Nobel-prizewinning experimental physicist Robert Millikan, of whom my old physics teacher told the following story (after much anticipatory twitching of the smiling muscles). The unit of self-importance is the Kan, but its value is so high that for practical purposes physicists employ the Millikan. ▪

A Scientific Apprenticeship

In September 1947 I enrolled as a graduate student in the physics department of Cornell University at Ithaca. I went there to learn how to do research in physics under the guidance of Hans Bethe. Bethe is not only a great physicist but also an outstanding trainer of students. When I arrived at Cornell and introduced myself to the great man, two things about him immediately impressed me. First, there was a lot of mud on his shoes. Second, the other students called him Hans. I had never seen anything like that in England. In England, professors were treated with respect and wore clean shoes.

Within a few days Hans found me a good problem to work on. He had an amazing ability to choose good problems, not too hard and not too easy, for students of widely varying skills and interests. He had eight or ten students doing research problems and never seemed to find it a strain to keep us busy and happy. He ate lunch with us at the cafeteria almost every day. After a few hours of conversation, he could judge accurately what each student was capable of doing. It had been arranged that I would only be at Cornell for nine months, and so he gave me a problem that he knew I could finish within that time. It worked out exactly as he said it would.

I was lucky to arrive at Cornell at that particular moment. Nineteen forty-seven was the year of the post-war flowering of physics, when new ideas and new experiments were sprouting everywhere from seeds that had lain dormant through the war. The scientists who had spent the war years at places like Bomber Command headquarters and Los Alamos came back to the universities impatient to get started again in pure science. They were in a hurry to make up for the years they had lost, and they went to work with energy and enthusiasm. Pure science in 1947 was starting to hum. And right in the middle of the renascence of pure physics was Hans Bethe.

At that time there was a single central unsolved problem that absorbed the attention of a large fraction of physicists. We called it the quantum electrodynamics problem. The problem was simply that there existed no accurate theory to describe the everyday behavior of atoms and electrons emitting and absorbing light. Quantum electrodynamics was the name of the missing theory. It was called quantum because it had to take into account the quantum nature of light, electro because it had to deal with electrons, and dynamics because it had to describe forces and motions. We had inherited from the prewar generation of physicists, Einstein and Bohr and Heisenberg and Dirac, the basic ideas for such a theory. But the basic ideas were not enough. The basic ideas could tell you roughly how an atom would behave. But we wanted to be able to calculate the behavior exactly. Of course it often happens in science that things are too complicated to be calculated exactly, so that one has to be content with a rough qualitative understanding. The strange thing in 1947 was that even the simplest and most elementary objects, hydrogen atoms and light quanta, could not be accurately understood. Hans Bethe was convinced that a correct and exact theory would emerge if we could figure out how to calculate consistently using the old pre-war ideas. He stood like Moses on the mountain showing us the promised land. It was for us students to move in and make ourselves at home there.

A few months before I arrived at Cornell, two important things had happened. First, there were some experiments at Columbia University in New York which measured the behaviour of an electron a thousand times more accurately than it had been measured before. This made the problem of creating an accurate theory far more urgent and gave the theorists some accurate numbers which they had to try to explain. Second, Hans Bethe himself did the first theoretical calculation that went substantially beyond what had been done before the war. He calculated the energy of an electron in an atom of hydrogen and found an answer agreeing fairly well with the Columbia measurement. This showed that he was on the right track. But his calculation was still a pastiche of old ideas held together by physical intuition. It had no firm mathematical basis. And it was not even consistent with Einstein's principle of relativity. That was how things stood in September when I joined Hans's group of students.

The problem that Hans gave me was to repeat his calculation of the electron energy with the minimum changes that were needed to make it consistent with Einstein. It was an ideal problem for somebody like me, who had a good mathematical background and little knowledge of physics. I plunged in and filled hundreds of pages with calculations, learning the physics as I went along. After a few months I had an answer, again agreeing near enough with Columbia. My calculation was still a pastiche. I had not improved on Hans's calculation in any fundamental sense. I came no closer than Hans had come to a basic understanding of the electron. But those winter months of calculation had given me skill and confidence. I had mastered the tools of my trade. I was now ready to start thinking.

As a relaxation from quantum electrodynamics, I was encouraged to spend a few hours a week in the student laboratory doing experiments. These were not real research experiments. We were just going through the motions, repeating famous old experiments, knowing beforehand what the answers ought to be. The other students grumbled at having to waste their time doing Mickey Mouse experiments. But I found the experiments fascinating. In all my time in England I had never been let loose in a laboratory. All these strange objects that I had read about, crystals and magnets and prisms and spectroscopes, were actually there and could be touched and handled. It seemed like a miracle when I measured the electric voltage produced by light of various colors falling on a metal surface and found that Einstein's law of the photoelectric effect is really true. Unfortunately I came to grief on the Millikan oil drop experiment. Millikan was a great physicist at the University of Chicago who first measured the electric charge of individual electrons. He made a mist of tiny drops of oil and watched them float around under his microscope while he pulled and pushed them with strong electric fields. The drops were so small that some of them carried a net electric charge of only one or two electrons. I had my oil drops floating nicely, and then I grabbed hold of the wrong knob to adjust the electric field. They found me stretched out on the floor, and that finished my career as an experimenter.

I never regretted my brief and almost fatal exposure to experiments. This experience brought home to me as nothing else could the truth of Einstein's remark, 'One may say the eternal mystery of the world is its comprehensibility.' Here was I, sitting at my desk for weeks on end, doing the most elaborate and sophisticated calculations to figure out how an electron should behave. And here was the electron on my little oil drop, knowing quite well how to behave without waiting for the result of my calculation. How could one seriously believe that the electron really cared about my calculation, one way or the other? And yet the experiments at Columbia showed that it did care. Somehow or other, all this complicated mathematics that I was scribbling established rules that the electron on the oil drop was bound to follow. We know that this is so. Why it is so, why the electron pays attention to our mathematics, is a mystery that even Einstein could not fathom.

J. Robert Oppenheimer

from 'WAR AND THE NATIONS'

J. Robert Oppenheimer, leader of the Los Alamos team of physicists that made the first atomic bombs, later in his agony coined the collective confessional '...the physicists have known sin' and he quoted the Bhagavad-Gita: 'I am become Death, the destroyer of worlds' (he was scholar enough to read it in the original Sanskrit). Oppenheimer was a victim of the McCarthy witch-hunts of the 1950s, and a leading light of the influential Pugwash conferences, along with my senior colleague the late Sir Rudolf Peierls who (together with Otto Frisch) did the calculations that first demonstrated the feasibility of an atomic bomb, and went on to work with Oppenheimer in Los Alamos. This extract is from a 1962 lecture in which Oppenheimer reflects on the moral and political fallout of the terrible weapons whose radioactive fallout had so blighted the cities of Hiroshima and Nagasaki. ■

———————

Nineteen thirty-nine was the year of fission and was also the year of the outbreak of the Second World War; a good many changes had come to all people, but also to physicists. Early in the 1920s up until the very early 1930s scientists from the Soviet Union were welcome and were frequently found in the great centres of learning in Europe and warm collegial relations were formed then between Russians, Englishmen, Germans, Scandinavians, many of which persist to this day. That was changed, too, in the 1930s. During the 1930s very many men of science, like very many other men, either had to leave or in conscience did leave Germany. Many of them came to Canada, many to the United Kingdom, and perhaps most of all to the United States. Some came from Italy as well. By 1939 the Western world was no longer a suburb of the scientific community, but a centre in its own right, and when fission was discovered the first analyses of what nuclei were involved and what prospects there were for its practical use for the release of energy were largely conducted in the United States. I remember that Uhlenbeck, who was still in Holland, thought it his duty to tell his government about this development; the Minister of Finance immediately ordered 50 tons of uranium ore from the Belgian mining company, and remarked: 'Clever, these physicists'.

Actually it was very largely the refugee scientists in England and in the United States who took the first steps to interest their governments in the making of atomic explosives and who took some steps, very primitive ones, in thinking out how this might be done and what might be involved in it. In fact, we all know that it was a letter from Einstein, written at the suggestion of Szilard, Wigner, and Teller, that first brought the matter to President Roosevelt's attention; in the United Kingdom I think it was Simon and Peierls who played this early part. Bohr remained in Denmark as long as it was humanly possible for him to do so. The governments were busy. They had a war on their hands and certainly any reasonable appraisal would have suggested that radar, probably the proximity fuse, and in principle if not in fact rockets would have very much more to do with the outcome of the war than would the atomic energy undertaking. It started slowly under crazy names like Tube Alloys in the United Kingdom, and Department of Substitute Materials in the United States. When I came into it my predecessor had the title Co-ordinator of Rapid Rupture.

There were really very many questions. Would a bomb work and what sort of a thing would it be, how much material would it need, what kind of energies would it release; would it ignite the atmosphere in nuclear reactions and end us all; could it be used to start fusion reactions? There was also the problem of producing, in industrial processes that had no previous analogue in human history, the very considerable number of pounds of the special materials, uranium and plutonium, of which the first bombs had to be made. By late 1941 an authorization for production was really given. There was an uneasy cooperation between the United Kingdom, Canada, and the United States, later substantially to improve, but never, I think, to become completely free of trouble, especially for our friends from the United Kingdom, though we learned much and gained much from all their help. There was also, of course, very much secrecy.

Late in 1942 we decided that we must get to work on how to make bombs themselves. On July 16th, 1945, early in the morning, the first bomb was exploded. It did a little better than we thought it might. One of the guards said: 'The long hairs have let it get away from them.' That day, the President of the United States, the Prime Minister of England, and Stalin were meeting in Potsdam. I believed, because I was told by Dr. Bush, that the President would take the occasion to discuss this development with Stalin, not in order to tell him how to make a bomb, which the President did not know, but to do something that seemed important at the time, to treat the Russians as allies in this undertaking and to start discussing with them how we were going to live with this rather altered situation in the world. It did not come off that way. The President said something, but it is completely unclear whether Stalin understood it or not. No one was present except Stalin's interpreter of the moment and the President, who does not know Russian. But it was a casual word and that was all.

The bombs were used against Japan. That had been foreseen and in principle approved by Roosevelt and Churchill when they met in Canada and again at Hyde Park. It was largely taken for granted; there were questions raised, but I believe there was very little deliberation and even less record of any deliberation there was. And I would like synoptically, briefly, on the basis of my memory of the time and of talk with many historians who have grappled with it, to tell you what little I think about this. I think first of all that we do not know and at the moment cannot know whether a political effort to end the war in the Far East could have been successful. The Japanese Government was deeply divided and stalemated in favour of war. The dissident part of the Government had made an overture through Moscow to the West. Moscow did nothing about it until Potsdam. Stalin told Truman about it. Stalin did not seem interested, Truman did not seem interested, and nothing happened. This was at the very time when the test bomb was successful and a couple of weeks before the bombing of Japan. The actual military plans at that time for the subjugation of Japan and the end of the war were clearly much more terrible in every way and for everyone concerned than the use of the bombs. There is no question about that; and these plans were discussed with us; they would have involved, it was thought, a half a million or a million casualties on the Allied side and twice that number on the Japanese side. Nevertheless, my own feeling is that if the bombs were to be used there could have been more effective warning and much less wanton killing than took place actually in the heat of battle and the confusion of the campaign. That is about all that I am clear about in hindsight. That, and one other thing: I am very glad that the bomb was not kept secret. I am glad that all of us knew, as a few of us already did, what was up and what readjustments in human life and in political institutions would be called for. Those are the days when we all drank one toast only: 'No more wars'.

When the war was over, the great men of physics spoke quite simply and eloquently, Einstein in advocacy of world government and Bohr, first to Roosevelt and to Churchill and to General Marshall and then finally quite openly, when nobody else listened but the public, of the need to work for a world which was completely open. He had in mind that we had some very great secrets and that we ought to be willing to relinquish them in exchange for the disappearance of secrecy from all countries and particularly from the secret-ridden communist societies. Stimson, who resigned as Secretary of War in September 1945, wrote: 'Mankind will not be able to live with the riven atom, without some government of the whole.' Among many reports that we in our innumerable commissions produced, I remember two. One of them, which remains, I think, to this day Top Secret, ended roughly: 'If this weapon does not persuade men of the need for international collaboration and the need to put an end to war, nothing that comes out of a laboratory ever will.' The other said: 'If there is to be any international action for the control of atomic energy there must be an international community of knowledge and of understanding.'

All of this was very deep and genuine and I think most of our community, and many other people also, believed it desirable. It was not exactly what Stalin wanted. And it really was not anything to which any government became very clearly or deeply or fully committed. In the absence of a practical way of getting there, the most that could be done was to put forward some tentative and not entirely disingenuous suggestions about the control of atomic energy which, if accepted, would have led in the direction of international collaboration and in the direction of a suitable beginning of world order. That is not how it has worked; and I remind you only of two obvious things. We are in an arms race of quite unparalleled deadliness—I think this is not the place to speak about the amount of devilment that is piled up on both sides, or about the precautions and the difficulties of making sure that it does not go off; on the other hand, we have lived sixteen and a half years without a nuclear war. In the balance, between the very great gravity of the risks we face and the obvious restraints that have seen us through this time, I have no counsel except that of sobriety and of some hope.

It may seem wrong to speak of this as an experience of physicists. It certainly is not an intellectual challenge like that out of which the theory of relativity was born or that which gave rise to the solution of the paradoxes of wave-particle duality and the quantum theory. I doubt if there is a certain specific right idea to be had in the field of how to remake the world to live with these armaments and to live with our other commitments and our other hopes. But it is true that we have been marked by our deep implication in this development, by the obvious fact that without physics it could not have happened, and by the heavy weight which has been laid on so many members of this community in counselling their government, in speaking publicly and in trying above all in the early phases to find a healthy direction. I do not think that even our young colleagues, tearing away at the new unsolved problems of fundamental physics, are as free of preoccupation for their relation to the good life and the good society, as we were, long ago, when we were their age.

There have been, as you know, many deep and painful conflicts among technical people, and I think one can pick up the paper almost any day and find examples of learned men calling their colleagues liars. We are torn by conflicts, and this, I think, was not openly and clearly true in 1945 and 1946. The arms race, the Cold War, the obduracy of the political conflict, and the immense and complex and terrifying scope of the technological enterprise are not a climate in which the simple discussion of physical problems finds very much place. But more than that, of course, these are not physical problems and they cannot be settled by the methods of science. The question of what our purpose is on earth, the question of how we may make a government that will represent these purposes, the question of what our own responsibility is, the question of what business it is of ours to think about these things, are not to be solved in any laboratory or settled by any equation or any mathematics. Part of the conflict among technical people is like the conflict among all people: it comes from conflicting assessments of what our antagonist's course may be, what his behaviour will be—a subject rich in mystery, even for the experts. Part of it comes because we are talking about a world in which there is no relevant previous experience. No world has ever faced a possibility of destruction—in a relevant sense annihilation—comparable to that which we face, nor a process of decision-making even remotely like that which is involved in this. Those of you who have been in battle know how tangled, unpredictable, and unamenable to prior planning the course of a battle often turns out to be, even when it was well planned. No one has any experience with warfare in the nuclear age. These are some of the reasons for acrimonious differences as to what fraction of a population may survive if you do this or do that, or what you may trust our antagonists to do and what you must suspect them of doing. In addition, the community of physicists is certainly no more than any other free of evil, free of vanity, or free of their own glory; we must expect rather ugly things to happen and they do.

But I would really think that on a few rather deep points which do not imply the answers to all the questions in which we could rightly be interested, we are as a community really rather clear as to what our duty is. It is, in the first place, to give an honest account of what we all know together, know in the way in which I know about the Lorentz contraction and wave-particle duality, know from deep scientific conviction and experience. We think that we should give that information openly whenever that is possible, that we should give it to our governments in secret when the governments ask for it, or, even if the governments do not ask for it, that they should be made aware of it, when we think it essential, as Einstein did in 1939. We all, I think, are aware that it is our duty to distinguish between knowledge in this rather special and proud, but therefore often abstract and irrelevant, sense, and our best guess, our most educated appraisal of proposals which rest on things that in the nature of the case cannot yet be known, like the little cost of some hundred million to build a certain kind of nuclear carrier. We think that it is even more important, and even more essential, to distinguish what we know in the vast regions of science where a great deal is known and more is coming to be known all the time, from all those other things of which we would like to speak and should speak in another context and in another way, those things for which we hope, those things which we value. Finally, I think we believe that whenever we see an opportunity, we have the duty to work for the growth of that international community of knowledge and understanding, of which I spoke earlier, with our colleagues in other lands, with our colleagues in competing, antagonistic, possibly hostile lands, with our colleagues and with others with whom we have any community of interest, any community of professional, of human, or of political concern.

We think of these activities as our contribution, not very different from those of anybody else, but with an emphasis conditioned by the experiences of growing, increasing understanding of the natural physical world, in an increasingly tangled, increasingly wonderful and unexpected situation. We think of this as our contribution to the making of a world which is varied and cherishes variety, which is free and cherishes freedom, and which is freely changing to adapt to the inevitable needs of change in the twentieth century and all centuries to come, but a world which, with all its variety, freedom, and change, is without nation states armed for war and above all, a world without war.

Max F. Perutz

'A PASSION FOR CRYSTALS'

■ X-ray crystallography is the subtle technique by which we know the three-dimensional structure of large biological molecules. Pioneered by William and Lawrence Bragg (the only father and son team to share a Nobel Prize), it was developed further by the physicist J. D. ('Sage') Bernal and then brought to triumphant fruition by a group of younger physicists turned biologist mentored by Bernal or by the younger Bragg. Max Perutz and Dorothy Hodgkin were two of the several Nobel-prizewinning crystallographers associated with Bernal. Here Perutz, in *I wish I'd made you angry earlier*, paints a picture of Dorothy Hodgkin, and of her science. Perutz, originally from Austria, was one of the leading lights of that extraordinary powerhouse of scientific achievement, which he helped to found, the MRC Laboratory of Molecular Biology at Cambridge. Dorothy Hodgkin was Oxford's most distinguished crystallographer. To my regret, I scarcely knew her, but I remember seeing her around the Oxford science labs, an awe-inspiring figure of serene dignity but with tragically arthritic hands. Perutz gives us a warm and affectionate portrait, as befits the gentle and reflective author and his equally gentle subject, both scientists of enormous distinction and becoming modesty. ■

In October 1964, the *Daily Mail* carried a headline 'Grandmother wins Nobel Prize'. Dorothy Hodgkin won it 'for her determination by X-ray techniques of the structures of biologically important molecules'.

She used a physical method first developed by W. L. Bragg, X-ray crystallography, to find the arrangements of the atoms in simple salts and minerals. She had the courage, skill, and sheer willpower to extend the method to compounds that were far more complex than anything attempted before. The most important of these were cholesterol, vitamin D, penicillin, and vitamin B_{12}. Later, she was most famous for her work on insulin, but this reached its climax only five years after she had won the prize.

In the early 1940s, when Howard Florey and Ernest Chain had isolated penicillin from Alexander Fleming's mould, some of the best chemists in Britain and the United States tried to find its chemical constitution. They were taken aback when a handsome young woman, using not chemistry but X-ray analysis, then still mistrusted as an upstart physical technique, had the face to tell them what it was. When Dorothy Hodgkin insisted that its core was a ring of three carbon atoms and a nitrogen which was believed to be too unstable to exist, one of the chemists, John Cornforth, exclaimed angrily, 'If that's the formula of penicillin, I'll give up chemistry and grow mushrooms'. Fortunately he swallowed his words and won the Chemistry Prize himself 30 years later. Hodgkin's formula proved right and was the starting-point for the synthesis of chemically modified penicillins that have saved many lives.

Pernicious anaemia used to be deadly until the early 1930s when it was discovered that it could be kept in check by liver extracts. In 1948, the active principle, vitamin B_{12}, was isolated from liver in crystalline form, and chemists began to wonder what its formula was. The first X-ray diffraction pictures showed that the vitamin contained over a thousand atoms, compared to penicillin's thirty-nine; it took Hodgkin and an army of helpers eight years to solve its structure. Like penicillin, vitamin B_{12} showed chemical features not encountered before, such as a strange ring of nitrogens and carbon atoms surrounding its central cobalt atom and a novel kind of bond from the cobalt atom to the carbon atoms of a sugar ring that provided the clue to the vitamin's biological function. The Nobel Prize was awarded to Hodgkin not just for determining the structures of several vitally important compounds, but also for extending the bounds of chemistry itself.

In 1935 Dorothy Crowfoot, as she then was, put a crystal of insulin in front of an X-ray beam and placed a photographic film behind it. That night, when she developed the film, she saw minute, regularly arranged spots forming a diffraction pattern that held out the prospect of solving insulin's structure. Later that night she wandered around the streets of Oxford, madly excited that she might be the first to determine the structure of a protein, but next morning she woke with a start: could she be sure that her crystals really were insulin rather than some trivial salt? She rushed back to the lab before breakfast. A simple spot test on a microscope slide showed that her crystals took up a stain characteristic for protein, which revived her hopes. She never imagined that it would take her thirty-four years to solve that complex structure, nor that once solved it would have practical application. It has recently enabled genetic engineers to change the chemistry of insulin in order to improve its benefits for diabetics.

Dorothy Crowfoot was born in Cairo in 1910. Her father, J. W. Crowfoot, was Education Officer in Khartoum and an archaeologist; her mother too was an archaeologist, with a particular interest in the history of weaving. When Dorothy was a child, they lived next door to the Sudan Government chemist, Dr A. F. Joseph. It was 'Uncle Joseph's' early encouragement that excited her interest in science. Later he introduced her to the Cambridge Professor of Physical Chemistry, T. Martin Lowry, who advised her to work with J. D. Bernal.

When Dorothy Crowfoot was 24 and working in Cambridge with Bernal on crystals of another protein, the digestive enzyme pepsin, Bernal made his crucial discovery of their rich X-ray diffraction patterns. But, on the day that he did, her parents had taken her to London to consult a specialist about persistent pains in her hands. He diagnosed the onset of the rheumatoid arthritis that was to cripple her hands and feet, but never slowed her determined pursuit of science.

At Oxford, Dorothy Hodgkin used to labour on the structure of life in a crypt-like room tucked away in a corner of Ruskin's Cathedral of Science, the Oxford Museum. Her Gothic window was high above, as in a monk's cell, and beneath it was a gallery reachable only by a ladder. Up there she would mount her crystals for X-ray analysis, and descend precariously, clutching her treasure with one hand and balancing herself on the ladder with the other. For all its gloomy setting, Hodgkin's lab was a jolly place. As Chemistry Tutor at Somerville College, she always had girls doing crystal structures for their fourth year and two or three research students of either sex working for their PhDs. They were a cheerful lot, not just because they were young, but because her gentle and affectionate guidance led most of them on to interesting results. Her best-known pupil, however, made her name in a career other than chemistry: Margaret Roberts, later Margaret Thatcher, worked as a fourth-year student on X-ray crystallography in Dorothy Hodgkin's laboratory.

In 1937, Dorothy had married the historian Thomas Hodgkin. Some women intellectuals regard their children as distracting impediments to their careers, but Dorothy radiated motherly warmth even while doing scientific work. Concentration came to her so easily that she could give all her attention to a child's chatter at one moment and switch to complex calculation the next.

She pursued her crystallographic studies, not for the sake of honours, but because this was what she liked to do. There was magic about her person. She had no enemies, not even among those whose scientific theories she demolished or whose political views she opposed. Just as her X-ray cameras bared the intrinsic beauty beneath the rough surface of things, so the warmth and gentleness of her approach to people uncovered in everyone, even the most hardened scientific crook, some hidden kernel of goodness. She was once asked in a BBC radio interview whether she felt handicapped in her career by being a woman. 'As a matter fact,' she replied gently, 'men were always particularly nice and helpful to me *because* I was a woman.' At scientific meetings she would seem lost in a dream, until she suddenly came out with some penetrating remark, usually made in a diffident tone of voice, and followed by a little laugh, as if wanting to excuse herself for having put everyone else to shame.

Dorothy Hodgkin's uncanny knack of solving difficult structures came from a combination of manual skill, mathematical ability, and profound knowledge of crystallography and chemistry. It often led her and her alone to recognise what the initially blurred maps emerging from X-ray analysis were trying to tell. She was a great chemist; a saintly, gentle, and tolerant lover of people; and a devoted protagonist of peace.

Barbara and George Gamow

'SAID RYLE TO HOYLE'

■   A couple of examples now of scientific comic verse, the first from *Mr Tompkins in Wonderland* by the cosmologist George Gamow, and penned with his wife Barbara. Gamow was enough of a 'character' to feature in the title of *Genes, Girls and Gamow*, Jim Watson's feministically challenged sequel to *The Double Helix*. It was Gamow's obsession with solving the genetic code that drew him and Watson together, in a sporadic correspondence about the genetic code, and they were among those who founded the in-groupish RNA Tie Club. But Gamow was more famous as a cosmologist and early champion of what was later to be called (scathingly, by its most vociferous opponent Fred Hoyle) the Big Bang. Hoyle and his colleagues Hermann Bondi and Thomas Gold had favoured the alternative Steady State theory, which held that the Universe had no beginning: new matter was continuously formed in the gaps left as galaxies drew apart. If the continuous creation of matter sounds improbable, Hoyle pointed out that it is no more so than the single spontaneous creation postulated in the Big Bang. Unfortunately (a good example of what T. H. Huxley had earlier described as 'The great tragedy of science—a beautiful hypothesis slain by an ugly fact') the Steady State hypothesis was decisively refuted by the radio telescope observations of Hoyle's Cambridge colleague Martin Ryle. The whole story is succinctly told in verse. ■

'Your years of toil,'
Said Ryle to Hoyle,
'Are wasted years, believe me.
The steady state
Is out of date.
Unless my eyes deceive me,
My telescope
Has dashed your hope;
Your tenets are refuted.
Let me be terse:

Our universe
Grows daily more diluted!'
Said Hoyle, 'You quote
Lemaître, I note,
And Gamow. Well, forget them!
That errant gang
And their Big Bang—
Why aid them and abet them?
You see, my friend,
It has no end
And there was no beginning,
As Bondi, Gold,
And I will hold
Until our hair is thinning!'
'Not so!' cried Ryle
With rising bile
And straining at the tether;
'*Far galaxies*
Are, as one sees,
More tightly packed together!'
'You make me boil!'
Exploded Hoyle,
His statement rearranging;
'*New matter's born*
Each night and morn.
The picture is unchanging!'
'Come off it, Hoyle!
I aim to foil
You yet' (The fun commences)
'And in a while,'
Continued Ryle,
'I'll bring you to your senses!'[1]

1. A fortnight before the publication date of the first printing of this book there appeared an article by F. Hoyle entitled: 'Recent Developments in Cosmology' (*Nature*, 9 Oct. 1965, p. iii). Hoyle writes: 'Ryle and his associates have counted radio sources... The indication of that radio count is that the Universe was more dense in the past than it is today.' The author has decided, however, not to change the lines of the arias of 'Cosmic Opera' since, once written, operas become classic. In fact, even today Desdemona sings a beautiful aria before she dies, after being strangled by Othello.

J. B. S. Haldane

'CANCER'S A FUNNY THING'

■ It might seem unusual to describe as 'comic' a poem about the poet's rectal carcinoma, but J. B. S. Haldane was an unusual man. Here are some rhymes worthy of Hilaire Belloc—or perhaps Tom Lehrer is a better comparison, given the subject matter—and they are all the funnier when they are scientific technical terms. Haldane ends on an up-beat note that is almost cheering. ■

I wish I had the voice of Homer
To sing of rectal carcinoma,
Which kills a lot more chaps, in fact,
Than were bumped off when Troy was sacked.
Yet, thanks to modern surgeon's skills,
It can be killed before it kills
Upon a scientific basis
In nineteen out of twenty cases.
I noticed I was passing blood
(Only a few drops, not a flood).
So pausing on my homeward way
From Tallahassee to Bombay
I asked a doctor, now my friend,
To peer into my hinder end,
To prove or to disprove the rumour
That I had a malignant tumour.
They pumped in $BaSO_4$.
Till I could really stand no more,
And, when sufficient had been pressed in,
They photographed my large intestine,
In order to decide the issue
They next scraped out some bits of tissue.
(Before they did so, some good pal
Had knocked me out with pentothal,
Whose action is extremely quick,

And does not leave me feeling sick.)
The microscope returned the answer
That I had certainly got cancer,
So I was wheeled into the theatre
Where holes were made to make me better.
One set is in my perineum
Where I can feel, but can't yet see 'em.
Another made me like a kipper
Or female prey of Jack the Ripper,
Through this incision, I don't doubt,
The neoplasm was taken out,
Along with colon, and lymph nodes
Where cancer cells might find abodes.
A third much smaller hole is meant
To function as a ventral vent:
So now I am like two-faced Janus
The only[1] god who sees his anus.
I'll swear, without the risk of perjury,
It was a snappy bit of surgery.
My rectum is a serious loss to me,
But I've a very neat colostomy,
And hope, as soon as I am able,
To make it keep a fixed time-table.
So do not wait for aches and pains
To have a surgeon mend your drains;
If he says 'cancer' you're a dunce
Unless you have it out at once,
For if you wait it's sure to swell,
And may have progeny as well.
My final word, before I'm done,
Is 'Cancer can be rather fun'.
Thanks to the nurses and Nye Bevan
The NHS is quite like heaven
Provided one confronts the tumour
With a sufficient sense of humour.
I know that cancer often kills,
But so do cars and sleeping pills;
And it can hurt one till one sweats,
So can bad teeth and unpaid debts.
A spot of laughter, I am sure,
Often accelerates one's cure;

So let us patients do our bit
To help the surgeons make us fit.

1. In India there are several more
With extra faces, up to four,
But both in Brahma and in Shiva
I own myself an unbeliever.

Jacob Bronowski

from THE IDENTITY OF MAN

■ Director of Research at the British National Coal Board, the mathematician Jacob Bronowski spoke no English when he immigrated from Poland, but he became a household name and a byword for polymathic wisdom, when he wrote and presented a thirteen-part documentary on BBC television called The Ascent of Man. Probably only Carl Sagan and David Attenborough rival Bronowski in the history of science documentaries. Bronowski's cultivated tones—slightly accented, wise, cerebral, learned, never talking down yet never obscure—are clearly audible to me when I read the written words on the page of this extract from *The Identity of Man*. ■

I have chosen to describe science as an account of the machinery of nature, not in engineering terms, but in linguistic ones. One persuasive reason is that I shall be talking in the next essay about literature, and whatever I have to say there by way of likeness or of contrast will be said more fairly if I use a common model of language in both places. But a more cogent reason, of course, is that language is a more telling and a better model for science than is any mechanism.

We receive experience from nature in a series of messages. From these messages we extract a content of information: that is, we decode the messages in some way. And from this code of information we then make a basic vocabulary of concepts and a basic grammar of laws, which jointly describe the inner organization that nature translates into the happenings and the appearances that we meet.

Somewhere in this decoding, the mind takes a critical step from the individual experience to the general law which embraces it. How do we guess the law and form the concepts that underlie it? How do we decide that there are, and how do we give properties to, such invisible things as atoms? That the atoms in their turn are composed of more fundamental particles? How do we convince ourselves that there is a universal quantity called energy, which is carried by single quanta, yet which spreads from place to place in a motion like a wave? And that the rearrangement of atoms, and still more fundamental particles, consumes or releases energy? How do we come to picture a living process in these dead terms?

Take as a concrete example again the structure of the eye, which Bishop Butler and Henri Bergson both thought too marvelous to be explained by mechanical evolution. After centuries of preparation, how do we come to conclude that the small rods and cones in my retina are sensitive to single quanta of light, that these quanta untwist the molecules of visual purple, that this chemical change is integrated electrically in my eye with others like it and signaled to my brain, and that the coloring of the picture that it evokes there has been fixed at my conception by the same fragments of my father's sperm and my mother's ovum that determined my sex?

I have only to describe this complex, farfetched and intricately connected sequence to make it evident that no simple set of observations will suffice to establish it. In the first place, it is a highly generalized account which could not be derived from single experiments, even at a single point in the sequence. We have to fit together many separate experiments to reach, for example, the plain conclusion that visual purple is bleached, or that this chemical change is signaled as an electric impulse. To say something persuasive about the optic nerve, we need the evidence of a host of other observations on a multitude of other nerves. And when we look beyond one of these generalizations along their whole connected sequence, we realize how they lock and engage with, how they are fixed and held in place by, all the generalizations of physics and chemistry and biology and the physiology of the nervous system.

The most modest research worker at his bench, pushing a probe into a neuron to measure the electric response when a light is flashed, is enmeshed in a huge and intertwined network of theories that he carries into his work from the whole field of science, all the way from Ohm's law to Avogadro's number. He is not alone; he is sustained and held and in some sense imprisoned by the state of scientific theory in every branch. And what he finds is not a single fact either: it adds a thread to the network, ties a knot here and another there, and by these connections at once binds and enlarges the whole system.

This is worth saying, even though it has always been so, because it is still neglected by philosophers. They see that science passes from fact to prediction, from instance to law, by a procedure of generalization—what is usually called induction. To reason in this direction, from the particular to the general, cannot be justified on logical grounds: David Hume showed that more than two hundred years ago. But more incisive than the question, What right have we to form inductions? is the question, How do we form them? Hume gave no explanation of this except habit—

> 'We are determined by CUSTOM alone to suppose the future conformable to the past'

—and philosophers have followed him ever since. Their theories are still dominated by their belief that science is an accumulation of facts, and that a generalization grows of itself from a heaping of single instances in one narrow field. They think that a scientist is persuaded that light arrives at the eye in a shower of quanta because he does an experiment, does it again, and repeats it to be sure.

Alas, this is not at all what any scientist does. He may indeed repeat an experiment two or three times, if its outcome strikes him as odd and unexpected. But even here, he means by odd and unexpected precisely that it conflicts with what other experiments in other fields have led him to believe. The suspicion with which all scientists treat the published evidence for extrasensory perception shows this. A set of results is odd and unexpected, in the end it is unbelievable, because it outrages the intricate network of connections that has been established between known phenomena.

Peter Medawar

▓ Sir Peter Medawar, Nobel-prizewinning zoologist and medical scientist, is surely the wittiest of all scientific writers. One extract cannot do him justice: the learning, the intelligence, the urbane and loftily confident erudition—which would be described as arrogance if he didn't effort-lessly get away with it. I have seen feminists bridle at the introductory sentence of *Pluto's Republic* ('A good many years ago, a neighbour whose sex chivalry forbids me to disclose...') but, however reluctantly, they had to struggle not to laugh at the patrician—I suppose they would say patriar-chal—wit. And what Frenchman would not bridle—but again struggle not to laugh—at 'Teilhard has accordingly resorted to that tipsy, euphoristic prose-poetry which is one of the more tiresome manifestations of the French spirit.' Before John Maynard Smith had met Medawar, he asked J. B. S. Haldane what his newly elected professor was like. Haldane invoked Shakespeare: 'He smiles and smiles, and is a villain.' Yes!

I could not resist giving Medawar more than his fair share of the pages of this book. The opening sentence of his Romanes Lecture on 'Science and Literature' ('I hope I shall not be thought ungracious if I say at the outset that nothing on earth would have induced me to attend the kind of lecture you may think I am about to give') prompted one literary critic to remark, 'This lecturer has never been thought ungracious in his life'. I have reproduced some passages from the lecture, which include increas-ingly needed attacks on wilful obscurantism. Then a brief extract from Medawar's historical sleuthing of the cause of Darwin's illness, with its splendidly Medawarlike put-down of Freudianism. As for his review of Teilhard de Chardin's *The Phenomenon of Man*, it has simply got to be the greatest negative book review ever written. You will not be satisfied with the brief extract here. The original can be found in *Pluto's Republic* and also in an earlier collection, *The Art of the Soluble*. Another book review, this time of Watson's *The Double Helix*, provoked enough discussion to justify a brief postscript. For Medawar, even a postscript is a work of art, and I have reproduced it here. Finally, to show that it isn't all controversy, I have chosen a fragment of Medawar's affectionate and respectful portrait of D'Arcy Thompson, whom we met earlier in this book. ▓

from 'SCIENCE AND LITERATURE'

Let me begin by discussing the character and interaction of imagination and critical reasoning in literature and in science. I shall use 'imagination' in a modern sense (modern on the literary time scale, I mean), or, at all events, in a sense fully differentiated from mere fancy or whimsical inventiveness. (It is worth remembering that when the phrase 'creative imagination' is used today, we are expected to look solemn and attentive, but in the eighteenth century we could as readily have looked contemptuous or even shocked.)

The official Romantic view is that Reason and the Imagination are antithetical, or at best that they provide alternative pathways leading to the truth, the pathway of Reason being long and winding and stopping short of the summit, so that while Reason is breathing heavily, there is Imagination capering lightly up the hill. It is true that Shelley[1] recognized a poetical element in science, though 'the poetry is concealed by the accumulation of facts and calculating processes'; true also that in one passage of his famous rhapsody, he was kind enough to say that poetry comprehends all science—though here, as he makes plain, he is using poetry in a general sense to stand for all exercises of the creative spirit, a sense that comprehends imaginative literature itself as one of its special instances. But in the ordinary usages to which I shall restrict myself, Reason and Imagination are antithetical. That was Shelley's view and Keat's, Wordsworth's, and Coleridge's; it was also Peacock's, for whom Reason was marching into territories formerly occupied by poets; and it was also the view of William Blake,[2] who came 'in the grandeur of Inspiration to cast off Rational Demonstration...to cast off Bacon, Locke, & Newton'; 'I will not Reason & Compare—my business is to create'.

This was not only the official view of the Romantic poets; it was also the official scientific view. When Newton wrote *Hypotheses non fingo*, he was taken to mean that he reprobated the exercise of the imagination in science. (He did not 'really' mean this, of course, but the importance of his disclaimer lies precisely in this misunderstanding of it.) Bacon too, and later on John Stuart Mill were taken as official spokesmen for the belief that there existed, or could be devised, a calculus of discovery, a formulary of intellectual behaviour, which could be relied upon to conduct the scientist towards the truth, and this new calculus was thought of almost as an antidote to the imagination, as it had been in Bacon's own day an antidote to what Macaulay[3] called the 'sterile exuberance' of scholastic thought. Even today this central canon of inductivism—that scientific thought is fully accountable to reason—is assumed quite unthinkingly to be true. 'Science is a matter of disinterested observation, unprejudiced insight and experimentation, and patient ratiocination within some system of logically correlated concepts'—an important opinion, for Aldous Huxley is a man thought to speak with equal authority for science and letters.

[...]

By the time of the New Philosophy, the competition or disputation between eloquence and wisdom, style and substance, medium and message had already been in progress for nearly 2,000 years, but as far as the New Philosophy was concerned, the Royal Society, with the formidable support of John Locke and Thomas Hobbes, may be thought to have settled the matter once and for all: scientific and philosophic writing were on no account to be made the subject of a literary spectacle and of exercises in the high rhetoric style.

This position has been threatened only during those two periods in which our native philosophic style (which is also a style of thinking) was obfuscated by influences from abroad. During the Gothic period of philosophic writing, which began before the middle of the nineteenth century and continued until the First World War, we were all oppressed and perhaps mildly stupefied by metaphysical profundities of German origin. But although those tuba notes from the depths of the Rhine filled us with thoughts of great solemnity and confusion, it was not as music, thank heavens, that we were expected to admire them. The style was not an object of admiration in itself. Today, though we are now much better armed against it, speculative metaphysics has given way to what might be called a salon philosophy as the chief exotic influence, and French writers enjoy the reverential attention that was at one time thought due to German. Style has now become an object of first importance, and what a style it is! For me it has a prancing, high-stepping quality, full of self-importance; elevated indeed, but in the balletic manner, and stopping from time to time in studied attitudes, as if awaiting an outburst of applause. It has had a deplorable influence on the quality of modern thought in philosophy and in the behavioural and 'human' sciences.

The style I am speaking of, like the one it superseded, is often marked by its lack of clarity, and hereabout we are apt to complain that it is sometimes very hard to follow. To say as much, however, may now be taken as a sign of eroded sensibilities. I could quote evidence of the beginnings of a whispering campaign against the virtues of clarity. A writer on structuralism in the *TLS* has recently suggested that thoughts which are confused and tortuous by reason of their profundity are most appropriately expressed in prose that is deliberately unclear. What a preposterously silly idea! I am reminded of an air-raid warden in wartime Oxford who, when bright moonlight seemed to be defeating the spirit of the blackout, exhorted us to wear dark glasses. He, however, was being funny on purpose.

I must not speak of obscurity as if it existed in just one species. A man may indeed write obscurely when he is struggling to resolve problems of great intrinsic difficulty. This was the obscurity of Kant, one of the greatest of all thinkers. There is no more moving or touching passage in his writings than that in which he confesses that he has no gift for lucid exposition, and expresses the hope that in due course others will help to make his intentions plain.[4]

[...]

The *rhetorical* use of obscurity is, however, a vice. It is often said—and it was said of Kant[5]—that the purpose of obscure or difficult writing is to create the illusion of profundity, and the accusation need not be thought an unjust one merely because it is trite. But in its more subtle usages, obscurity can be used to create the illusion of a deeply reasoned discourse. Suppose we read a text with a closely reasoned argument which is complex and hard to follow. We struggle with it, and as we go along we may say, 'I don't see how he makes that out', or 'I can see now what he's getting at', and in the end we shall probably get there, and either agree with what the author says or find reasons for taking a different view. But suppose there is no argument; suppose that the text is asseverative in manner, perhaps because analytical reasoning has been repudiated in favour of reasoning of some higher kind. If now the text is made hard to follow because of *non sequiturs*, digressions, paradoxes, impressive-sounding references to Gödel, Wittgenstein, and topology, 'in' jokes, trollopy metaphors, and a general determination to keep all vulgar sensibilities at bay, then again we shall have great difficulty in finding out what the author intends us to understand. We shall have to reason it out therefore, much as we reasoned out Latin unseens or a passage in some language we don't fully understand. In both texts some pretty strenuous reasoning may be interposed between the author's conceptions and our understanding of them, and it is strangely easy to forget that in one case the reasoning was the author's but in the other case our own. We have thus been the victims of a confidence trick.

Let me end this section with a declaration of my own. In all territories of thought which science or philosophy can lay claim to, including those upon which literature has also a proper claim, no one who has something original or important to say will willingly run the risk of being misunderstood; people who write obscurely are either unskilled in writing or up to mischief. The writers I am speaking of are, however, in a purely literary sense, extremely skilled.

1. Percy Bysshe Shelley, *A Defence of Poetry* (1821).
2. William Blake, *Milton* (1804), book 2, pl. 41; and *Jerusalem* (1804), ch. 1, pl. 10.
3. Thomas Babington Macaulay, *Lord Bacon* (1837), an extended review of Montagu's edition of Bacon's works that first appeared in the *Edinburgh Review*.
4. Immanuel Kant, *Critique of Pure Reason*, introduction to the second edition (1787).
5. See Kant's preface to *The Metaphysic of Morals* (1797).

* * *

from 'DARWIN'S ILLNESS'

Kempf believed that Darwin's forty years' disabling illness was a neurotic manifestation of a conflict between his sense of duty towards a rather domineering father and a sexual attachment to his mother, who died when he was eight. His mother, a gentle and latterly an ailing creature, fond of flowers and pets, had propounded a riddle which it was Darwin's life-work to resolve: How, by looking inside a flower, might its name be discovered? Kempf wrote in 1918 with an arch delicacy that sometimes obscures his meaning, but Good's more recent interpretation leaves us in no doubt. For Good, 'there is a wealth of evidence that unmistakably points' to the idea that Darwin's illness was 'a distorted expression of the aggression, hate, and resentment felt, at an unconscious level, by Darwin towards his tyrannical father'. These deep and terrible feelings found outward expression in Darwin's touching reverence toward his father and his father's memory, and in his describing his father as the kindest and wisest man he ever knew: clear evidence, if evidence were needed, of how deeply his true inner sentiments had been repressed.

* * *

from 'THE PHENOMENON OF MAN'

Everything does not happen continuously at any one moment in the universe. Neither does everything happen everywhere in it.

There are no summits without abysses.

When the end of the world is mentioned, the idea that leaps into our minds is always one of catastrophe.

Life was born and propagates itself on the earth as a solitary pulsation.

In the last analysis the best guarantee that a thing should happen is that it appears to us as vitally necessary.

This little bouquet of aphorisms, each one thought sufficiently important by its author to deserve a paragraph to itself, is taken from Père Teilhard's *The Phenomenon of Man*. It is a book widely held to be of the utmost profundity and significance; it created something like a sensation upon its publication in France, and some reviewers hereabouts called it the Book of the Year—one, the Book of the Century. Yet the greater part of it, I shall show, is nonsense, tricked out with a variety of metaphysical conceits, and its author can be excused of dishonesty only on the grounds that before deceiving others he has taken great pains to deceive himself. *The Phenomenon of Man* cannot be read without a feeling of suffocation, a gasping and flailing around for sense. There is an argument in it, to be sure—a feeble argument, abominably expressed—and this I shall expound in due course; but consider first the style, because it is the style that creates the illusion of content, and which is a cause as well as merely a symptom of Teilhard's alarming apocalyptic seizures.

The Phenomenon of Man stands square in the tradition of *Naturphilosophie*, a philosophical indoor pastime of German origin which does not seem even by accident (though there is a great deal of it) to have contributed anything of permanent value to the storehouse of human thought. French is not a language that lends itself naturally to the opaque and ponderous idiom of nature-philosophy, and Teilhard has accordingly resorted to the use of that tipsy, euphoristic prose-poetry which is one of the more tiresome manifestations of the French spirit.

[...]

How have people come to be taken in by *The Phenomenon of Man*? We must not underestimate the size of the market for works of this kind, for philosophy-fiction. Just as compulsory primary education created a market catered for by cheap dailies and weeklies, so the spread of secondary and latterly of tertiary education has created a large population of people, often with well-developed literary and scholarly tastes, who have been educated far beyond their capacity to undertake analytical thought.

* * *

from the postscript to 'LUCKY JIM'

'Lucky Jim' was a defence of Watson against the storm of outraged criticism that burst out after the publication of *The Double Helix*. Nothing has occurred to shake my belief that the discovery of the structure and biological functions of the nucleic acids is the greatest achievement of science in the twentieth century. In defending Watson I felt much as advocates must feel when defending a client who is unmistakably guilty of many of the charges brought against him: I have in mind particularly his lack of adequate acknowledgement of the work of scientists such as Chargaff who made really important contributions to the elucidation of the problem which he and Crick finally solved. I showed Francis Crick my review before it appeared in the *New York Review of Books* and was very pleased when he said of my parallel with Alice's Wonderland and the Mad Hatter's tea-party: 'That's quite right, you know, it was exactly like that.'

One passage in this review of *The Double Helix* came in for a lot of criticism. I see it struck W. H. Auden too, unfavourably I should guess, The passage runs:

> It just so happens that during the 1950s, the first great age of molecular biology, the English schools of Oxford and particularly of Cambridge produced more than a score of graduates of quite outstanding ability—much more brilliant, inventive, articulate and dialectically skilful than most young scientists; right up in the Jim Watson class. But Watson had one towering advantage over all of them: in addition to being extremely clever he had something important to be clever *about*.

Surely, I was asked, you don't intend to imply that Shakespeare and Tolstoy etc. are not important and that it is hardly possible to be clever about them? Of *course* this is not what I intended. I had it in mind that many of the brilliant contemporaries of Jim Watson and many of the brightest literary students of the later 1950s entered the advertising or entertainment industries or contented themselves with petty literary pursuits. The widely prevalent opinion that almost any literary work, even if it amounts to no more than writing advertising copy or a book review, not to mention that Ph.D. thesis on 'Some little known laundry bills of George Moore', is intrinsically superior to almost any scientific activity is not one with which a scientist can be expected to sympathize.

* * *

from 'D'ARCY THOMPSON AND
GROWTH AND FORM'

D'Arcy Wentworth Thompson was an aristocrat of learning whose intellectual endowments are not likely ever again to be combined within one man. He was a classicist of sufficient distinction to have become President of the Classical Associations of England and Wales and of Scotland; a mathematician good enough to have had an entirely mathematical paper accepted for publication by the Royal Society; and a naturalist who held important chairs for sixty-four years, that is, for all but the length of time into which we must nowadays squeeze the whole of our lives from birth until professional retirement. He was a famous conversationalist and lecturer (the two are often thought to go together, but seldom do), and the author of a work which, considered as literature, is the equal of anything of Pater's or Logan Pearsall Smith's in its complete mastery of the *bel canto* style. Add to all this that he was over six feet tall, with the build and carriage of a Viking and with the pride of bearing that comes from good looks known to be possessed.

D'Arcy Thompson (he was always called that, or D'Arcy) had not merely the makings but the actual accomplishments of three scholars. All three were eminent, even if, judged by the standards which he himself would have applied to them, none could strictly be called great. If the three scholars had merely been added together in D'Arcy Thompson, each working independently of the others, then I think we should find it hard to repudiate the idea that he was an amateur, though a patrician among amateurs; we should say, perhaps, that great as were his accomplishments, he lacked that deep sense of engagement that marks the professional scholar of the present day. But they were not merely added together; they were integrally—Clifford Dobell said chemically—combined. I am trying to say that he was not one of those who have made two or more separate and somewhat incongruous reputations, like a composer-chemist or politician-novelist, or like the one man who has both ridden in the Grand National and become an FRS; but that he was a man who comprehended many things with an undivided mind. In the range and quality of his learning, the uses to which he put it, and the style in which he made it known I see not an amateur, but, in the proper sense of that term, a natural philosopher—though one dare not call him so without a hurried qualification, for fear he might be thought to have practised what the Germans call *Naturphilosophie*.

Jonathan Kingdon

from SELF-MADE MAN

▇ Peter Medawar said of D'Arcy Thompson that his three roles—zoologist, mathematician and classicist—were integrally, 'chemically' combined. Something similar could be said of Jonathan Kingdon the zoologist and Jonathan Kingdon the artist (in a different way, one could make the same point about Desmond Morris). Kingdon wrote and illustrated the definitive volumes on the mammals of Africa. As an anatomical artist, he has been compared to Leonardo da Vinci. His anthropological musings are thoughtful and provocative. This extract is from *Self-Made Man*, a scientifically informed reverie on human origins, about which I wrote in a review:

> A gentle wisdom comes out of Africa, a timeless vision that looks through and beyond the effete faddishness, the forgettable ephemerality of contemporary culture and preoccupation. With the eyes of an artist and the mind of a scientist and polymath, Jonathan Kingdon gazes deep into the past and if you look deeply enough into our human past you come inevitably to the home continent of Africa. He was born there and so, as it happens, was I, although we'd both be classed under that ludicrous name (which he rightly ridicules) 'Caucasian'. But, as Kingdon reminds us, wherever we were individually born and whatever our 'race', we are all Africans. ▇

Before the Wise Men

Three-and-a-half million years ago at Laetoli in northern Tanzania there was a local disaster. Carbonatite ash and tiny globules of lava had rained down as the Sadiman-Lemagrut volcano erupted and the Rift Valley slopes below were powdered with a sort of raw cement. Then it rained. Such events would have been common enough along the Great Rift and fatal, or at least frightening, for those who witnessed them. We know there were witnesses, because before the mushy cement set into a hard pavement a female hominid, a southern ape woman, or *Australopithecus*, and her youngster trudged through it, probably seeking to escape a suddenly poisoned homeland. After their passage a three-toed horse went by and a rather confused hare dithered in the noxious mud.

All along the tortuous path that leads back from us to ever earlier ancestors were people who had to fill their stomachs and with exquisite spasms of sexual chemistry pass on their genes. Time and again they faltered, as drought, poisoned ash, and a multitude of hazards conspired to destroy their frail substance. Fossil bones and footsteps and ruined homes are the solid facts of history, but the surest hints, the most enduring signs, lie in those minuscule genes. For a moment we protect them with our lives, then like relay runners with a baton, we pass them on to be carried by our descendants. There is a poetry in genetics which is more difficult to discern in broken bones, and genes are the only unbroken living thread that weaves back and forth through all those boneyards.

African prehistory and palaeontology are new sciences. At a time when the value of fossils is taken for granted it is easy to forget that human fossils remained virtually unnoticed until Darwin created the scientific and philosophical framework in which they could find relevance. Before the interior of Africa had been explored and with no relevant fossils to hand, Darwin wrote:

> It is probable that Africa was formerly inhabited by extinct apes closely allied to the gorilla and chimpanzee and that these two species are now man's nearest allies; it is somewhat more probable that our early progenitors lived on the African continent than elsewhere.

This farsighted prediction was the result of the framework in which Darwin ordered his observations and deductions. That framework is now the basis for biological teaching and for our understanding of human prehistory. There are now many more hard facts to learn but Darwinian leaps are still necessary to bridge the unknown spaces in between.

Histories cannot be entirely taught. To perceive history as in any sense a living past, rather than a procession of learnt facts, requires feats of imagination that are essentially private and voluntary. For imagination to be more than fantasy, our experience of the living world must offer some sense of continuity to help bridge the chasm between that poisoned day at Laetoli and the present. Three-toed hipparions have gone but a barking zebra signals some sort of equine continuity. The sun that rises over a now extinct Lemagrut will never again illuminate the tread of an *Australopithecus* family but the genes that could build two flat feet like theirs are not extinct. Toe by toe and heel by heel there are countless feet being built in countless wombs today that at the right age could retrace those trails across volcanic mud.

Richard Leakey and Roger Lewin

from ORIGINS RECONSIDERED

■ The discovery of a major hominid fossil rightly hits headlines all around the world. If you see such a headline, it is a reasonable bet that a member of the Leakey family will turn out to be responsible. Of Richard Leakey I once wrote that he is

> ...a robust hero of a man, who actually lives up to the cliché, 'a big man in every sense of the word'. Like other big men he is loved by many, feared by some, and not over-preoccupied with the judgments of any.

Through turbulent years in and out of Kenyan politics, on two separate occasions running the Kenya Wildlife Service (which he revitalized into a crack fighting force against the elephant poachers), he is still best known for his work on hominid fossils. Here, writing with Roger Lewin, he describes the dramatic discovery, by his co-worker Kamoya Kimeu, of the Turkana Boy, the most complete specimen of *Homo erectus* (*ergaster* in the terminology of some authorities), the species that is the immediate ancestor of *Homo sapiens*. ■

A Giant Lake

'Kamoya has found a small piece of hominid frontal, about 1.5 by 2 inches, in good condition', Alan recorded in his field diary on August 23, 1984. 'It was on a slope on the bank opposite the camp. The slope itself is covered with black lava pebbles. How he found it, I'll never know.'

Kamoya's skill at finding hominid fossils is legendary. A fossil hunter needs sharp eyes and a keen search image, a mental template that subconsciously evaluates everything he sees in his search for telltale clues. A kind of mental radar works even if he isn't concentrating hard. A fossil mollusk expert has a mollusk search image. A fossil antelope expert has an antelope search image. Kamoya is a fossil hominid expert, and there is no one better at finding fossilized remains of our ancestors. Yet even when one has a good internal radar, the search is incredibly more difficult than it sounds. Not only are the fossils often the same color as the rocks among which they are found, so they blend in with the background; they are also usually broken into odd-shaped fragments. The search image has to accommodate this complication.

In our business, we don't expect to find a whole skull lying on the surface, staring up at us. The typical find is a small piece of petrified bone. The fossil hunter's search image therefore has to have an infinite number of dimensions, matching every conceivable angle of every shape of fragment of every bone in the human body. Often Kamoya can spot a hominid fossil fragment on a rock-strewn sediment slope from a dozen paces; someone else on his hands and knees staring right at it might fail to see it.

I met Kamoya in 1964, on my first serious foray into the hominid fossil business. He was part of a team of workers on an expedition to Lake Natron, just over the southwest border with Tanzania. We immediately struck up a friendship and professional relationship that has continued ever since. He demonstrated his skill even then, by finding a fossil hominid jaw of the same species as *Zinjanthropus*, which my mother had discovered five years earlier at Olduvai Gorge. Kamoya's find was the only known lower jaw of this hominid species, so I was very impressed. Particularly so as Kamoya spotted the fossil barely protruding from a cliff face, not two feet from where I had been looking for fossils a little earlier. Part of Kamoya's secret is that, although he's a stockily built man with a great sense of calm about him, he is always on the move, restless, rarely idle. So it was when he found the piece of hominid skull that had brought Alan and me on our trip to west Turkana.

'We had our camp by the Nariokotome River', explains Kamoya. 'It's dry most of the time, but about a hundred yards upstream from the camp you can dig down and find water, two feet down if there has been rain recently, maybe ten feet if it's been very dry. But you can always find water.' Kamoya and his team were on their way from the northern part of the western shore to some areas in the south, where we knew there were some promising fossil deposits. The geologist Frank Brown and the paleontologist John Harris were part of this north-to-south sweep, the final stages of a four-year survey of likely fossil localities on the west side. We had decided that 1984 would be the year serious work began in the search for hominid fossils there. And we had reason to be optimistic, because a couple of small fragments had been discovered early in the survey.

Nariokotome had been the site of a camp the previous year, so Kamoya knew that shade and water could be found there. 'We arrived about midday, dirty and tired', he recalls. 'The first thing we did was to look for water. Yes, it was there, just like last year, except we had to dig a little deeper this year.' Their bodies and clothes washed, lunch eaten, the men declared the rest of the day was a holiday. Not for Kamoya. He thought he would take a look at a gully across the dry riverbed, just three hundred yards away.

'I don't know what it was about that gully that attracted Kamoya to it', says Frank Brown, who had been with Kamoya the previous three seasons of the west-side survey. 'We passed by in 1981, the second year of the survey, and he took a look then but found nothing. It was the same the next year. Nothing. And then this year, 1984, bingo! He finds a hominid.' Kamoya's explanation is typically enigmatic: 'It just looked interesting.' I count myself a fairly skilled hunter of fossil hominids too, and I occasionally get a sense—nothing tangible—that I'm going to find something, so I understand what Kamoya means. But even to my eye this gully looked unpromising, a scatter of pebbles on a slope, a goat track snaking by a ragged thorn bush, the dry bed of a stream that cuts the gully, and a local dirt road just a few yards away, running north to south.

'The soil's a light color in the gully,' explains Kamoya, 'and the stones are black, pieces of lava. The fossil is a little lighter than the lava, so it was easy to see. I found what I was looking for.' The fragment was not much bigger than a couple of postage stamps put together, but nevertheless it was diagnostic. A flattish piece of bone with a slight curvature indicated skull, and a skull from a big-brained animal. In addition, the impression of the brain on the inner surface was very faint. Together, these clues triggered Kamoya's search image to say *hominid skull*. A similar piece of bone, thinner, with a tighter curvature and with deeper brain impressions on the inner surface might have indicated an antelope, for instance.

It wasn't immediately obvious where on the hominid skull Kamoya's fossil fragment had come from, but it turned out to be a part of the frontal region. Kamoya did know that the skull was more than a million years old—1.6 million years, according to Frank Brown's calculation—so he guessed he had found a *Homo erectus*, the hominid species directly ancestral to *Homo sapiens*.

The earliest member of the hominid family evolved somewhere between five and ten million years ago, according to current estimates. A good average date, therefore, is 7.5 million years ago for the origin of the first hominid species. One of the defining characteristics of hominids is the mode of locomotion: we and all our immediate ancestors walked erect on two legs, or bipedally. Although the earliest members of the family were bipedal, and therefore had their hands free from the immediate business of locomotion, the making of stone tools and the expansion of the brain came relatively late in our history, beginning about 2.5 million years ago. There is some debate about it, but I am convinced that the making of stone tools is a characteristic of our own branch of the human family, the *Homo* lineage, and that it is closely associated with the expansion of the brain. The evolutionary increments in these respects were small at first but became significant with the appearance of *Homo erectus*. As we shall see throughout this book, the origin of *Homo erectus* represents a major turning point in human history. From the vantage point of today, it speaks of leaving an essentially apelike past and embarking on a distinctly humanlike future. For this reason, Kamoya's discovery was potentially very important.

'I called my people over,' says Kamoya, 'and we searched the ground surface. We found one more piece, but that was all. So we built a pile of stones, a cairn, to mark exactly where the fossils had come from.' It was too late that evening to call me in Nairobi, so Kamoya had waited until the following morning to give the news. In fact, the news was twofold, because a little earlier John Harris also had found a piece of skull, which he thought might be a hominid or a large monkey. This one was about two million years old, again according to Frank's initial estimate. So when Kamoya and Peter met Alan and me at the airstrip, there was a lot to talk about, plans to make for further exploration of the two fossil discoveries—and, of course, camp gossip.

[…]

'We have many bones to show you', promised Kamoya as we unpacked the belly of the plane. 'You will like the hominids.' I knew I would. 'Skeletons?' I joked, and we all laughed at the improbable prospect. With evening upon us, we drank beer by the mess tent; the darkness fell quickly, as it always does this close to the equator.

Over dinner we discussed our plans, the thought of new hominid fossils uppermost in our minds. I proposed that our first visit the next day be to John's site to see the fragment that might be hominid or a large monkey. If the fossil really was hominid, and if it really was two million years old, it could be very important. The hominid story around two million years ago is unclear but is crucial to the origins of the big-brained creatures that eventually became us, so any new fossil is potentially illuminating. I had the feeling that one day we would be in for some real surprises in this slice of our prehistory, two million years ago. Perhaps John's fossil would provide it. On the other hand, I was not optimistic about the prospects of Kamoya's *Homo erectus* site. 'Seldom have I seen anything less hopeful', I recorded in my diary before turning in that night, tired but happy to be at the lake.

[...]

Over lunch [the next day] we decided that the rest of the Hominid Gang would start sieving operations at Kamoya's site, which was close to camp. Alan, John, and I would join them. 'We sieved for about two hours,' Alan recorded in his field diary that evening, 'picking off the boulder lag and screening. It is very dusty and the stones are black.' It wasn't pleasant, and I knew it would get worse.

[...]

After two hours of scooping dry earth, shaking it through mesh, and finding absolutely nothing of any interest, we found our enthusiasm waning, and Frank asked whether we'd like to see some fossil stromatolites he'd found. Needing no more of a pretext, Alan and I excused ourselves from the sieving and set off. John came too. We all thought nothing more would come of the work at hand.

[...]

We visited several other fossil localities as John told us more of what he'd found in the weeks before we arrived, more of our ancestors' environmental setting. We headed back to camp, giving little thought to the sieving task we had left behind at Kamoya's hominid site. But as we neared the shade of the Nariokotome camp we heard people shouting: 'We've found more bone! Lots of skull!'

We ran to where Kamoya was sitting, his treasure arrayed before him, like jewels plucked from the dry earth. 'The right temporal, left and right parietals, and bits of frontals of a beautifully preserved (if broken) *Homo erectus*', is how Alan described the find in his field diary. 'That's a lesson', I later noted in mine. 'The most unpromising site, as Kamoya's surely was, can sometimes surprise us.' Like everyone else, I was elated. There was great excitement, joking, and laughter. Here, beginning to take shape before our eyes, was part of the front and sides of the cranium of a human ancestor, *Homo erectus*, upright man.

Donald C. Johanson and Maitland A. Edey

from LUCY

If there is one fossil that is even more celebrated than the Turkana Boy, it is Lucy, *Australopithecus afarensis*. And if there is one among paleoan-thropologists who rivals Richard Leakey as the alpha male of the tribe, it is Donald Johanson. The peerless joy of scientific discovery rings through Johanson's description, written with Maitland Edey, of the sensational finding of AL 288-1, and of her being named after 'Lucy in the sky with diamonds' as the camp tape recorder belted the Beatles out at full volume into the night sky: 'We were sky-high, you must remember, from finding her.' 'The camp was rocking with excitement. That first night we never went to bed at all.'

Any science can be like that, if you understand it properly.

Mornings are not my favorite time. I am a slow starter and much prefer evenings and nights. At Hadar I feel best just as the sun is going down. I like to walk up one of the exposed ridges near the camp, feel the first stirrings of evening air and watch the hills turn purple. There I can sit alone for a while, think about the work of the day just ended, plan the next, and ponder the larger questions that have brought me to Ethiopia. Dry silent places are intensifiers of thought, and have been known to be since early Christian anchorites went into the desert to face God and their own souls.

Tom Gray joined me for coffee. Tom was an American graduate student who had come out to Hadar to study the fossil animals and plants of the region, to reconstruct as accurately as possible the kinds and frequencies and relationships of what had lived there at various times in the remote past and what the climate had been like. My own target—the reason for our expedition—was hominid fossils: the bones of extinct human ancestors and their close relatives. I was interested in the evidence for human evolution. But to understand that, to interpret any hominid fossils we might find, we had to have the supporting work of other specialists like Tom.

'So, what's up for today?' I asked.

Tom said he was busy marking fossil sites on a map.

'When are you going to mark in Locality 162?'

'I'm not sure where 162 is', he said.

'Then I guess I'll have to show you.' I wasn't eager to go out with Gray that morning. I had a tremendous amount of work to catch up on. We had had a number of visitors to the camp recently. Richard and Mary Leakey, two well-known experts on hominid fossils from Kenya, had left only the day before. During their stay I had not done any paperwork, any cataloging. I had not written any letters or done detailed descriptions of any fossils. I *should* have stayed in camp that morning—but I didn't. I felt a strong subconscious urge to go with Tom, and I obeyed it. I wrote a note to myself in my daily diary: *Nov. 30, 1974. To Locality 162 with Gray in AM. Feel good.*

As a paleoanthropologist—one who studies the fossils of human ancestors—I am superstitious. Many of us are, because the work we do depends a great deal on luck. The fossils we study are extremely rare, and quite a few distinguished paleoanthropologists have gone a lifetime without finding a single one. I am one of the more fortunate. This was only my third year in the field at Hadar, and I had already found several. I know I am lucky, and I don't try to hide it. That is why I wrote 'feel good' in my diary. When I got up that morning I felt it was one of those days when you should press your luck. One of those days when something terrific might happen.

Throughout most of that morning, nothing did. Gray and I got into one of the expedition's four Land-Rovers and slowly jounced our way to Locality 162. This was one of several hundred sites that were in the process of being plotted on a master map of the Hadar area, with detailed information about geology and fossils being entered on it as fast as it was obtained. Although the spot we were headed for was only about four miles from camp, it took us half an hour to get there because of the rough terrain. When we arrived it was already beginning to get hot.

At Hadar, which is a wasteland of bare rock, gravel and sand, the fossils that one finds are almost all exposed on the surface of the ground. Hadar is in the center of the Afar desert, an ancient lake bed now dry and filled with sediments that record the history of past geological events. You can trace volcanic-ash falls there, deposits of mud and silt washed down from distant mountains, episodes of volcanic dust, more mud, and so on. Those events reveal themselves like layers in a slice of cake in the gullies of new young rivers that recently have cut through the lake bed here and there. It seldom rains at Hadar, but when it does it comes in an overpowering gush—six months' worth overnight. The soil, which is bare of vegetation, cannot hold all that water. It roars down the gullies, cutting back their sides and bringing more fossils into view.

Gray and I parked the Land-Rover on the slope of one of those gullies. We were careful to face it in such a way that the canvas water bag that was hanging from the side mirror was in the shade. Gray plotted the locality on the map. Then we got out and began doing what most members of the expedition spent a great deal of their time doing: we began surveying, walking slowly about, looking for exposed fossils.

Some people are good at finding fossils. Others are hopelessly bad at it. It's a matter of practice, of training your eye to see what you need to see.

I will never be as good as some of the Afar people. They spend all their time wandering around in the rocks and sand. They have to be sharp-eyed; their lives depend on it. Anything the least bit unusual they notice. One quick educated look at all those stones and pebbles, and they'll spot a couple of things a person not acquainted with the desert would miss.

Tom and I surveyed for a couple of hours. It was now close to noon, and the temperature was approaching 110. We hadn't found much: a few teeth of the small extinct horse *Hipparion*; part of the skull of an extinct pig; some antelope molars; a bit of a monkey jaw. We had large collections of all these things already, but Tom insisted on taking these also as added pieces in the overall jigsaw puzzle of what went where.

'I've had it,' said Tom. 'When do we head back to camp?'

'Right now. But let's go back this way and survey the bottom of that little gully over there.'

The gully in question was just over the crest of the rise where we had been working all morning. It had been thoroughly checked out at least twice before by other workers, who had found nothing interesting. Nevertheless, conscious of the 'lucky' feeling that had been with me since I woke, I decided to make that small final detour. There was virtually no bone in the gully. But as we turned to leave, I noticed something lying on the ground partway up the slope.

'That's a bit of a hominid arm,' I said.

'Can't be. It's too small. Has to be a monkey of some kind.'

We knelt to examine it.

'Much too small,' said Gray again.

I shook my head. 'Hominid.'

'What makes you so sure?' he said.

'That piece right next to your hand. That's hominid too.'

'Jesus Christ,' said Gray. He picked it up. It was the back of a small skull. A few feet away was part of a femur: a thighbone. 'Jesus Christ,' he said again. We stood up, and began to see other bits of bone on the slope: a couple of vertebrae, part of a pelvis—all of them hominid. An unbelievable, impermissible thought flickered through my mind. Suppose all these fitted together? Could they be parts of a single, extremely primitive skeleton? No such skeleton had ever been found—anywhere.

'Look at that', said Gray. 'Ribs.'

A single individual

'I can't believe it,' I said. 'I just can't believe it.'

'By God, you'd better believe it!' shouted Gray. 'Here it is, Right here!' His voice went up into a howl. I joined him. In that 110-degree heat we began jumping up and down. With nobody to share our feelings, we hugged each other, sweaty and smelly, howling and hugging in the heat-shimmering gravel, the small brown remains of what now seemed almost certain to be parts of a single hominid skeleton lying all around us.

'We've got to stop jumping around,' I finally said. 'We may step on something. Also, we've got to make sure.'

'Aren't you sure, for Christ's sake?'

'I mean, suppose we find two left legs. There may be several individuals here, all mixed up. Let's play it cool until we can come back and make absolutely sure that it all fits together.'

We collected a couple of pieces of jaw, marked the spot exactly and got into the blistering Land-Rover for the run back to camp. On the way we picked up two expedition geologists who were loaded down with rock samples they had been gathering.

'Something big,' Gray kept saying to them. 'Something big. Something *big.*'

'Cool it,' I said.

But about a quarter of a mile from camp, Gray could not cool it. He pressed his thumb on the Land-Rover's horn, and the long blast brought a scurry of scientists who had been bathing in the river. 'We've got it,' he yelled. 'Oh, Jesus, we've got it. We've got The Whole Thing!'

That afternoon everyone in camp was at the gully, sectioning off the site and preparing for a massive collecting job that ultimately took three weeks. When it was done, we had recovered several hundred pieces of bone (many of them fragments) representing about forty percent of the skeleton of a single individual. Tom's and my original hunch had been right. There was no bone duplication.

But a single individual of what? On preliminary examination it was very hard to say, for nothing quite like it had ever been discovered. The camp was rocking with excitement. That first night we never went to bed at all. We talked and talked. We drank beer after beer. There was a tape recorder in the camp, and a tape of the Beatles' song 'Lucy in the Sky with Diamonds' went belting out into the night sky, and was played at full volume over and over again out of sheer exuberance. At some point during that unforgettable evening—I no longer remember exactly when—the new fossil picked up the name of Lucy, and has been so known ever since, although its proper name—its acquisition number in the Hadar collection—is AL 288-1.

Stephen Jay Gould

'WORM FOR A CENTURY, AND ALL SEASONS'

The American palaeontologist Stephen Jay Gould was my exact contemporary and we enjoyed—or suffered—a kind of love/hate relationship, on opposite sides of the Atlantic and opposite sides of several schisms in the broad church of Darwinian theory. We disagreed about much, but each respected the other as a writer. A good case could be made that Gould was our generation's finest exponent of the scientific short story, and in choosing one for this collection I had an embarrassment of riches. After much dithering, I finally went for his essay on Darwin's 'worm book', which shows Gould at his best in so many ways. There is the love of history, and in particular the love of Darwin. There is the artful extraction of the general from the particular, of the overarching principle from minutely and lovingly dissected detail. Finally, there is the moving conclusion, where Gould unites the worms of Darwin's wise old age to the coral reefs of his brilliant youth:

> Was Darwin really conscious of what he had done as he wrote his last professional lines, or did he proceed intuitively, as men of his genius sometimes do? Then I came to the very last paragraph and I shook with the joy of insight. Clever old man; he knew full well. In his last words, he looked back to his beginning, compared those worms with his first corals, and completed his life's work in both the large and the small.

Gould shook with the joy of insight. When I first read those lines, I shook vicariously with him. What an exquisite piece of writing. ■

In the preface to his last book, an elderly Charles Darwin wrote: 'The subject may appear an insignificant one, but we shall see that it possesses some interest; and the maxim "de minimis lex non curat" [the law is not concerned with trifles] does not apply to science.'

Trifles may matter in nature, but they are unconventional subjects for last books. Most eminent graybeards sum up their life's thought and offer a few pompous suggestions for reconstituting the future. Charles Darwin wrote about worms—*The Formation of Vegetable Mould, Through the Action of Worms, With Observations on Their Habits* (1881).

This month[1] marks the one-hundredth anniversary of Darwin's death—and celebrations are under way throughout the world. Most symposiums and books are taking the usual high road of broad implication—Darwin and modern life, or Darwin and evolutionary thought. For my personal tribute, I shall take an ostensibly minimalist stance and discuss Darwin's 'worm book'. But I do this to argue that Darwin justly reversed the venerable maxim of his legal colleagues.

Darwin was a crafty man. He liked worms well enough, but his last book, although superficially about nothing else, is (in many ways) a covert summation of the principles of reasoning that he had laboured a lifetime to identify and use in the greatest transformation of nature ever wrought by a single man. In analysing his concern with worms, we may grasp the sources of Darwin's general success.

The book has usually been interpreted as a curiosity, a harmless work of little importance by a great naturalist in his dotage. Some authors have even used it to support a common myth about Darwin that recent scholarship has extinguished. Darwin, his detractors argued, was a man of mediocre ability who became famous by the good fortune of his situation in place and time. His revolution was 'in the air' anyway, and Darwin simply had the patience and pertinacity to develop the evident implications. He was, Jacques Barzun once wrote (in perhaps the most inaccurate epitome I have ever read), 'a great assembler of facts and a poor joiner of ideas…a man who does not belong with the great thinkers'.

To argue that Darwin was merely a competent naturalist mired in trivial detail, these detractors pointed out that most of his books are about minutiae or funny little problems—the habits of climbing plants, why flowers of different form are sometimes found on the same plant, how orchids are fertilized by insects, four volumes on the taxonomy of barnacles, and finally, how worms churn the soil. Yet all these books have both a manifest and a deeper or implicit theme—and detractors missed the second (probably because they didn't read the books and drew conclusions from the titles alone). In each case, the deeper subject is evolution itself or a larger research programme for analysing history in a scientific way.

Why is it, we may ask at this centenary of his passing, that Darwin is still so central a figure in scientific thought? Why must we continue to read his books and grasp his vision if we are to be competent natural historians? Why do scientists, despite their notorious unconcern with history, continue to ponder and debate his works? Three arguments might be offered for Darwin's continuing relevance to scientists.

We might honor him first as the man who 'discovered' evolution. Although popular opinion may grant Darwin this status, such an accolade is surely misplaced, for several illustrious predecessors shared his conviction that organisms are linked by ties of physical descent. In nineteenth-century biology, evolution was a common enough heresy.

As a second attempt, we might locate Darwin's primary claim upon continued scientific attention in the extraordinarily broad and radical implications of his proffered evolutionary mechanism—natural selection. Indeed, I have pushed this theme relentlessly in my two previous books, focusing upon three arguments: natural selection as a theory of local adaptation, not inexorable progress; the claim that order in nature arises as a coincidental by-product of struggle among individuals; and the materialistic character of Darwin's theory, particularly his denial of any causal role to spiritual forces, energies, or powers. I do not now abjure this theme, but I have come to realize that it cannot represent the major reason for Darwin's continued *scientific* relevance, though it does account for his impact upon the world at large. For it is too grandiose, and working scientists rarely traffic in such abstract generality.

Everyone appreciates a nifty idea or an abstraction that makes a person sit up, blink hard several times to clear the intellectual cobwebs, and reverse a cherished opinion. But science deals in the workable and soluble, the idea that can be fruitfully embodied in concrete objects suitable for poking, squeezing, manipulating, and extracting. The idea that counts in science must lead to fruitful work, not only to speculation that does not engender empirical test, no matter how much it stretches the mind.

I therefore wish to emphasize a third argument for Darwin's continued importance, and to claim that his greatest achievement lay in establishing principles of *useful* reason for sciences (like evolution) that attempt to reconstruct history. The special problems of historical science (as contrasted, for example, with experimental physics) are many, but one stands out most prominently: science must identify processes that yield observed results. The results of history lie strewn around us, but we cannot, in principle, directly observe the processes that produced them. How then can we be scientific about the past?

As a general answer, we must develop criteria for inferring the processes we cannot see from results that have been preserved. This is the quintessential problem of evolutionary theory: How do we use the anatomy, physiology, behaviour, variation, and geographic distribution of modern organisms, and the fossil remains in our geological record, to infer the pathways of history?

Thus, we come to the covert theme of Darwin's worm book, for it is both a treatise on the habits of earthworms and an exploration of how we can approach history in a scientific way.

Darwin's mentor, the great geologist Charles Lyell, had been obsessed with the same problem. He argued, though not with full justice, that his predecessors had failed to construct a science of geology because they had not developed procedures for inferring an unobservable past from a surrounding present and had therefore indulged in unprovable reverie and speculation. 'We see,' he wrote in his incomparable prose, 'the ancient spirit of speculation revived and a desire manifestly shown to cut, rather than patiently to untie, the Gordian Knot.' His solution, an aspect of the complex world view later called uniformitarianism, was to observe the work of present processes and to extrapolate their rates and effects into the past. Here Lyell faced a problem. Many results of the past—the Grand Canyon for example—are extensive and spectacular, but most of what goes on about us every day doesn't amount to much—a bit of erosion here or deposition there. Even a Stromboli or a Vesuvius will cause only local devastation. If modern forces do too little, then we must invoke more cataclysmic processes, now expired or dormant, to explain the past. And we are in catch-22: if past processes were effective and different from present processes, we might explain the past in principle, but we could not be scientific about it because we have no modern analogue in what we can observe. If we rely only upon present processes, we lack sufficient oomph to render the past.

Lyell sought salvation in the great theme of geology: time. He argued that the vast age of our earth provides ample time to render all observed results, however spectacular, by the simple summing of small changes over immense periods. Our failure lay, not with the earth, but with our habits of mind: we had been previously unwilling to recognize how much work the most insignificant processes can accomplish with enough time.

Darwin approached evolution in the same way. The present becomes relevant, and the past therefore becomes scientific, only if we can sum the small effects of present processes to produce observed results. Creationists did not use this principle and therefore failed to understand the relevance of small-scale variation that pervades the biological world (from breeds of dogs to geographical variation in butterflies). Minor variations are the stuff of evolution (not merely a set of accidental excursions around a created ideal type), but we recognize this only when we are prepared to sum small effects through long periods of time.

Darwin recognized that this principle, as a basic mode of reasoning in historical science, must extend beyond evolution. Thus, late in his life, he decided to abstract and exemplify his historical method by applying it to a problem apparently quite different from evolution—a project broad enough to cap an illustrious career. He chose earthworms and the soil. Darwin's refutation of the legal maxim 'de minimis lex non curat' was a conscious double-entendre. Worms are both humble and interesting, and a worm's work, when summed over all worms and long periods of time, can shape our landscape and form our soils.

Thus, Darwin wrote at the close of his preface, refuting the opinions of a certain Mr Fish who denied that worms could account for much 'considering their weakness and their size':

> Here we have an instance of that inability to sum up the effects of a continually recurrent cause, which has often retarded the progress of science, as formerly in the case of geology, and more recently in that of the principle of evolution.

Darwin had chosen well to illustrate his generality. What better than worms: the most ordinary, commonplace, and humble objects of our daily observation and dismissal. If they, working constantly beneath our notice, can form much of our soil and shape our landscape, then what event of magnitude cannot arise from the summation of small effects. Darwin had not abandoned evolution for earthworms; rather, he was using worms to illustrate the general method that had validated evolution as well. Nature's mills, like God's, grind both slowly and exceedingly small.

Darwin made two major claims for worms. First, in shaping the land, their effects are directional. They triturate particles of rock into ever smaller fragments (in passing them through their gut while churning the soil), and they denude the land by loosening and disaggregating the soil as they churn it; gravity and erosive agents then move the soil more easily from high to low ground, thus leveling the landscape. The low, rolling character of topography in areas inhabited by worms is, in large part, a testimony to their slow but persistent work.

Second, in forming and churning the soil, they maintain a steady state amidst constant change. As the primary theme of his book (and the source of its title), Darwin set out to prove that worms form the soil's upper layer, the so-called vegetable mold. He describes it in the opening paragraph:

> The share which worms have taken in the formation of the layer of vegetable mould, which covers the whole surface of the land in every moderately humid country, is the subject of the present volume. This mould is generally of a blackish color and a few inches in thickness. In different districts it differs but little in appearance, although it may rest on various subsoils.

The uniform fineness of the particles of which it is composed is one of its chief characteristic features.

Darwin argues that earthworms form vegetable mold by bringing 'a large quantity of fine earth' to the surface and depositing it there in the form of castings. (Worms continually pass soil through their intestinal canals, extract anything they can use for food, and 'cast' the rest; the rejected material is not feces but primarily soil particles, reduced in average size by trituration and with some organic matter removed.) The castings, originally spiral in form and composed of fine particles, are then disaggregated by wind and water, and spread out to form vegetable mold. 'I was thus led to conclude,' Darwin writes, 'that all the vegetable mould over the whole country has passed many times through, and will again pass many times through, the intestinal canals of worms.'

The mold doesn't continually thicken after its formation, for it is compacted by pressure into more solid layers a few inches below the surface. Darwin's theme here is not directional alteration, but continuous change within apparent constancy. Vegetable mold is always the same, yet always changing. Each particle cycles through the system, beginning at the surface in a casting, spreading out, and then working its way down as worms deposit new castings above; but the mold itself is not altered. It may retain the same thickness and character while all its particles cycle. Thus, a system that seems to us stable, perhaps even immutable, is maintained by constant turmoil. We who lack an appreciation of history and have so little feel for the aggregated importance of small but continuous change scarcely realize that the very ground is being swept from beneath our feet; it is alive and constantly churning.

Darwin uses two major types of arguments to convince us that worms form the vegetable mold. He first proves that worms are sufficiently numerous and widely spread in space and depth to do the job. He demonstrates 'what a vast number of worms live unseen by us beneath our feet'—some 53,707 per acre (or 356 pounds of worms) in good British soil. He then gathers evidence from informants throughout the world to argue that worms are far more widely distributed, and in a greater range of apparently unfavorable environments, than we usually imagine. He digs to see how deeply they extend into the soil, and cuts one in two at fifty-five inches, although others report worms at eight feet down or more.

With plausibility established, he now seeks direct evidence for constant cycling of vegetable mold at the earth's surface. Considering both sides of the issue, he studies the foundering of objects into the soil as new castings pile up above them, and he collects and weighs the castings themselves to determine the rate of cycling.

Darwin was particularly impressed by the evenness and uniformity of foundering for objects that had once lain together at the surface. He sought fields that, twenty years or more before, had been strewn with objects of substantial size—burned coals, rubble from the demolition of a building, rocks collected from the ploughing of a neighbouring field. He trenched these fields and found, to his delight, that the objects still formed a clear layer, parallel to the surface but now several inches below it and covered with vegetable mold made entirely of fine particles. 'The straightness and regularity of the lines formed by the embedded objects, and their parallelism with the surface of the land, are the most striking features of the case', he wrote. Nothing could beat worms for a slow and meticulous uniformity of action.

Darwin studied the sinking of 'Druidical stones' at Stonehenge and the foundering of Roman bathhouses, but he found his most persuasive example at home, in his own field, last plowed in 1841:

> For several years it was clothed with an extremely scant vegetation, and was so thickly covered with small and large flints (some of them half as large as a child's head) that the field was always called by my sons 'the stony field'. When they ran down the slope the stones clattered together. I remember doubting whether I should live to see these larger flints covered with vegetable mould and turf. But the smaller stones disappeared before many years had elapsed, as did every one of the larger ones after a time; so that after thirty years (1871) a horse could gallop over the compact turf from one end of the field to the other, and not strike a single stone with his shoes. To anyone who remembered the appearance of the field in 1842, the transformation was wonderful. This was certainly the work of the worms.

In 1871, he cut a trench in his field and found 2.5 inches of vegetable mold, entirely free from flints: 'Beneath this lay coarse clayey earth full of flints, like that in any of the neighbouring ploughed fields.... The average rate of accumulation of the mould during the whole thirty years was only .088 inch per year (i.e., nearly one inch in twelve years).'

In various attempts to collect and weigh castings directly, Darwin estimated from 7.6 to 18.1 tons per acre per year. Spread out evenly upon the surface, he calculated that from 0.8 to 2.2 inches of mold would form anew every ten years. In gathering these figures, Darwin relied upon that great, unsung, and so characteristically British institution—the corps of zealous amateurs in natural history, ready to endure any privation for a precious fact. I was particularly impressed by one anonymous contributor: 'A lady,' Darwin tells us, 'on whose accuracy I can implicitly rely, offered to collect during a year all the castings thrown up on two separate square yards, near Leith Hill Place, in Surrey.' Was she the analogue of a modern Park Avenue woman of means, carefully scraping up after her dog: one bag for a cleaner New York, the other for Science with a capital S?

The pleasure of reading Darwin's worm book lies not only in recognizing its larger point but also in the charm of detail that Darwin provides about worms themselves. I would rather peruse 30 pages of Darwin on worms than slog through 300 pages of eternal verities explicitly preached by many writers. The worm book is a labor of love and intimate, meticulous detail. In the book's other major section, Darwin spends 100 pages describing experiments to determine which ends of leaves (and triangular paper cutouts, or abstract 'leaves') worms pull into their burrows first. Here we also find an overt and an underlying theme, in this case leaves and burrows versus the evolution of instinct and intelligence, Darwin's concern with establishing a usable definition of intelligence, and his discovery (under that definition) that intelligence pervades 'lower' animals as well. All great science is a fruitful marriage of detail and generality, exultation and explanation. Both Darwin and his beloved worms left no stone unturned.

I have argued that Darwin's last book is a work on two levels—an explicit treatise on worms and the soil and a covert discussion of how to learn about the past by studying the present. But was Darwin consciously concerned with establishing a methodology for historical science, as I have argued, or did he merely stumble into such generality in his last book? I believe that his worm book follows the pattern of all his other works, from first to last: every compendium on minutiae is also a treatise on historical reasoning—and each book elucidates a different principle.

Consider his first book on a specific subject, *The Structure and Distribution of Coral-Reefs* (1842). In it, he proposed a theory for the formation of atolls, 'those singular rings of coral-land which rise abruptly out of the unfathomable ocean,' that won universal acceptance after a century of subsequent debate. He argued that coral reefs should be classified into three categories—fringing reefs that abut an island or continent, barrier reefs separated from island or continent by a lagoon, and atolls, or rings of reefs, with no platform in sight. He linked all three categories with his 'subsidence theory', rendering them as three stages of a single process: the subsidence of an island or continental platform beneath the waves as living coral continues to grow upward. Initially, reefs grow right next to the platform (fringing reefs). As the platform sinks, reefs grow up and outward, leaving a separation between sinking platform and living coral (a barrier reef). Finally the platform sinks entirely, and a ring of coral expresses its former shape (an atoll). Darwin found the forms of modern reefs 'inexplicable, excepting on the theory that their rocky bases slowly and successively sank beneath the level of the sea, whilst the corals continued to grow upwards'.

This book is about coral, but it is also about historical reasoning. Vegetable mold formed fast enough to measure its rate directly; we capture the past by summing effects of small and observable present causes. But what if rates are too slow, or scales too large, to render history by direct observation of present processes? For such cases, we must develop a different method. Since large-scale processes begin at different times and proceed at diverse rates, the varied stages of different examples should exist simultaneously in the present. To establish history in such cases, we must construct a theory that will explain a series of present phenomena as stages of a single historical process. The method is quite general. Darwin used it to explain the formation of coral reefs. We invoke it today to infer the history of stars. Darwin also employed it to establish organic evolution itself. Some species are just beginning to split from their ancestors, others are midway through the process, still others are on the verge of completing it.

But what if evidence is limited to the static object itself? What if we can neither watch part of its formation nor find several stages of the process that produced it? How can we infer history from a lion? Darwin treated this problem in his treatise on the fertilization of orchids by insects (1862); the book that directly followed the *Origin of Species*. I have discussed his solution in several essays and will not dwell on it here: we infer history from imperfections that record constraints of descent. The 'various contrivances' that orchids use to attract insects and attach pollen to them are the highly altered parts of ordinary flowers, evolved in ancestors for other purposes. Orchids work well enough, but they are jury-rigged to succeed because flowers are not optimally constructed for modification to these altered roles. If God wanted to make insect attractors and pollen stickers from scratch, he would certainly have built differently.

Thus, we have three principles for increasing adequacy of data: if you must work with a single object, look for imperfections that record historical descent; if several objects are available, try to render them as stages of a single historical process; if processes can be directly observed, sum up their effects through time. One may discuss these principles directly or recognize the 'little problems' that Darwin used to exemplify them: orchids, coral reefs, and worms—the middle book, the first, and the last.

Darwin was not a conscious philosopher. He did not, like Huxley and Lyell, write explicit treatises on methodology. Yet I do not think he was unaware of what he was doing, as he cleverly composed a series of books at two levels, thus expressing his love for nature in the small and his ardent desire to establish both evolution and the principles of historical science. I was musing on this issue as I completed the worm book two weeks ago. Was Darwin really conscious of what he had done as he wrote his last professional lines, or did he proceed intuitively, as men of his genius sometimes do? Then I came to the very last paragraph, and I shook with the joy of insight. Clever old man; he knew full well. In his last words, he looked back to his beginning, compared those worms with his first corals, and completed his life's work in both the large and the small:

> The plough is one of the most ancient and most valuable of man's inventions; but long before he existed the land was in fact regularly ploughed, and still continues to be thus ploughed by earthworms. It may be doubted whether there are many other animals which have played so important a part in the history of the world, as have these lowly organized creatures. Some other

animals, however, still more lowly organized, namely corals, have done more conspicuous work in having constructed innumerable reefs and islands in the great oceans; but these are almost confined to the tropical zones.

At the risk of unwarranted ghoulishness, I cannot suppress a final irony. A year after publishing his worm book, Darwin died on 19 April 1882. He wished to be buried in the soil of his adopted village, where he would have made a final and corporeal gift to his beloved worms. But the sentiments (and politicking) of fellow scientists and men of learning secured a guarded place for his body within the well-mortared floor of Westminster Abbey. Ultimately the worms will not be cheated, for there is no permanence in history, even for cathedrals. But ideas and methods have all the immortality of reason itself. Darwin has been gone for a century, yet he is with us whenever we choose to think about time.

1. Darwin died on 19 April 1882 and this column first appeared in *Natural History* in April 1982.

John Tyler Bonner

from LIFE CYCLES

Biologists have learned that if they want to understand how some great feat of evolution happened it is an enormous help if it happened more than once, because this gives us a basis for comparison. The genetic code, unfortunately, seems to have evolved only once (or if there were rival codes, perhaps resembling those devised by Crick and Gamow, they are no longer with us). If we are to get a comparative perspective on how genetic codes evolve, we'll have to wait until extraterrestrial life is discovered, and that probably means waiting forever. Multicellularity—the coming together of cells in bodies—evolved several times. One of these independent evolutions is so alien and strange that it might as well have evolved on Mars; and the other-worldly perspective that it affords us might illuminate our own more familiar multicellularity. Nobody knows more about slime moulds than John Tyler Bonner, and he tells us about them here. I have never met him, but I have long felt an affinity with his genial and imaginative writing style. ▪

Beginnings

I have devoted my life to slime molds. This may seem a peculiar occupation—narrow at best, slightly revolting at its worst—but let me explain why they captivated me and how they opened my eyes so that I wanted to understand not only what made them tick, but how they fit into the general pattern of living things and what the principles are that integrate all of life.

Slime molds are an extremely common organism, widespread all over the world. Yet because they are microscopic and live mostly in the darkness of the soil, they are hard to see, and for that reason they have been little known until recent years. However, if one takes a small bit of topsoil or humus from almost anywhere and brings it into the laboratory, one can easily grow them on small petri dishes containing transparent agar culture medium. There, through the low powers of a microscope, it is possible to follow their life cycle, which to me has always been a sight of great beauty.

The molds begin as encapsulated spores which split open, and out of each spore emerges a single amoeba. This amoeba immediately begins to feed on the bacteria that are supplied as food, and after about three hours of eating they divide in two. At this rate it does not take long for them to eat all the bacteria on the agar surface—usually about two days. Next comes the magic. After a few hours of starvation, these totally independent cells stream into aggregation centers to form sausage-shaped masses of cells, each of which now acts as an organized multicellular organism. It can crawl towards light, orient in heat gradients, and show an organized unity in various other ways. It looks like a small, translucent slug about a millimeter long (indeed, this migrating mass of amoebae is now commonly called a 'slug'). It has clear front and hind ends, and its body is sheathed in a very delicate coating of slime which it leaves behind as it moves, looking like a microscopic, collapsed sausage casing.

After a period of migration whose length depends very much on the conditions of the slug's immediate environment, the slug stops, points up into the air, and slowly transforms itself into a fruiting body consisting of a delicately tapered stalk one or more millimeters high, with a terminal globe of spores at its tip. This wonderful metamorphosis is achieved first by the anterior cells of the slug, which will become the stalk cells. They form a small, internal cellulose cylinder that is continuously extended at the tip. As this is occurring, the anterior cells around the top of the newly created cylinder pour into the cylinder, like a fountain flowing in reverse. The result is that the tip of the cylinder (which is the stalk) rises up into the air. As it does, the mass of posterior cells, which are to become the spores, adheres to the rising tip, and in this way the spore mass is lifted upward. During this process each amoeba in the spore mass becomes a spore, imprisoned in a thick-walled, capsule-shaped coat, ready to begin the next generation. The stalk cells inside the thin, tapering cellulose cylinder become large with huge, internal vacuoles; during this process they die, using up their last supplies of energy to build thick cellulose walls. It is a remarkable fact that the anterior cells, on the other hand—the leaders in the crawling slug—die, while the laggard cells in the hind region turn into spores, any one of which can start a new generation. Slime molds seem to support the old army principle of never going out in front—never volunteer for anything.

This entire life cycle (which happens to be asexual) takes about four days in the laboratory. The organisms are very easy to grow, and in many ways ideal for experimental work. The species I have described is only one of about fifty species, making comparative studies possible. Today, in this modern, technical world, one can view one's experiments with extraordinary ease. For instance, I have in my laboratory a video camera on my microscope, and on the screen I can follow the results of any operation I might perform on the migrating slug. If I follow it for two hours, I can immediately play back the changes on time lapse, so the two hours can be speeded up to two minutes. The possibilities make going to the laboratory each day a delight of anticipation. The life of an experimental biologist is one of minute and often humdrum detail involving endless, frustrating experiments that do not work, but the rewards, albeit rare,

are great. Suddenly—and how exciting it is when it happens—something will go right and give one a flash of insight into how things work.

A few years ago an old friend who happened to be a veterinarian was sick in the hospital recovering from an operation. While I was visiting him, his surgeon came by and my friend introduced me as 'Dr Bonner'. The surgeon asked me, 'Are you a small animal or a large animal man?' Without thinking, and somewhat to his alarm, I replied that I was a 'teensy-weensy animal man'. I have often thought of this episode in the context of the many years I have helped students revive their sick cultures into healthy and thriving ones. One of my main roles in life, then, has been that of a slime mold veterinarian.

Oliver Sacks

from UNCLE TUNGSTEN

■ Oliver Sacks is best known as a sympathetic clinical neurologist with a wonderful portfolio of unsettling case histories and a gift for recounting and drawing lessons from them. But he himself was first drawn to science by a love of chemistry, awakened by his mother's brother, 'Uncle Tungsten', surely the best sort of uncle any child could wish for. Sacks is still fascinated by chemistry, and this extract from his *Uncle Tungsten* well conveys the romantic pull that science can exert on an intelligent young mind. John Maynard Smith and I were once being shown around the Panama jungle as honoured guests by a young American researcher, and Maynard Smith whispered to me, 'Isn't it nice to listen to a man who really loves his animals.' The 'animals' in this case were trees. In Oliver Sacks's case they are elements, but the principle remains. ■

We had called him Uncle Tungsten for as long as I could remember, because he manufactured lightbulbs with filaments of fine tungsten wire.

His firm was called Tungstalite, and I often visited him in the old factory in Farringdon and watched him at work, in a wing collar, with his shirtsleeves rolled up. The heavy, dark tungsten powder would be pressed, hammered, sintered at red heat, then drawn into finer and finer wire for the filaments. Uncle's hands were seamed with the black powder, beyond the power of any washing to get out (he would have to have the whole thickness of epidermis removed, and even this, one suspected, would not have been enough). After thirty years of working with tungsten, I imagined, the heavy element was in his lungs and bones, in every vessel and viscus, every tissue of his body. I thought of this as a wonder, not a curse—his body invigorated and fortified by the mighty element, given a strength and enduringness almost more than human.

Whenever I visited the factory, he would take me around the machines, or have his foreman do so. (The foreman was a short, muscular man, a Popeye with enormous forearms, a palpable testament to the benefits of working with tungsten.) I never tired of the ingenious machines, always beautifully clean and sleek and oiled, or the furnace where the black powder was compacted from a powdery incoherence into dense, hard bars with a grey sheen.

During my visits to the factory, and sometimes at home, Uncle Dave would teach me about metals with little experiments. I knew that mercury, that strange liquid metal, was incredibly heavy and dense. Even lead floated on it, as my uncle showed me by floating a lead bullet in a bowl of quicksilver. But then he pulled out a small grey bar from his pocket, and to my amazement, this sank immediately to the bottom. That, he said, was *his* metal, tungsten.

Uncle loved the density of the tungsten he made, and its refractoriness, its great chemical stability. He loved to handle it—the wire, the powder, but the massy little bars and ingots most of all. He caressed them, balanced them (tenderly, it seemed to me) in his hands. 'Feel it, Oliver,' he would say, thrusting a bar at me. 'Nothing in the world feels like sintered tungsten.' He would tap the little bars and they would emit a deep clink. 'The sound of tungsten,' Uncle Dave would say, 'nothing like it.' I did not know whether this was true, but I never questioned it.

[...]

Uncle Dave loved handling the metals and minerals in his cabinet, allowing me to handle them, expatiating on their wonders. He saw the whole earth, I think, as a gigantic natural laboratory, where heat and pressure caused not only vast geologic movements, but innumerable chemical miracles too. 'Look at these diamonds,' he would say, showing me a specimen from the famous Kimberley mine. 'They are almost as old as the earth. They were formed thousands of millions of years ago, deep in the earth, under unimaginable pressures. Then they were brought to the surface in this kimberlite, tracking hundreds of miles from the earth's mantle, and then through the crust, till they finally reached the surface. We may never see the interior of the earth directly, but this kimberlite and its diamonds are a sample of what it is like. People have tried to manufacture diamonds,' he added, 'but we cannot match the temperatures and pressures that are necessary.'[1]

On one visit, Uncle Dave showed me a large bar of aluminum. After the dense platinum metals, I was amazed at how light it was, scarcely heavier than a piece of wood. 'I'll show you something interesting', he said. He took a smaller lump of aluminum, with a smooth, shiny surface, and smeared it with mercury. All of a sudden—it was like some terrible disease—the surface broke down, and a white substance like a fungus rapidly grew out of it, until it was a quarter of an inch high, then half an inch high, and it kept growing and growing until the aluminum was completely eaten up. 'You've seen iron rust—oxidizing, combining with the oxygen in the air', Uncle said. 'But here, with the aluminum, it's a million times faster. That big bar is still quite shiny, because it's covered by a fine layer of oxide, and that protects it from further change. But rubbing it with mercury destroys the surface layer, so then the aluminum has no protection, and it combines with the oxygen in seconds.'

I found this magical, astounding, but also a little frightening—to see a bright and shiny metal reduced so quickly to a crumbling mass of oxide. It made me think of a curse or a spell, the sort of disintegration I sometimes saw in my dreams. It made me think of mercury as evil, as a destroyer of metals. Would it do this to every sort of metal?

'Don't worry,' Uncle answered, 'the metals we use here, they're perfectly safe. If I put this little bar of tungsten in the mercury, it would not be affected at all. If I put it away for a million years, it would be just as bright and shiny as it is now.' The tungsten, at least, was stable in a precarious world.

'You've seen,' Uncle Dave went on, 'that when the surface layer is broken, the aluminum combines very rapidly with oxygen in the air to form this white oxide, which is called alumina. It is similar with iron as it rusts; rust is an iron oxide. Some metals are so avid for oxygen that they will combine with it, tarnishing, forming an oxide, the moment they are exposed to the air. Some will even pull the oxygen out of water, so one has to keep them in a sealed tube or under oil.' Uncle showed me some chunks of metal with a whitish surface, in a bottle of oil. He fished out a chunk and cut it with his penknife. I was amazed at how soft it was; I had never seen a metal cut like this. The cut surface had a brilliant, silvery luster. This was calcium, Uncle said, and it was so active that it never occurred in nature as the pure metal, but only as compounds or minerals from which it had to be extracted. The white cliffs of Dover, he said, were chalk; others were made of limestone—these were different forms of calcium carbonate, a major component in the crust of the earth. The calcium metal, as we spoke, had oxidized completely, its bright surface now a dull, chalky white. 'It's turning into lime,' Uncle said, 'calcium oxide.'

But sooner or later Uncle's soliloquies and demonstrations before the cabinet all returned to *his* metal. 'Tungsten,' he said. 'No one realized at first how perfect a metal it was. It has the highest melting point of any metal, it is tougher than steel, and it keeps its strength at high temperatures—an ideal metal!'

Uncle had a variety of tungsten bars and ingots in his office. Some he used as paperweights, but others had no discernible function whatever, except to give pleasure to their owner and maker. And indeed, by comparison, steel bars and even lead felt light and somehow porous, tenuous. 'These lumps of tungsten have an extraordinary concentration of mass,' he would say. 'They would be deadly as weapons—far deadlier than lead.'

They had tried to make tungsten cannonballs at the beginning of the century, he added, but found the metal too hard to work—though they used it sometimes for the bobs of pendulums. If one wanted to weigh the earth, Uncle Dave suggested, and to use a very dense, compact mass to 'balance' against it, one could do no better than to use a huge sphere of tungsten. A ball only two feet across, he calculated, would weigh five thousand pounds.

One of tungsten's mineral ores, scheelite, Uncle Dave told me, was named after the great Swedish chemist Carl Wilhelm Scheele, who was the first to show that it contained a new element. The ore was so dense that miners called it 'heavy stone' or *tung sten*, the name subsequently given to the element itself. Scheelite was found in beautiful orange crystals that fluoresced bright blue in ultraviolet light. Uncle Dave kept specimens of scheelite and other fluorescent minerals in a special cabinet in his office. The dim light of Farringdon Road on a November evening, it seemed to me, would be transformed when he turned on his Wood's lamp and the luminous chunks in the cabinet suddenly glowed orange, turquoise, crimson, green.

Though scheelite was the largest source of tungsten, the metal had first been obtained from a different mineral, called wolframite. Indeed, tungsten was sometimes called wolfram, and still retained the chemical symbol W. This thrilled me, because my own middle name was Wolf. Heavy seams of the tungsten ores were often found with tin ore, and the tungsten made it more difficult to isolate the tin. This was why, my uncle continued, they had originally called the metal wolfram—for, like a hungry animal, it 'stole' the tin. I liked the name *wolfram*, its sharp, animal quality, its evocation of a ravening, mystical wolf—and thought of it as a tie between Uncle Tungsten, Uncle Wolfram, and myself, O. Wolf Sacks.

[...]

Scheele was one of Uncle Dave's great heroes. Not only had he discovered tungstic acid and molybdic acid (from which the new element molybdenum was made), but hydrofluoric acid, hydrogen sulfide, arsine, and prussic acid, and a dozen organic acids, too. All this, Uncle Dave said, he did by himself, with no assistants, no funds, no university position or salary, but working alone, trying to make ends meet as an apothecary in a small provincial Swedish town. He had discovered oxygen, not by a fluke, but by making it in several different ways; he had discovered chlorine; and he had pointed the way to the discovery of manganese, of barium, of a dozen other things.

Scheele, Uncle Dave would say, was wholly dedicated to his work, caring nothing for fame or money and sharing his knowledge, whatever he had, with anyone and everyone. I was impressed by Scheele's generosity, no less than his resourcefulness, by the way in which (in effect) he gave the actual discovery of elements to his students and friends—the discovery of manganese to Johan Gahn, the discovery of molybdenum to Peter Hjelm, and the discovery of tungsten itself to the d'Elhuyar brothers.

Scheele, it was said, never forgot anything if it had to do with chemistry. He never forgot the look, the feel, the smell of a substance, or the way it was transformed in chemical reactions, never forgot anything he read, or was told, about the phenomena of chemistry. He seemed indifferent, or inattentive, to most things else, being wholly dedicated to his single passion, chemistry. It was this pure and passionate absorption in phenomena—noticing everything, forgetting nothing—that constituted Scheele's special strength.

Scheele epitomized for me the romance of science. There seemed to me an integrity, an essential goodness, about a life in science, a lifelong love affair. I had never given much thought to what I might be when I was 'grown up'—growing up was hardly imaginable—but now I knew: I wanted to be a chemist. A chemist like Scheele, an eighteenth-century chemist coming fresh to the field, looking at the whole undiscovered world of natural substances and minerals, analyzing them, plumbing their secrets, finding the wonder of unknown and new metals.

1. There were many attempts to manufacture diamonds in the nineteenth century, the most famous being those of Henri Moissan, the French chemist who first isolated fluorine and invented the electrical furnace. Whether Moissan actually got any diamonds is doubtful—the tiny, hard crystals he took for diamond were probably silicon carbide (which is now called moissanite). The atmosphere of this early diamond-making, with its excitements, its dangers, its wild ambitions, is vividly conveyed in H. G. Wells's story 'The Diamond Maker'.

Lewis Thomas

'SEVEN WONDERS'

I have long admired the writings of Lewis Thomas, whom I bracketed in my mind with that other fine American writer of science, Loren Eiseley, and I was especially delighted when Rockefeller University, in 2007, awarded me the Lewis Thomas Prize. The prize recognizes scientists who give us 'not merely new information but cause for reflection, even revelation, as in a poem or painting'. I immediately disclaim any such accolade for me, but the description is obviously, and correctly, intended to apply to Lewis Thomas himself, a distinguished medical scientist who carried on the long and hon-ourable tradition of literary doctors from Chekhov to Somerset Maugham. Even by the high standards of literary doctors, however, Lewis Thomas is an outstanding stylist. Here, he considers his personal 'seven wonders'. By the way, the scrapie agent is now known to be a protein—a prion. ◼

A while ago I received a letter from a magazine editor inviting me to join six other people at dinner to make a list of the Seven Wonders of the Modern World, to replace the seven old, out-of-date Wonders. I replied that I couldn't manage it, not on short order anyway, but still the question keeps hanging around in the lobby of my mind. I had to look up the old biodegradable Wonders, the Hanging Gardens of Babylon and all the rest, and then I had to look up that word 'wonder' to make sure I understood what it meant. It occurred to me that if the magazine could get any seven people to agree on a list of any such seven things you'd have the modern Seven Wonders right there at the dinner table.

Wonder is a word to wonder about. It contains a mixture of messages: something marvellous and miraculous, surprising, raising unanswer-able questions about itself, making the observer wonder, even raising sceptical questions like, 'I *wonder* about that'. Miraculous and marvel-lous are clues; both words come from an ancient Indo-European root meaning simply to smile or to laugh. Anything wonderful is something to smile in the presence of, in admiration (which, by the way, comes from the same root, along with, of all telling words, 'mirror').

I decided to try making a list, not for the magazine's dinner party but for this occasion: seven things I wonder about the most.

I shall hold the first for the last, and move along.

My Number Two Wonder is a bacterial species never seen on the face of the earth until 1982, creatures never dreamed of before, living viola-tion of what we used to regard as the laws of nature, things literally straight out of Hell. Or anyway what we used to think of as Hell, the hot unlivable interior of the earth. Such regions have recently come into scientific view from the research submarines designed to descend twenty-five hundred meters or more to the edge of deep holes in the sea bottom, where open vents spew superheated seawater in plumes from chimneys in the earth's crust, known to oceanographic scientists as 'black smokers'. This is not just hot water, or steam, or even steam under pressure as exists in a laboratory autoclave (which we have relied upon for decades as the surest way to destroy all microbial life). This is extremely hot water under extremely high pressure, with temperatures in excess of 300 degrees centigrade. At such heat, the existence of life as we know it would be simply inconceivable. Proteins and DNA would fall apart, enzymes would melt away, anything alive would die instantaneously. We have long since ruled out the possibility of life on Venus because of that planet's comparable temperature; we have ruled out the possibility of life in the earliest years of this planet, four billion or so years ago, on the same ground.

B. J. A. Baross and J. W. Deming have recently discovered the presence of thriving colonies of bacteria in water fished directly from these deep-sea vents. Moreover, when brought to the surface, encased in titanium syringes and sealed in pressurized chambers heated to 250 degrees centigrade, the bacteria not only survive but reproduce themselves enthusiastically. They can be killed only by chilling them down in boiling water.

And yet they look just like ordinary bacteria. Under the electron microscope they have the same essential structure—cell walls, ribosomes, and all. If they were, as is now being suggested, the original archebacteria, ancestors of us all, how did they or their progeny ever learn to cool down? I cannot think of a more wonderful trick.

My Number Three Wonder is *oncideres*, a species of beetle encountered by a pathologist friend of mine who lives in Houston and has a lot of mimosa trees in his backyard. This beetle is not new, but it qualifies as a Modern Wonder because of the exceedingly modern questions raised for evolutionary biologists about the three consecutive things on the mind of the female of the species. Her first thought is for a mimosa tree, which she finds and climbs, ignoring all other kinds of trees in the vicinity. Her second thought is for the laying of eggs, which she does by crawling out on a limb, cutting a longitudinal slit with her mandible and depositing her eggs beneath the slit. Her third and last thought concerns the welfare of her offspring; beetle larvae cannot survive in live wood, so she backs up a foot or so and cuts a neat circular girdle all around the limb, through the bark and down into the cambium. It takes her eight hours to finish this cabinetwork. Then she leaves and where she goes I do not know. The limb dies from the girdling, falls to the ground in the next breeze, the larvae feed and grow into the next generation, and the questions lie there unanswered. How on earth did these three linked thoughts in her mind evolve together in evolution? How could any one of the three become fixed as beetle behavior by itself, without the other two? What are the odds favoring three totally separate bits of behavior—liking a particular tree, cutting a slit for eggs, and then girdling the limb—happening together by random chance among a beetle's genes? Does this smart beetle know what she is doing? And how did the mimosa tree enter the picture in its evolution? Left to themselves, unpruned, mimosa trees have a life expectancy of twenty-five to thirty years. Pruned each year, which is what the beetle's girdling labor accomplishes, the tree can flourish for a century. The mimosa-beetle relationship is an elegant example of symbiotic partnership, a phenomenon now recognized as pervasive in nature. It is good for us to have around on our intellectual mantelpiece such creatures as this insect and its friend the tree, for they keep reminding us how little we know about nature.

The Fourth Wonder on my list is an infectious agent known as the scrapie virus, which causes a fatal disease of the brain in sheep, goats, and several laboratory animals. A close cousin of scrapie is the C-J virus, the cause of some cases of senile dementia in human beings. These are called 'slow viruses', for the excellent reason that an animal exposed to infection today will not become ill until a year and a half or two years from today. The agent, whatever it is, can propagate itself in abundance from a few infectious units today to more than a billion next year. I use the phrase 'whatever it is' advisedly. Nobody has yet been able to find any DNA or RNA in the scrapie or C-J viruses. It may be there, but if so it exists in amounts too small to detect. Meanwhile, there is plenty of protein, leading to a serious proposal that the virus may indeed be *all* protein. But protein, so far as we know, does not replicate itself all by itself, not on this planet anyway. Looked at this way, the scrapie agent seems the strangest thing in all biology and, until someone in some laboratory figures out what it is, a candidate for Modern Wonder.

My Fifth Wonder is the olfactory receptor cell, located in the epithelial tissue high in the nose, sniffing the air for clues to the environment, the fragrance of friends, the smell of leaf smoke, breakfast, nighttime and bedtime, and a rose, even, it is said, the odor of sanctity. The cell that does all these things, firing off urgent messages into the deepest parts of the brain, switching on one strange unaccountable memory after another, is itself a proper brain cell, a certified neuron belonging to the brain but miles away out in the open air, nosing around the world. How it manages to make sense of what it senses, discriminating between jasmine and anything else non-jasmine with infallibility, is one of the deep secrets of neurobiology. This would be wonder enough, but there is more. This population of brain cells, unlike any other neurons of the vertebrate central nervous system, turns itself over every few weeks; cells wear out, die, and are replaced by brand-new cells rewired to the same deep centers miles back in the brain, sensing and remembering the same wonderful smells. If and when we reach an understanding of these cells and their functions, including the moods and whims under their governance, we will know a lot more about the mind than we do now, a world away.

Sixth on my list is, I hesitate to say, another insect, the termite. This time, though, it is not the single insect that is the Wonder, it is the collectivity. There is nothing at all wonderful about a single, solitary termite, indeed there is really no such creature, functionally speaking, as a lone termite, any more than we can imagine a genuinely solitary human being; no such thing. Two or three termites gathered together on a dish are not much better; they may move about and touch each other nervously, but nothing happens. But keep adding more termites until they reach a critical mass, and then the miracle begins. As though they had suddenly received a piece of extraordinary news, they organize in platoons and begin stacking up pellets to precisely the right height, then turning the arches to connect the columns, constructing the cathedral and its chambers in which the colony will live out its life for the decades ahead, air-conditioned and humidity-controlled, following the chemical blueprint coded in their genes, flawlessly, stone-blind. They are not the dense mass of individual insects they appear to be; they are an organism, a thoughtful, meditative brain on a million legs. All we really know about this new thing is that it does its architecture and engineering by a complex system of chemical signals.

The Seventh Wonder of the modern world is a human child, any child. I used to wonder about childhood and the evolution of our species. It seemed to me unparsimonious to keep expending all that energy on such a long period of vulnerability and defenselessness, with nothing to show for it, in biological terms, beyond the feckless, irresponsible pleasure of childhood. After all, I used to think, it is one sixth of a whole human life span! Why didn't our evolution take care of that, allowing us to jump catlike from our juvenile to our adult (and, as I thought) productive stage of life? I had forgotten about language, the single human trait that marks us out as specifically human, the property that enables our survival as the most compulsively, biologically, obsessively social of all creatures on earth, more interdependent and interconnected even than the famous social insects. I had forgotten that, and forgotten that children *do* that in childhood. Language is what childhood is for.

There is another related but different creature, nothing like so wonderful as a human child, nothing like so hopeful, something to worry about all day and all night. It is *us*, aggregated together in our collective, critical masses. So far, we have learned how to be useful to each other only when we collect in small groups—families, circles of friends, once in a while (although still rarely) committees. The drive to be useful is encoded in our genes. But when we gather in very large numbers, as in the modern nation-state, we seem capable of levels of folly and self-destruction to be found nowhere else in all of Nature.

As a species, taking all in all, we are still too young, too juvenile, to be trusted. We have spread across the face of the earth in just a few thousand years, no time at all as evolution clocks time, covering all livable parts of the planet, endangering other forms of life, and now threatening ourselves. As a species, we have everything in the world to learn about living, but we may be running out of time. Provisionally, but only provisionally, we are a Wonder.

And now the first on my list, the one I put off at the beginning of making a list, the first of all Wonders of the modern world. To name this one, you have to redefine the world as it has indeed been redefined in this most scientific of all centuries. We named the place we live in the *world* long ago, from the Indo-European root *wiros*, which meant man. We now live in the whole universe, that stupefying piece of expanding geometry. Our suburbs are the local solar system, into which, sooner or later, we will spread life, and then, likely, beyond into the galaxy. Of all celestial bodies within reach or view, as far as we can see, out to the edge, the most wonderful and marvellous and mysterious is turning out to be our own planet earth. There is nothing to match it anywhere, not yet anyway.

It is a living system, an immense organism, still developing, regulating itself, making its own oxygen, maintaining its own temperature, keeping all its infinite living parts connected and interdependent, including us. It is the strangest of all places, and there is everything in the world to learn about it. It can keep us awake and jubilant with questions for millennia ahead, if we can learn not to meddle and not to destroy. Our great hope is in being such a young species, thinking in language only a short while, still learning, still growing up.

We are not like the social insects. They have only the one way of doing things and they will do it forever, coded for that way. We are coded differently, not just for binary choices, *go* or *no-go*. We can go four ways at once, depending on how the air feels: *go, no-go*, but also *maybe*, plus *what the hell let's give it a try*. We are in for one surprise after another if we keep at it and keep alive. We can build structures for human society never seen before, thoughts never thought before, music never heard before.

Provided we do not kill ourselves off, and provided we can connect ourselves by the affection and respect for which I believe our genes are also coded, there is no end to what we might do on or off this planet.

At this early stage in our evolution, now through our infancy and into our childhood and then, with luck, our growing up, what our species needs most of all, right now, is simply a future.

James Watson

from AVOID BORING PEOPLE

Peter Medawar characteristically went too far when he said, of the molecular biology revolution ushered in by Watson and Crick, 'It is simply not worth arguing with anyone so obtuse as not to realise that this complex of discoveries is the greatest achievement of science in the twentieth century.' Not worth arguing? How about relativity? Quantum mechanics? Nevertheless, I would go so far as to say that there are times when to label a scientist as a Nobel Prize winner sounds like an understatement, and Watson and Crick constitute the best example I know. For me, the greatest achievement of Watson and Crick was to turn genetics from a branch of wet and squishy physiology into a branch of information technology, in the process slaying, as I suggested above, the ghost of vitalism.

Both Watson and Crick have written fascinating autobiographies. I felt that *The Double Helix* was too well known to need anthologizing here and I chose, instead, some diverting paragraphs from Watson's later memoir, *Avoid Boring People*. This work, with its calculatedly ambiguous title and perhaps not entirely calculated tactlessness, is a kind of sequel to *The Double Helix*. Each chapter concludes with a list of 'Remembered Lessons', which can be read as homilies, rather in the manner of Medawar's *Advice to a Young Scientist*. I have chosen five of these lessons, but it is entertaining just to read down the list. Here are some more: Put lots of spin on balls (G. H. Hardy would have loved that: he was obsessed with 'spin' but it seems to have been a subtle concept, related to cricket but not in any simple way). Work on Sundays. Don't take up golf.

Remembered Lessons

NEVER BE THE BRIGHTEST PERSON IN A ROOM

Getting out of intellectual ruts more often than not requires unexpected intellectual jousts. Nothing can replace the company of others who have the background to catch errors in your reasoning or provide facts that may either prove or disprove your argument of the moment. And the sharper those around you, the sharper you will become. It's contrary to human nature, and especially to human male nature, but being the top dog in the pack can work against greater accomplishments. Much better to be the least accomplished chemist in a super chemistry department than the superstar in a less lustrous department. By the early 1950s, Linus Pauling's scientific interactions with fellow scientists were effectively monologues instead of dialogues. He then wanted adoration, not criticism.

[...]

WORK WITH A TEAMMATE WHO IS YOUR INTELLECTUAL EQUAL

Two scientists acting together usually accomplish more than two loners each going their own way. The best scientific pairings are marriages of convenience in that they bring together the complementary talents of those involved. Given, for example, Francis's penchant for high-level crystallographic theory, there was no need for me also to master it. All I needed were its implications for interpreting DNA X-ray photographs. The possibility, of course, existed that Francis might err in some fashion I couldn't spot, but keeping good relations with others in the field outside our partnership meant that he would always have his ideas checked by others with even greater crystallographic talents. For my part, I brought to our two-man team a deep understanding of biology and a compulsive enthusiasm for solving what proved to be a fundamental problem of life.

An intelligent teammate can shorten your flirtation with a bad idea. For all too long I kept trying to build DNA models with the sugar phosphate backbone in the center, convinced that if I put the backbone on the outside, there would be no stereochemical restriction on how it could fold up into a regular helix. Francis's scorn for this assertion made me reverse course much sooner than I would have otherwise. Soon I too realized that my past argument had been lousy and, in fact, the stereochemistry of the sugar-phosphate groups would of course move them to outer positions of helices that use approximately ten nucleotides to make a complete turn.

In general, a scientific team of more than two is a crowded affair. Once you have three people working on a common objective, either one member effectively becomes the leader or the third eventually feels a less-than-equal partner and resents not being around when key decisions are made. Three-person operations also make it hard to assign credit. People naturally believe in the equal partnerships of successful duos—Rodgers and Hammerstein, Lewis and Clark. Most don't believe in the equal contributions of three-person crews.

[…]

AVOID GATHERINGS OF MORE THAN TWO NOBEL PRIZE WINNERS

All too often some well-intentioned person gathers together Nobel laureates to enhance an event promoting his or her university or city. The host does so convinced that these special guests will exude genius and incandescent or at least brilliantly eccentric personalities. The fact is that many years pass between the awarding of a prize and the work it acknowledges, so even recently awarded Nobelists have likely seen better days. The honorarium, no matter how hefty, will not compensate you for the realization that you probably look and act as old and tired as the other laureates, whose conversation is boring you perhaps as much as yours is boring them. The best way to remain lively is to restrict your professional contact to young, not yet famous colleagues. Though they likely will beat you at tennis, they will also keep your brain moving.

[…]

SIT IN THE FRONT ROW WHEN A SEMINAR'S TITLE INTRIGUES YOU

By far the best way to profit from seminars that interest you is to sit in the front row. Not being bored, you do not risk the embarrassment of falling asleep in front of everybody's eyes. If you cannot follow the speaker's train of thought from where you are, you are in a good place to interrupt. Chances are you are not alone in being lost and most everyone in the audience will silently applaud. Your prodding may in fact reveal whether the speaker indeed has a take-home message or has simply deluded himself into believing he does. Waiting until a seminar is over to ask questions is pathologically polite. You will probably forget where you got lost and start questioning results you actually understood.

Now, if you have suspicions that a seminar will bore you but are not sure enough to risk skipping it, sit in the back row. There a dull, glazed expression will not be conspicuous, and if you walk out, your departure may be thought temporary and compelled by the call of nature. Szilard did not follow this advice, habitually sitting in a front row and getting up abruptly in the middle of talks when he'd had too much of too little. Those outside his close circle of friends were relieved when his inherent restlessness made him move on to a potentially more exciting domicile.

[...]

EXTEND YOURSELF INTELLECTUALLY THROUGH COURSES THAT INITIALLY FRIGHTEN YOU

All through my undergraduate days, I worried that my limited mathematical talents might keep me from being more than a naturalist. In deciding to go for the gene, whose essence was surely in its molecular properties, there seemed no choice but to tackle my weakness head-on. Not only was math at the heart of virtually all physics, but the forces at work in three-dimensional molecular structures could not be described except with math. Only by taking higher math courses would I develop sufficient comfort to work at the leading edge of my field, even if I never got near the leading edge of math. And so my Bs in two genuinely tough math courses were worth far more in confidence capital than any A I would likely have received in a biology course, no matter how demanding. Though I would never use the full extent of analytical methods I had learned, the Poisson distribution analyses needed to do most phage experiments soon became rather satisfying, even in the age of slide rules, instead of a source of crippling anxiety.

Francis Crick

from WHAT MAD PURSUIT

Francis Crick's autobiography is called *What Mad Pursuit*. I wrote the following for its jacket blurb:

Francis Crick's is the dominant intellect from the heroic age of molecular biology when authors-per-paper could be counted on one hand and heroic individual intelligence could still dominate. We expect brilliance from his book and we get it, together with mature wisdom. What we may not expect—but also get—is a generous and charming modesty that belies the famous opening sentence of an alternative volume. This modesty is personal and does not preclude a justified pride, almost arrogance, on behalf of a discipline—molecular biology—that earned the right to be arrogant by cutting the philosophical claptrap, getting its head down, and in short order solving many of the outstanding problems of life. Francis Crick seems to epitomize the ruthlessly successful science that he did so much to found.

The 'alternative volume' I had in mind was, of course, *The Double Helix*. Here I have reprinted Crick's own personal response to the discovery of the double helix and to its aftermath, ending with a delightful anecdote about Jim Watson trying to explain DNA after dinner.

What was it like to live with the double helix? I think we realized almost immediately that we had stumbled onto something important. According to Jim, I went into the Eagle, the pub across the road where we lunched every day, and told everyone that we'd discovered the secret of life. Of that I have no recollection, but I do recall going home and telling Odile that we seemed to have made a big discovery. Years later she told me that she hadn't believed a word of it. 'You were always coming home and saying things like that,' she said, 'so naturally I thought nothing of it.' Bragg was in bed with flu at the time, but as soon as he saw the model and grasped the basic idea he was immediately enthusiastic. All past differences were forgiven and he became one of our strongest supporters. We had a constant stream of visitors, a contingent from Oxford that included Sydney Brenner, so that Jim soon began to tire of my repetitious enthusiasm. In fact at times he had cold feet, thinking that perhaps it was all a pipe dream, but the experimental data from King's College, when we finally saw them, were a great encouragement. By summer most of our doubts had vanished and we were able to take a long cool look at the structure, sorting out its accidental features (which were somewhat inaccurate) from its really fundamental properties, which time has shown to be correct.

For a number of years after that, things were fairly quiet. I named my family's Cambridge house in Portugal Place 'The Golden Helix' and eventually erected a simple brass helix on the front of it, though it was a single helix rather than a double one. It was supposed to symbolize not DNA but the basic idea of a helix. I called it golden in the same way that Apuleius called his story 'The Golden Ass', meaning beautiful. People have often asked me whether I intend to gild it, but we never got further than painting it yellow.

Finally one should perhaps ask the personal question—am I glad that it happened as it did? I can only answer that I enjoyed every moment of it, the downs as well as the ups. It certainly helped me in my subsequent propaganda for the genetic code. But to convey my own feelings, I cannot do better than quote from a brilliant and perceptive lecture I heard years ago in Cambridge by the painter John Minton in which he said of his own artistic creations, 'The important thing is to be there when the picture is painted'. And this, it seems to me, is partly a matter of luck and partly good judgement, inspiration, and persistent application.

There was in the early fifties a small, somewhat exclusive biophysics club at Cambridge, called the Hardy Club, named after a Cambridge zoologist of a previous generation who had turned physical chemist. The list of those early members now has an illustrious ring, replete with Nobel laureates and Fellows of the Royal Society, but in those days we were all fairly young and most of us not particularly well known. We boasted only one F.R.S.—Alan Hodgkin—and one member of the House of Lords—Victor Rothschild. Jim was asked to give an evening talk to this select gathering. The speaker was customarily given dinner first at Peterhouse. The food there was always good but the speaker was also plied with sherry before dinner, wine with it, and, if he was so rash as to accept them, drinks after dinner as well. I have seen more than one speaker struggling to find his way into his topic through a haze of alcohol. Jim was no exception. In spite of it all he managed to give a fairly adequate description of the main points of the structure and the evidence supporting it, but when he came to sum up he was quite overcome and at a loss for words. He gazed at the model, slightly bleary-eyed. All he could manage to say was 'It's so beautiful, you see, so beautiful!' But then, of course, it was.

Lewis Wolpert

from THE UNNATURAL NATURE OF SCIENCE

■ Lewis Wolpert is a distinguished embryologist and writer about biology generally. He is never one to duck controversy and his powerful cannon, while not always tightly bolted to the deck, is capable more often than not of landing a good broadside on the right target. Wolpert persuasively emphasizes the distinction between science and technology, and one of the differences is that science is often counter-intuitive. Wolpert's point is most strongly made by the paradoxes and mysteries of modern physics but, as the following paragraphs from *The Unnatural Nature of Science* show, we don't have to venture beyond classical physics and biology to cast doubt on T. H. Huxley's opinion that

> Science is nothing but trained and organized common sense, differing from the latter only as a veteran may differ from a raw recruit: and its methods differ from those of common sense only as far as the guardsman's cut and thrust differ from the manner in which a savage wields his club.

The strangeness of scientific theory is one of its appeals, and it is not incompatible with Huxley's view that the practice of science is organized common sense. ■

––––––––

The physics of motion provides one of the clearest examples of the counter-intuitive and unexpected nature of science. Most people not trained in physics have some sort of vague ideas about motion and use these to predict how an object will move. For example, when students are presented with problems requiring them to predict where an object—a bomb, say—will land if dropped from an aircraft, they often get the answer wrong. The correct answer—that the bomb will hit that point on the ground more or less directly below the point at which the aircraft has arrived at the moment of impact—is often rejected. The underlying confusion partly comes from not recognizing that the bomb continues to move forward when released and this is not affected by its downwards fall. This point is made even more dramatically by another example. Imagine being in the centre of a very large flat field. If one bullet is dropped from your hand and another is fired horizontally from a gun at exactly the same time, which will hit the ground first? They will, in fact, hit the ground at the same time, because the bullet's rate of fall is quite independent of its horizontal motion. That the bullet which is fired is travelling horizontally has no effect on how fast it falls under the action of gravity.

Another surprising feature of motion is that the most natural state for an object is movement at constant speed—not, as most of us think, being stationary. A body in motion will continue to move forever unless there is a force that stops it. This was a revolutionary idea first proposed by Galileo in the early seventeenth century and was quite different from Aristotle's more common-sense view, from the fourth century BC, that the motion of an object required the continuous action of a force. Galileo's argument is as follows. Imagine a perfectly flat plane and a perfectly round ball. If the plane is slightly inclined the ball will roll down it and go on and on and on. But a ball going up a slope with a slight incline will have its velocity retarded. From this it follows that motion along a horizontal plane is perpetual, 'for if the velocity be uniform it cannot be diminished or slackened, much less destroyed'. So, on a flat slope, with no resistance, an initial impetus will keep the ball moving forever, even though there is no force. Thus the natural state of a physical object is motion along a straight line at constant speed, and this has come to be known as Newton's first law of motion. That a real ball will in fact stop is due to the opposing force provided by friction between a real ball and a real plane. The enormous conceptual change that the thinking of Galileo required shows that science is not just about accounting for the 'unfamiliar' in terms of the familiar. Quite the contrary: *science often explains the familiar in terms of the unfamiliar.*

Julian Huxley

from ESSAYS OF A BIOLOGIST

■   I have long thought that science should inspire great poetry, but scientists have published disappointingly few poems. That versatile and enlightened doctor Erasmus Darwin was highly rated as a poet in the eighteenth century, but his rhyming couplets of epic science do not suit modern tastes. Charles Darwin, Erasmus's intellectual as well as his biological grandson, enjoyed poetry as a young man but in later life, I am sorry to say, he found even Shakespeare 'nauseating'. Not so Julian Huxley, the biological grandson of Darwin's Bulldog and one of the leading intellectual grandchildren of Charles Darwin himself. For some reason that I can't now reconstruct, as an undergraduate I identified with the young Julian Huxley. I knew that like me he had read Zoology at Balliol College, Oxford, where he was a contemporary of my Dawkins grandfather, but I didn't know then that I was to be Huxley's successor (not immediate) as Tutor in Zoology at New College. Huxley was said to have been the last Oxford tutor capable of teaching the whole range of Zoology, and he was indeed versatile, though not in the Erasmus class. Huxley's essays influenced me, but I now find myself admiring them less than I did. Nevertheless, for old times' sake, I wanted this anthology to include something by him, even if short. I also wanted at least one poem, and I remembered that each of the *Essays of a Biologist* is introduced by a short poem, of which this is the best. Huxley was not a great poet but perhaps the following lines will serve to alert others to the poetic inspiration that a deep understanding of science—and nobody would deny him that—can provide. ■

God and Man

> The world of things entered your infant mind
> To populate that crystal cabinet.
> Within its walls the strangest partners met,
> And things turned thoughts did propagate their kind.
> For, once within, corporeal fact could find
> A spirit. Fact and you in mutual debt

Built there your little microcosm—which yet
Had hugest tasks to its small self assigned.

Dead men can live there, and converse with stars:
Equator speaks with pole, and night with day:
Spirit dissolves the world's material bars—
A million isolations burn away.
The Universe can live and work and plan,
At last made God within the mind of man.

Albert Einstein

'RELIGION AND SCIENCE'

■ The greatest scientist of the age, he has been called: many would say the greatest scientist of all time, and few would leave him out of the top three, with Newton and Darwin. Not an experimentalist, and only an average mathematician, Einstein's supreme gift was his unprecedented, unparalleled imagination, guided by a kind of scientifically disciplined aesthetic. Great scientists look toward the far horizon and see that what is 'obvious' to common sense can be wrong. If you make a wildly counter-intuitive assumption and follow it through to its conclusion, you can—if you are a genius like Einstein—arrive at a wholly new kind of 'obvious'. Nobody has ever done this kind of thing better than Einstein. We will encounter his thoughts on relativity later in the book. Here I have chosen a meditation on religion—in the special Einsteinian sense of the word which, despite his notorious fondness for figures of speech such as 'God' or 'The Old One', is probably best characterized as 'atheistic pantheism'. ■

Everything that the human race has done and thought is concerned with the satisfaction of deeply felt needs and the assuagement of pain. One has to keep this constantly in mind if one wishes to understand spiritual movements and their development. Feeling and longing are the motive force behind all human endeavor and human creation, in however exalted a guise the latter may present themselves to us. Now what are the feelings and needs that have led men to religious thought and belief in the widest sense of the words? A little consideration will suffice to show us that the most varying emotions preside over the birth of religious thought and experience. With primitive man it is above all fear that evokes religious notions—fear of hunger, wild beasts, sickness, death. Since at this stage of existence understanding of causal connections is usually poorly developed, the human mind creates illusory beings more or less analogous to itself on whose wills and actions these fearful happenings depend. Thus one tries to secure the favor of these beings by carrying out actions and offering sacrifices which, according to the tradition handed down from generation to generation, propitiate them or make them well disposed toward a mortal. In this sense I am speaking of a religion of fear. This, though not created, is in an important degree stabilized by the formation of a special priestly caste which sets itself up as a mediator between the people and the beings they fear, and erects a hegemony on this basis. In many cases a leader or ruler or a privileged class whose position rests on other factors combines priestly functions with its secular authority in order to make the latter more secure; or the political rulers and the priestly caste make common cause in their own interests.

The social impulses are another source of the crystallization of religion. Fathers and mothers and the leaders of larger human communities are mortal and fallible. The desire for guidance, love, and support prompts men to form the social or moral conception of God. This is the God of Providence, who protects, disposes, rewards, and punishes; the God who, according to the limits of the believer's outlook, loves and cherishes the life of the tribe or of the human race, or even life itself; the comforter in sorrow and unsatisfied longing; he who preserves the souls of the dead. This is the social or moral conception of God.

The Jewish scriptures admirably illustrate the development from the religion of fear to moral religion, a development continued in the New Testament. The religions of all civilized peoples, especially the peoples of the Orient, are primarily moral religions. The development from a religion of fear to moral religion is a great step in peoples' lives. And yet, that primitive religions are based entirely on fear and the religions of civilized peoples purely on morality is a prejudice against which we must be on our guard. The truth is that all religions are a varying blend of both types, with this differentiation: that on the higher levels of social life the religion of morality predominates.

Common to all these types is the anthropomorphic character of their conception of God. In general, only individuals of exceptional endowments, and exceptionally high-minded communities, rise to any considerable extent above this level. But there is a third stage of religious experience which belongs to all of them, even though it is rarely found in a pure form: I shall call it cosmic religious feeling. It is very difficult to elucidate this feeling to anyone who is entirely without it, especially as there is no anthropomorphic conception of God corresponding to it.

The individual feels the futility of human desires and aims and the sublimity and marvelous order which reveal themselves both in nature and in the world of thought. Individual existence impresses him as a sort of prison and he wants to experience the universe as a single significant whole. The beginnings of cosmic religious feeling already appear at an early stage of development, e.g., in many of the Psalms of David and in some of the Prophets. Buddhism, as we have learned especially from the wonderful writings of Schopenhauer, contains a much stronger element of this.

The religious geniuses of all ages have been distinguished by this kind of religious feeling, which knows no dogma and no God conceived in man's image; so that there can be no church whose central teachings are based on it. Hence it is precisely among the heretics of every age that we find men who were filled with this highest kind of religious feeling and were in many cases regarded by their contemporaries as atheists, sometimes also as saints. Looked at in this light, men like Democritus, Francis of Assisi, and Spinoza are closely akin to one another.

How can cosmic religious feeling be communicated from one person to another, if it can give rise to no definite notion of a God and no theology? In my view, it is the most important function of art and science to awaken this feeling and keep it alive in those who are receptive to it.

We thus arrive at a conception of the relation of science to religion very different from the usual one. When one views the matter historically, one is inclined to look upon science and religion as irreconcilable antagonists, and for a very obvious reason. The man who is thoroughly convinced of the universal operation of the law of causation cannot for a moment entertain the idea of a being who interferes in the course of events—provided, of course, that he takes the hypothesis of causality really seriously. He has no use for the religion of fear and equally little for social or moral religion. A God who rewards and punishes is inconceivable to him for the simple reason that a man's actions are determined by necessity, external and internal, so that in God's eyes he cannot be responsible, any more than an inanimate object is responsible for the motions it undergoes. Science has therefore been charged with undermining morality, but the charge is unjust. A man's ethical behavior should be based effectually on sympathy, education, and social ties and needs; no religious basis is necessary. Man would indeed be in a poor way if he had to be restrained by fear of punishment and hope of reward after death.

It is therefore easy to see why the churches have always fought science and persecuted its devotees. On the other hand, I maintain that the cosmic religious feeling is the strongest and noblest motive for scientific research. Only those who realize the immense efforts and, above all, the devotion without which pioneer work in theoretical science cannot be achieved are able to grasp the strength of the emotion out of which alone such work, remote as it is from the immediate realities of life, can issue. What a deep conviction of the rationality of the universe and what a yearning to understand, were it but a feeble reflection of the mind revealed in this world, Kepler and Newton must have had to enable them to spend years of solitary labor in disentangling the principles of celestial mechanics! Those whose acquaintance with scientific research is derived chiefly from its practical results easily develop a completely false notion of the mentality of the men who, surrounded by a skeptical world, have shown the way to kindred spirits scattered wide through the world and the centuries. Only one who has devoted his life to similar ends can have a vivid realization of what has inspired these men and given them the strength to remain true to their purpose in spite of countless failures. It is cosmic religious feeling that gives a man such strength. A contemporary has said, not unjustly, that in this materialistic age of ours the serious scientific workers are the only profoundly religious people.

Carl Sagan

from THE DEMON-HAUNTED WORLD

■ Carl Sagan inspired a whole generation of young scientists, especially in America, and his death from cancer in 1996 was a grievous loss to science and the whole world of reality-based thinking. Open any one of his books and you need go no further than the Table of Contents to experience the tingling of the poetic nerve endings that will continue throughout the book: The shores of the cosmic ocean...One voice in the cosmic fugue...The harmony of worlds...The backbone of night...The edge of forever...Who speaks for Earth? Carl Sagan himself would be a good candidate for the answer to the last question. Quite apart from his contributions to public understanding and appreciation of science, Sagan's own research contributions to planetary science would have been fully enough to ensure his election to the National Academy of Sciences, and it is widely believed that envy at his massive success in communicating science to the millions was the direct cause of his being blackballed for election to the Academy. Parallel to his poetic evocations of the universe, Sagan was also an influential voice against superstition and paranormal mumbo jumbo of all kinds. Debunking is often thought to be a killjoy activity: unsexy, necessary but poor box-office. I have never understood this attitude although I have often encountered it. Carl Sagan eloquently belies it in his marvellous book *The Demon-Haunted World* from which the following excerpt is taken. ■

[Science] is more than a body of knowledge; it is a way of thinking. I have a foreboding of an America in my children's or grandchildren's time—when the United States is a service and information economy; when nearly all the key manufacturing industries have slipped away to other countries; when awesome technological powers are in the hands of a very few, and no one representing the public interest can even grasp the issues; when the people have lost the ability to set their own agendas or knowledgeably question those in authority; when, clutching our crystals and nervously consulting our horoscopes, our critical faculties in decline, unable to distinguish between what feels good and what's true, we slide, almost without noticing, back into superstition and darkness. The dumbing down of America is most evident in the slow decay of substantive content in the enormously influential media, the 30-second sound bites (now down to 10 seconds or less), lowest common denominator programming, credulous presentations on pseudoscience and superstition, but especially a kind of celebration of ignorance.

[...]

We've arranged a global civilization in which most crucial elements—transportation, communications, and all other industries; agriculture, medicine, education, entertainment, protecting the environment; and even the key democratic institution of voting—profoundly depend on science and technology. We have also arranged things so that almost no one understands science and technology. This is a prescription for disaster. We might get away with it for a while, but sooner or later this combustible mixture of ignorance and power is going to blow up in our faces.

A Candle in the Dark is the title of a courageous, largely Biblically based, book by Thomas Ady, published in London in 1656, attacking the witch-hunts then in progress as a scam 'to delude the people'. Any illness or storm, anything out of the ordinary, was popularly attributed to witchcraft. Witches must exist, Ady quoted the 'witchmongers' as arguing, 'else how should these things be, or come to pass?' For much of our history, we were so fearful of the outside world, with its unpredictable dangers, that we gladly embraced anything that promised to soften or explain away the terror. Science is an attempt, largely successful, to understand the world, to get a grip on things, to get hold of ourselves, to steer a safe course. Microbiology and meteorology now explain what only a few centuries ago was considered sufficient cause to burn women to death.

Ady also warned of the danger that 'the Nations [will] perish for lack of knowledge'. Avoidable human misery is more often caused not so much by stupidity as by ignorance, particularly our ignorance about ourselves. I worry that, especially as the millennium edges nearer, pseudoscience and superstition will seem year by year more tempting, the siren song of unreason more sonorous and attractive. Where have we heard it before? Whenever our ethnic or national prejudices are aroused, in times of scarcity, during challenges to national self-esteem or nerve, when we agonize about our diminished cosmic place and purpose, or when fanaticism is bubbling up around us—then, habits of thought familiar from ages past reach for the controls.

The candle flame gutters. Its little pool of light trembles. Darkness gathers. The demons begin to stir.

There is much that science doesn't understand, many mysteries still to be resolved. In a Universe tens of billions of light years across and some ten or fifteen billion years old, this may be the case forever. We are constantly stumbling on surprises. Yet some New Age and religious writers assert that scientists believe that 'what they find is all there is'. Scientists may reject mystic revelations for which there is no evidence except somebody's say-so, but they hardly believe their knowledge of Nature to be complete.

Science is far from a perfect instrument of knowledge. It's just the best we have. In this respect, as in many others, it's like democracy. Science by itself cannot advocate courses of human action, but it can certainly illuminate the possible consequences of alternative courses of action.

The scientific way of thinking is at once imaginative and disciplined. This is central to its success. Science invites us to let the facts in, even when they don't conform to our preconceptions. It counsels us to carry alternative hypotheses in our heads and see which best fit the facts. It urges on us a delicate balance between no-holds-barred openness to new ideas, however heretical, and the most rigorous sceptical scrutiny of everything—new ideas and established wisdom. This kind of thinking is also an essential tool for a democracy in an age of change.

One of the reasons for its success is that science has built-in, error-correcting machinery at its very heart. Some may consider this an overbroad characterization, but to me every time we exercise self-criticism, every time we test our ideas against the outside world, we are doing science. When we are self-indulgent and uncritical, when we confuse hopes and facts, we slide into pseudoscience and superstition.

Every time a scientific paper presents a bit of data, it's accompanied by an error bar—a quiet but insistent reminder that no knowledge is complete or perfect. It's a calibration of how much we trust what we think we know. If the error bars are small, the accuracy of our empirical knowledge is high; if the error bars are large, then so is the uncertainty in our knowledge. Except in pure mathematics nothing is known for certain (although much is certainly false).

Moreover, scientists are usually careful to characterize the veridical status of their attempts to understand the world—ranging from conjectures and hypotheses, which are highly tentative, all the way up to laws of Nature which are repeatedly and systematically confirmed through many interrogations of how the world works. But even laws of Nature are not absolutely certain. There may be new circumstances never before examined—inside black holes, say, or within the electron, or close to the speed of light—where even our vaunted laws of Nature break down and, however valid they may be in ordinary circumstances, need correction.

Humans may crave absolute certainty; they may aspire to it; they may pretend, as partisans of certain religions do, to have attained it. But the history of science—by far the most successful claim to knowledge accessible to humans—teaches that the most we can hope for is successive improvement in our understanding, learning from our mistakes, an asymptotic approach to the Universe, but with the proviso that absolute certainty will always elude us.

We will always be mired in error. The most each generation can hope for is to reduce the error bars a little, and to add to the body of data to which error bars apply. The error bar is a pervasive, visible self-assessment of the reliability of our knowledge. You often see error bars in public opinion polls ('an uncertainty of plus or minus three per cent', say). Imagine a society in which every speech in the *Congressional Record*, every television commercial, every sermon had an accompanying error bar or its equivalent.

One of the great commandments of science is, 'Mistrust arguments from authority'. (Scientists, being primates, and thus given to dominance hierarchies, of course do not always follow this commandment.) Too many such arguments have proved too painfully wrong. Authorities must prove their contentions like everybody else. This independence of science, its occasional unwillingness to accept conventional wisdom, makes it dangerous to doctrines less self-critical, or with pretensions to certitude.

Because science carries us toward an understanding of how the world is, rather than how we would wish it to be, its findings may not in all cases be immediately comprehensible or satisfying. It may take a little work to restructure our mindsets. Some of science is very simple. When it gets complicated, that's usually because the world is complicated—or because *we're* complicated. When we shy away from it because it seems too difficult (or because we've been taught so poorly), we surrender the ability to take charge of our future. We are disenfranchised. Our self-confidence erodes.

But when we pass beyond the barrier, when the findings and methods of science get through to us, when we understand and put this knowledge to use, many feel deep satisfaction. This is true for everyone, but especially for children—born with a zest for knowledge, aware that they must live in a future moulded by science, but so often convinced in their adolescence that science is not for them. I know personally, both from having science explained to me and from my attempts to explain it to others, how gratifying it is when we get it, when obscure terms suddenly take on meaning, when we grasp what all the fuss is about, when deep wonders are revealed.

In its encounter with Nature, science invariably elicits a sense of reverence and awe. The very act of understanding is a celebration of joining, merging, even if on a very modest scale, with the magnificence of the Cosmos. And the cumulative worldwide build-up of knowledge over time converts science into something only a little short of a trans-national, trans-generational meta-mind.

'Spirit' comes from the Latin word 'to breathe'. What we breathe is air, which is certainly matter, however thin. Despite usage to the contrary, there is no necessary implication in the word 'spiritual' that we are talking of anything other than matter (including the matter of which the brain is made), or anything outside the realm of science. On occasion, I will feel free to use the word. Science is not only compatible with spirituality; it is a profound source of spirituality. When we recognize our place in an immensity of light years and in the passage of ages, when we grasp the intricacy, beauty and subtlety of life, then that soaring feeling, that sense of elation and humility combined, is surely spiritual. So are our emotions in the presence of great art or music or literature, or of acts of exemplary selfless courage such as those of Mohandas Gandhi or Martin Luther King Jr. The notion that science and spirituality are somehow mutually exclusive does a disservice to both.

PART III

WHAT SCIENTISTS THINK

Richard Feynman

from THE CHARACTER OF PHYSICAL LAW

■ Richard Feynman, the great Nobel Prizewinning theoretical physicist, is better known as a lecturer of genius than as a writer, and he was deeply revered by all who knew him or even just heard him. I never did, and must sadly be content with the written word. I have already quoted his evocation of the wonder of a flower. The following extracts from *The Character of Physical Law* give a flavour of what it must have been like to attend one of his legendary lectures. ■

It is odd, but on the infrequent occasions when I have been called upon in a formal place to play the bongo drums, the introducer never seems to find it necessary to mention that I also do theoretical physics. I believe that is probably because we respect the arts more than the sciences. The artists of the Renaissance said that man's main concern should be for man, and yet there are other things of interest in the world. Even the artists appreciate sunsets, and the ocean waves, and the march of the stars across the heavens. There is then some reason to talk of other things sometimes. As we look into these things we get an aesthetic pleasure from them directly on observation. There is also a rhythm and a pattern between the phenomena of nature which is not apparent to the eye, but only to the eye of analysis; and it is these rhythms and patterns which we call Physical Laws.

[...]

The conservation of energy would let us think that we have as much energy as we want. Nature never loses or gains energy. Yet the energy of the sea, for example, the thermal motion of all the atoms in the sea, is practically unavailable to us. In order to get that energy organized, herded, to make it available for use, we have to have a difference in temperature, or else we shall find that although the energy is there we cannot make use of it. There is a great difference between energy and availability of energy. The energy of the sea is a large amount, but it is not available to us.

The conservation of energy means that the total energy in the world is kept the same. But in the irregular jigglings that energy can be spread about so uniformly that, in certain circumstances, there is no way to make more go one way than the other—there is no way to control it any more.

I think that by an analogy I can give some idea of the difficulty, in this way. I do not know if you have ever had the experience—I have—of sitting on the beach with several towels, and suddenly a tremendous downpour comes. You pick up the towels as quickly as you can, and run into the bathhouse. Then you start to dry yourself, and you find that this towel is a little wet, but it is drier than you are. You keep drying with this one until you find it is too wet—it is wetting you as much as drying you—and you try another one; and pretty soon you discover a horrible thing—that all the towels are damp and so are you. There is no way to get any drier, even though you have many towels, because there is no difference in some sense between the wetness of the towels and the wetness of yourself. I could invent a kind of quantity which I could call 'ease of removing water'. The towel has the same ease of removing water from it as you have, so when you touch yourself with the towel, as much water comes off the towel on to you as comes from you to the towel. It does not mean there is the same amount of water in the towel as there is on you—a big towel will have more water in it than a little towel—but they have the same dampness. When things get to the same dampness then there is nothing you can do any longer.

Now the water is like the energy, because the total amount of water is not changing. (If the bathhouse door is open and you can run into the sun and get dried out, or find another towel, then you're saved, but suppose everything is closed, and you can't get away from these towels or get any new towels.) In the same way if you imagine a part of the world that is closed, and wait long enough, in the accidents of the world the energy, like the water, will be distributed over all of the parts evenly until there is nothing left of one-way-ness, nothing left of the real interest of the world as we experience it.

Erwin Schrödinger

from WHAT IS LIFE?

P 415

■ The Austrian physicist Erwin Schrödinger (he of the probabilistic cat) made major contributions to quantum theory, but biologists know him too as an important influence on several key figures of the molecular biology revolution, including Maurice Wilkins and Francis Crick. Indeed, Matt Ridley, in his biography of Crick, says that Schrödinger's little book, *What is Life?*, which was based on a series of lectures delivered in Trinity College, Dublin, in 1943, influenced a whole generation of physicists to go into biology. What follows is an excerpt from that book, in which he considers entropy and develops the idea that living things suck order out of their surroundings. ■

Living Matter Evades the Decay to Equilibrium

What is the characteristic feature of life? When is a piece of matter said to be alive? When it goes on 'doing something', moving, exchanging material with its environment, and so forth, and that for a much longer period than we would expect an inanimate piece of matter to 'keep going' under similar circumstances. When a system that is not alive is isolated or placed in a uniform environment, all motion usually comes to a standstill very soon as a result of various kinds of friction; differences of electric or chemical potential are equalized, substances which tend to form a chemical compound do so, temperature becomes uniform by heat conduction. After that the whole system fades away into a dead, inert lump of matter. A permanent state is reached, in which no observable events occur. The physicist calls this the state of thermodynamical equilibrium, or of 'maximum entropy'. *= cha - ds*

Practically, a state of this kind is usually reached very rapidly. Theoretically, it is very often not yet an absolute equilibrium, not yet the true maximum of entropy. But then the final approach to equilibrium is very slow. It could take anything between hours, years, centuries...To give an example—one in which the approach is still fairly rapid: if a glass filled with pure water and a second one filled with sugared water are placed together in a hermetically closed case at constant temperature, it appears at first that nothing happens, and the impression of complete equilibrium is created. But after a day or so it is noticed that the pure water, owing to its higher vapour pressure, slowly evaporates and condenses on the solution. The latter overflows. Only after the pure water has totally evaporated has the sugar reached its aim of being equally distributed among all the liquid water available.

These ultimate slow approaches to equilibrium could never be mistaken for life, and we may disregard them here. I have referred to them in order to clear myself of a charge of inaccuracy.

It Feeds On 'Negative Entropy' *Resembling an enigma*

It is by avoiding the rapid decay into the inert state of 'equilibrium' that an organism appears so enigmatic; so much so, that from the earliest times of human thought some special non-physical or supernatural force (*vis viva*, entelechy) was claimed to be operative in the organism, and in some quarters is still claimed.

How does the living organism avoid decay? The obvious answer is: by eating, drinking, breathing and (in the case of plants) assimilating. The technical term is *metabolism*. The Greek word (μεταβάλλειν) means change or exchange. Exchange of what? Originally the underlying idea is, no doubt, exchange of material (e.g. the German for metabolism is *Stoffwechsel*.) That the exchange of material should be the essential thing is absurd. Any atom of nitrogen, oxygen, sulphur, etc. is as good as any other of its kind; what could be gained by exchanging them? For a while in the past our curiosity was silenced by being told that we feed upon energy. In some very advanced country (I don't remember whether it was Germany or the USA or both) you could find menu cards in restaurants indicating, in addition to the price, the energy content of every dish. Needless to say, taken literally, this is just as absurd. For an adult organism the energy content is as stationary as the material content. Since, surely, any calorie is worth as much as any other calorie, one cannot see how a mere exchange could help.

What then is that precious something contained in our food which keeps us from death? That is easily answered. Every process, event, happening—call it what you will; in a word, everything that is going on in Nature means an increase of the entropy of the part of the world where it is going on. Thus a living organism continually increases its entropy—or, as you may say, produces positive entropy—and thus tends to approach the dangerous state of maximum entropy, which is death. It can only keep aloof from it, i.e. alive, by continually drawing from its environment negative entropy—which is something very positive as we shall immediately see. What an organism feeds upon is negative entropy. Or, to put it less paradoxically, the essential thing in metabolism is that the organism succeeds in freeing itself from all the entropy it cannot help producing while alive.

What is Entropy?

What is entropy? Let me first emphasize that it is not a hazy concept or idea, but a measurable physical quantity just like the length of a rod, the temperature at any point of a body, the heat of fusion of a given crystal or the specific heat of any given substance. At the absolute zero point of temperature (roughly $-273°C$) the entropy of any substance is zero. When you bring the substance into any other state by slow, reversible little steps (even if thereby the substance changes its physical or chemical nature or splits up into two or more parts of different physical or chemical nature) the entropy increases by an amount which is computed by dividing every little portion of heat you had to supply in that procedure by the absolute temperature at which it was supplied—and by summing up all these small contributions. To give an example, when you melt a solid, its entropy increases by the amount of the heat of fusion divided by the temperature at the melting-point. You see from this, that the unit in which entropy is measured is cal./°C (just as the calorie is the unit of heat or the centimetre the unit of length).

The Statistical Meaning of Entropy

I have mentioned this technical definition simply in order to remove entropy from the atmosphere of hazy mystery that frequently veils it.

Much more important for us here is the bearing on the statistical concept of order and disorder, a connection that was revealed by the investigations of Boltzmann and Gibbs in statistical physics. This too is an exact quantitative connection, and is expressed by

$$\text{entropy} = k \log D,$$

where k is the so-called Boltzmann constant ($= 3.2983 \cdot 10^{-24}$ cal./°C), and D a quantitative measure of the atomistic disorder of the body in question. To give an exact explanation of this quantity D in brief non-technical terms is well-nigh impossible. The disorder it indicates is partly that of heat motion, partly that which consists in different kinds of atoms or molecules being mixed at random, instead of being neatly separated, e.g. the sugar and water molecules in the example quoted above. Boltzmann's equation is well illustrated by that example. The gradual 'spreading out' of the sugar over all the water available increases the disorder D, and hence (since the logarithm of D increases with D) the entropy. It is also pretty clear that any supply of heat increases the turmoil of heat motion, that is to say, increases D and thus increases the entropy; it is particularly clear that this should be so when you melt a crystal, since you thereby destroy the neat and permanent arrangement of the atoms or molecules and turn the crystal lattice into a continually changing random distribution.

An isolated system or a system in a uniform environment (which for the present consideration we do best to include as a part of the system we contemplate) increases its entropy and more or less rapidly approaches the inert state of maximum entropy. We now recognize this fundamental law of physics to be just the natural tendency of things to approach the chaotic state (the same tendency that the books of a library or the piles of papers and manuscripts on a writing desk display) unless we obviate it. (The analogue of irregular heat motion, in this case, is our handling those objects now and again without troubling to put them back in their proper places.)

Organization Maintained by Extracting 'Order' From the Environment

How would we express in terms of the statistical theory the marvellous faculty of a living organism, by which it delays the decay into thermodynamical equilibrium (death)? We said before: 'It feeds upon negative entropy', attracting, as it were, a stream of negative entropy upon itself, to compensate the entropy increase it produces by living and thus to maintain itself on a stationary and fairly low entropy level.

If D is a measure of disorder, its reciprocal, $1/D$, can be regarded as a direct measure of order. Since the logarithm of $1/D$ is just minus the logarithm of D, we can write Boltzmann's equation thus:

$$-(\text{entropy}) = k \log(1/D).$$

Hence the awkward expression 'negative entropy' can be replaced by a better one: entropy, taken with the negative sign, is itself a measure of order. Thus the device by which an organism maintains itself stationary at a fairly high level of orderliness (= fairly low level of entropy) really consists in continually sucking orderliness from its environment. This conclusion is less paradoxical than it appears at first sight. Rather could it be blamed for triviality. Indeed, in the case of higher animals we know the kind of orderliness they feed upon well enough, viz. the extremely well-ordered state of matter in more or less complicated organic compounds, which serve them as foodstuffs. After utilizing it they return it in a very much degraded form—not entirely degraded, however, for plants can still make use of it. (These, of course, have their most powerful supply of 'negative entropy' in the sunlight.)

NOTE TO CHAPTER 6

The remarks on *negative entropy* have met with doubt and opposition from physicist colleagues. Let me say first, that if I had been catering for them alone I should have let the discussion turn on *free energy* instead. It is the more familiar notion in this context. But this highly technical term seemed linguistically too near to *energy* for making the average reader alive to the contrast between the two things. He is likely to take *free* as more or less an *epitheton ornans* without much relevance, while actually the concept is a rather intricate one, whose relation to Boltzmann's order–disorder principle is less easy to trace than for entropy and 'entropy taken with a negative sign', which by the way is not my invention. It happens to be precisely the thing on which Boltzmann's original argument turned.

But F. Simon has very pertinently pointed out to me that my simple thermodynamical considerations cannot account for our having to feed on matter 'in the extremely well ordered state of more or less complicated organic compounds' rather than on charcoal or diamond pulp. He is right. But to the lay reader I must explain that a piece of un-burnt coal or diamond, together with the amount of oxygen needed for its combustion, is also in an extremely well-ordered state, as the physicist understands it. Witness to this: if you allow the reaction, the burning of the coal, to take place, a great amount of heat is produced. By giving it off to the surroundings, the system disposes of the very considerable entropy increase entailed by the reaction, and reaches a state in which it has, in point of fact, roughly the same entropy as before.

Yet we could not feed on the carbon dioxide that results from the reaction. And so Simon is quite right in pointing out to me, as he did, that actually the energy content of our food *does* matter; so my mocking at the menu cards that indicate it was out of place. Energy is needed to replace not only the mechanical energy of our bodily exertions, but also the heat we continually give off to the environment. And that we give off heat is not accidental, but essential. For this is precisely the manner in which we dispose of the surplus entropy we continually produce in our physical life process.

This seems to suggest that the higher temperature of the warm-blooded animal includes the advantage of enabling it to get rid of its entropy at a quicker rate, so that it can afford a more intense life process. I am not sure how much truth there is in this argument (for which I am responsible, not Simon). One may hold against it, that on the other hand many warm-blooders are *protected* against the rapid loss of heat by coats of fur or feathers. So the parallelism between body temperature and 'intensity of life', which I believe to exist, may have to be accounted for more directly by van't Hoff's law. The higher temperature itself speeds up the chemical reactions involved in living. (That it actually does, has been confirmed experimentally in species which take the temperature of the surroundings.)

Daniel Dennett

■   Daniel Dennett is a professional philosopher, and an extremely distinguished one who would find a place in any anthology of modern philosophical writings. He is here in this science anthology too because, of all modern philosophers, he is the scientist's philosopher. He thinks like a scientist, sounds like a scientist, and reads scientific journals more assiduously than most scientists do. He is a big man with a big intellect and a warm and jovial presence. His literary style is sometimes criticized as discursive, but his books well recompense the time you need to set aside for them. The first excerpt is from *Darwin's Dangerous Idea* and introduces his vivid image of 'universal acid'. The second is from *Consciousness Explained* but, as it happens, this particular passage is not about consciousness but is again a musing on Darwinian evolution. Like everything Dennett writes, both these excerpts make you think, and he makes it a pleasure to do so. ■

from DARWIN'S DANGEROUS IDEA

Did you ever hear of universal acid? This fantasy used to amuse me and some of my schoolboy friends—I have no idea whether we invented or inherited it, along with Spanish fly and saltpetre, as a part of underground youth culture. Universal acid is a liquid so corrosive that it will eat through *anything*! The problem is: what do you keep it in? It dissolves glass bottles and stainless-steel canisters as readily as paper bags. What would happen if you somehow came upon or created a dollop of universal acid? Would the whole planet eventually be destroyed? What would it leave in its wake? After everything had been transformed by its encounter with universal acid, what would the world look like? Little did I realize that in a few years I would encounter an idea—Darwin's idea—bearing an unmistakable likeness to universal acid: it eats through just about every traditional concept, and leaves in its wake a revolutionized world-view, with most of the old landmarks still recognizable, but transformed in fundamental ways.

Darwin's idea had been born as an answer to questions in biology, but it threatened to leak out, offering answers—welcome or not—to questions in cosmology (going in one direction) and psychology (going in the other direction). If *re*design could be a mindless, algorithmic process of evolution, why couldn't that whole process itself be the product of evolution, and so forth, *all the way down*? And if mindless evolution could account for the breathtakingly clever artifacts of the biosphere, how could the products of our own 'real' minds be exempt from an evolutionary explanation? Darwin's idea thus also threatened to spread *all the way up*, dissolving the illusion of our own authorship, our own divine spark of creativity and understanding.

Much of the controversy and anxiety that has enveloped Darwin's idea ever since can be understood as a series of failed campaigns in the struggle to contain Darwin's idea within some acceptably 'safe' and merely partial revolution. Cede some or all of modern biology to Darwin, perhaps, but hold the line there! Keep Darwinian thinking out of cosmology, out of psychology, out of human culture, out of ethics, politics, and religion! In these campaigns, many battles have been won by the forces of containment: flawed applications of Darwin's idea have been exposed and discredited, beaten back by the champions of the pre-Darwinian tradition. But new waves of Darwinian thinking keep coming. They seem to be improved versions, not vulnerable to the refutations that defeated their predecessors, but are they sound extensions of the unquestionably sound Darwinian core idea, or might they, too, be perversions of it, and even more virulent, more dangerous, than the abuses of Darwin already refuted?

* * *

from CONSCIOUSNESS EXPLAINED

The Birth of Boundaries and Reasons

In the beginning, there were no reasons; there were only causes. Nothing had a purpose, nothing had so much as a function; there was no teleology in the world at all. The explanation for this is simple: there was nothing that had interests. But after millennia there happened to emerge simple *replicators*. While *they* had no inkling of their interests, and perhaps properly speaking had no interests, we, peering back from our godlike vantage point at their early days, can nonarbitrarily assign them certain interests—generated by their defining 'interest' in self-replication. That is, maybe it really made no difference, was a matter of no concern, didn't matter to anyone or anything whether or not they succeeded in replicating (though it does seem that we can be grateful that they did), but at least we can assign them interests conditionally. *If* these simple replicators are to survive and replicate, thus persisting in the face of increasing entropy, their environment must meet certain conditions: conditions conducive to replication must be present or at least frequent.

Put more anthropomorphically, if these simple replicators want to continue to replicate, they should hope and strive for various things; they should avoid the 'bad' things and seek the 'good' things. When an entity arrives on the scene capable of behavior that staves off, however primitively, its own dissolution and decomposition, it brings with it into the world its 'good'. That is to say, it creates a point of view from which the world's events can be roughly partitioned into the favorable, the unfavorable, and the neutral. And its own innate proclivities to seek the first, shun the second, and ignore the third contribute essentially to the definition of the three classes. As the creature thus comes to have interests, the world and its events begin creating *reasons* for it— whether or not the creature can fully recognize them. The first reasons preexisted their own recognition. Indeed, the first problem faced by the first problem-facers was to learn how to recognize and act on the reasons that their very existence brought into existence.

As soon as something gets into the business of self-preservation, boundaries become important, for if you are setting out to preserve yourself, you don't want to squander effort trying to preserve the whole world: you draw the line. You become, in a word, *selfish*. This primordial form of selfishness (which, as a primordial form, lacks most of the flavors of our brand of selfishness) is one of the marks of life. Where one bit of granite ends and the next bit begins is a matter of slight moment; the fracture boundary may be real enough, but nothing works to protect the territory, to push back the frontier or retreat. 'Me against the world'— this distinction between everything on the inside of a closed boundary and everything in the external world—is at the heart of all biological processes, not just ingestion and excretion, respiration and transpiration. Consider, for instance, the immune system, with its millions of different antibodies arrayed in defense of the body against millions of different alien intruders. This army must solve the fundamental problem of recognition: telling one's self (and one's friends) from everything else. And the problem has been solved in much the way human nations, and their armies, have solved the counterpart problem: by standardized, mechanized identification routines—the passports and customs officers in miniature are molecular shapes and shape detectors. It is important to recognize that this army of antibodies has no generals, no GHQ with a battle plan, or even a description of the enemy: the antibodies represent their enemies only in the way a million locks represent the keys that open them.

We should note several other facts that are already evident at this earliest stage. Whereas evolution depends on history, Mother Nature is no snob, and origins cut no ice with her. It does not matter where or how an organism acquired its prowess; handsome is as handsome does. So far as we know, of course, the pedigrees of the early replicators were all pretty much the same: they were each of them the product of one blind, dumb-luck series of selections or another. But had some time-traveling hyperengineer inserted a *robot-replicator* into the milieu, and if its prowess was equal or better than the prowess of its natural-grown competition, its descendants might now be among us—might even be us!

Natural selection cannot tell how a system got the way it got, but that doesn't mean there might not be profound differences between systems 'designed' by natural selection and those designed by intelligent engineers. For instance, human designers, being farsighted but blinkered, tend to find their designs thwarted by unforeseen side effects and interactions, so they try to guard against them by giving each element in the system a single function, and insulating it from all the other elements. In contrast, Mother Nature (the process of natural selection) is famously myopic and lacking in goals. Since she doesn't foresee at all, she has no way of worrying about unforeseen side effects. Not 'trying' to avoid them, she tries out designs in which many such side effects occur; most such designs are terrible (ask any engineer), but every now and then there is a *serendipitous side effect*: two or more unrelated functional systems interact to produce a bonus: multiple functions for single elements. Multiple functions are not unknown in human-engineered artifacts, but they are relatively rare; in nature they are everywhere, and…one of the reasons theorists have had such a hard time finding plausible designs for consciousness in the brain is that they have tended to think of brain elements as serving just one function each.

Ernst Mayr

from THE GROWTH OF BIOLOGICAL THOUGHT

■ I gratefully admire those distinguished scientists whose native language is not English but who write it better than many native speakers. We have already met Dobzhansky (Russian), Tinbergen (Dutch), and Bronowski (Polish). Now here is Ernst Mayr (German). Mayr died in 2005 at the age of 100, the last surviving giant of the neo-Darwinian synthesis. His classic book is *Animal Species and Evolution* but I have chosen an extract from a later and more philosophical/historical book, *The Growth of Biological Thought*. This explores a perennial obsession of Mayr, essentialism, and the distinction from what he calls 'population thinking'. Many have wondered why it took so long for Darwin and Wallace to arrive on the scene. Nobody could deny that Darwin was outstandingly clever, but the problems solved by, for example, Newton and Galileo seem on the face of it harder than Darwin's problem. Yet Darwin lived two centuries later than Newton. What held humanity back so long, after the spectacular successes of Newton and other physicists? For Mayr, the answer was essentialism—what has been called the dead hand of Plato. ■

Generalizations in biology are almost invariably of a probabilistic nature. As one wit has formulated it, there is only one universal law in biology: 'All biological laws have exceptions.' This probabilistic conceptualization contrasts strikingly with the view during the early period of the scientific revolution that causation in nature is regulated by laws that can be stated in mathematical terms. Actually, this idea occurred apparently first to Pythagoras. It has remained a dominant idea, particularly in the physical sciences, up to the present day. Again and again it was made the basis of some comprehensive philosophy, but taking very different forms in the hands of various authors. With Plato it gave rise to essentialism, with Galileo to a mechanistic world picture, and with Descartes to the deductive method. All three philosophies had a fundamental impact on biology.

Plato's thinking was that of a student of geometry: a triangle, no matter what combination of angles it has, always has the *form* of a triangle, and is thus discontinuously different from a quadrangle or any other polygon. For Plato, the variable world of phenomena in an analogous manner was nothing but the reflection of a limited number of fixed and unchanging forms, *eide* (as Plato called them) or *essences* as they were called by the Thomists in the Middle Ages. These essences are what is real and important in this world. As ideas they can exist independent of any objects. Constancy and discontinuity are the points of special emphasis for the essentialists. Variation is attributed to the imperfect manifestation of the underlying essences. This conceptualization was the basis not only of the realism of the Thomists but also of so-called idealism or of the positivism of later philosophers, up to the twentieth century. Whitehead, who was a peculiar mixture of a mathematician and a mystic (perhaps one should call him a Pythagorean), once stated: 'The safest general characterization of the European philosophical tradition is that it consists in a series of footnotes to Plato.' No doubt, this was meant as praise, but it really was a condemnation, so far as it was true at all. What it really says is that European philosophy through all the centuries was unable to free itself from the strait jacket of Plato's essentialism. Essentialism, with its emphasis on discontinuity, constancy, and typical values ('typology'), dominated the thinking of the Western world to a degree that is still not yet fully appreciated by the historians of ideas. Darwin, one of the first thinkers to reject essentialism (at least in part), was not at all understood by the contemporary philosophers (all of whom were essentialists), and his concept of evolution through natural selection was therefore found unacceptable. Genuine change, according to essentialism, is possible only through the saltational origin of new essences. Because evolution as explained by Darwin, is by necessity gradual, it is quite incompatible with essentialism. However, the philosophy of essentialism fitted well with the thinking of the physical scientists, whose 'classes' consist of identical entities, be they sodium atoms, protons, or pi-mesons.

[. . .]

Western thinking for more than two thousand years after Plato was dominated by essentialism. It was not until the nineteenth century that a new and different way of thinking about nature began to spread, so-called population thinking. What is population thinking and how does it differ from essentialism? Population thinkers stress the uniqueness of everything in the organic world. What is important for them is the individual, not the type. They emphasize that every individual in sexually reproducing species is uniquely different from all others, with much individuality even existing in uniparentally reproducing ones. There is no 'typical' individual, and mean values are abstractions. Much of what in the past has been designated in biology as 'classes' are populations consisting of unique individuals.

There was a potential for population thinking in Leibniz's theory of monads, for Leibniz postulated that each monad was individualistically different from every other monad, a major departure from essentialism. But essentialism had such a strong hold in Germany that Leibniz's suggestion did not result in any population thinking. When it finally developed elsewhere, it had two roots; one consisted of the British animal breeders (Bakewell, Sebright, and many others) who had come to realize that every individual in their herds had different heritable characteristics, on the basis of which they selected the sires and dams of the next generation. The other root was systematics. All practising naturalists were struck by the observation that when collecting a 'series' of specimens of a single species they found that no two specimens were ever completely alike. Not only did Darwin stress this in his barnacle work, but even Darwin's critics concurred on this point. Wollaston, for instance, wrote 'amongst the millions of people who have been born into the world, we are certain that no two have ever been precisely alike in every respect; and in a similar manner it is not too much to affirm the same of all living creatures (however alike some of them may seem to our uneducated eyes) that have ever existed.' Similar statements were made by many mid nineteenth century taxonomists. This uniqueness is true not only for individuals but even for stages in the life cycle of any individual, and for aggregations of individuals whether they be demes, species, or plant and animal associations. Considering the large number of genes that are either turned on or turned off in a given cell, it is quite possible that not even any two cells in the body are completely identical. This uniqueness of biological individuals means that we must approach groups of biological entities in a very different spirit from the way we deal with groups of identical inorganic entities. This is the basic meaning of population thinking. The differences between biological individuals are real, while the mean values which we may calculate in the comparison of groups of individuals (species, for example) are man-made inferences. This fundamental difference between the classes of the physical scientists and the populations of the biologist has various consequences. For instance, he who does not understand the uniqueness of individuals is unable to understand the working of natural selection.

The statistics of the essentialist are quite different from those of the populationist. When we measure a physical constant—for instance, the speed of light—we know that under equivalent circumstances it is constant and that any variation in the observational results is due to inaccuracy of measurement, the statistics simply indicating the degree of reliability of our results. The early statistics from Petty and Graunt to Quetelet was essentialistic statistics, attempting to arrive at true values in order to overcome the confusing effects of variation. Quetelet, a follower of Laplace, was interested in deterministic laws. He hoped by his method to be able to calculate the characteristics of the 'average man', that is, to discover the 'essence' of man. Variation was nothing but 'errors' around the mean values.

Francis Galton was perhaps the first to realize fully that the mean value of variable biological populations is a construct. Differences in height among a group of people are real and not the result of inaccuracies of measurement. The most interesting parameter in the statistics of natural populations is the actual variation, its amount, and its nature. The amount of variation is different from character to character and from species to species. Darwin could not have arrived at a theory of natural selection if he had not adopted populational thinking. The sweeping statements in the racist literature, on the other hand, are almost invariably based on essentialistic (typological) thinking.

Garrett Hardin

from 'THE TRAGEDY OF THE COMMONS'

■ The American ecologist Garrett Hardin is responsible for introducing into our language the phrase 'The tragedy of the commons', and the important idea that it signifies. Here is an extract from the landmark paper in which he did so. It can go by other names, but everybody needs to understand the concept. ■

We can make little progress in working toward optimum population size until we explicitly exorcize the spirit of Adam Smith in the field of practical demography. In economic affairs, *The Wealth of Nations* (1776) popularized the 'invisible hand', the idea that an individual who 'intends only his own gain', is, as it were, 'led by an invisible hand to promote...the public interest'. Adam Smith did not assert that this was invariably true, and perhaps neither did any of his followers. But he contributed to a dominant tendency of thought that has ever since interfered with positive action based on rational analysis, namely, the tendency to assume that decisions reached individually will, in fact, be the best decisions for an entire society. If this assumption is correct it justifies the continuance of our present policy of laissez-faire in reproduction. If it is correct we can assume that men will control their individual fecundity so as to produce the optimum population. If the assumption is not correct, we need to re-examine our individual freedoms to see which ones are defensible.

Tragedy of Freedom in a Commons

The rebuttal to the invisible hand in population control is to be found in a scenario first sketched in a little-known pamphlet in 1833 by a mathematical amateur named William Forster Lloyd (1794–1852). We may well call it 'the tragedy of the commons', using the word 'tragedy' as the philosopher Whitehead used it: 'The essence of dramatic tragedy is not unhappiness. It resides in the solemnity of the remorseless working of things.' He then goes on to say, 'This inevitableness of destiny can only be illustrated in terms of human life by incidents which in fact involve unhappiness. For it is only by them that the futility of escape can be made evident in the drama.'

The tragedy of the commons develops in this way. Picture a pasture open to all. It is to be expected that each herdsman will try to keep as many cattle as possible on the commons. Such an arrangement may work reasonably satisfactorily for centuries because tribal wars, poaching, and disease keep the numbers of both man and beast well below the carrying capacity of the land. Finally, however, comes the day of reckoning, that is, the day when the long-desired goal of social stability becomes a reality. At this point, the inherent logic of the commons remorselessly generates tragedy.

As a rational being, each herdsman seeks to maximize his gain. Explicitly or implicitly, more or less consciously, he asks, 'What is the utility to me of adding one more animal to my herd?' This utility has one negative and one positive component.

1) The positive component is a function of the increment of one animal. Since the herdsman receives all the proceeds from the sale of the additional animal, the positive utility is nearly +1.

2) The negative component is a function of the additional overgrazing created by one more animal. Since, however, the effects of over-grazing are shared by all the herdsmen, the negative utility for any particular decision-making herdsman is only a fraction of 1.

Adding together the component partial utilities, the rational herdsman concludes that the only sensible course for him to pursue is to add another animal to his herd. And another; and another....But this is the conclusion reached by each and every rational herdsman sharing a commons. Therein is the tragedy. Each man is locked into a system that compels him to increase his herd without limit—in a world that is limited. Ruin is the destination toward which all men rush, each pursuing his own best interest in a society that believes in the freedom of the commons. Freedom in a commons brings ruin to all.

Some would say that this is a platitude. Would that it were! In a sense, it was learned thousands of years ago, but natural selection favors the forces of psychological denial. The individual benefits as an individual from his ability to deny the truth even though society as a whole, of which he is a part, suffers.

Education can counteract the natural tendency to do the wrong thing, but the inexorable succession of generations requires that the basis for this knowledge be constantly refreshed.

A simple incident that occurred a few years ago in Leominster, Massachusetts, shows how perishable the knowledge is. During the Christmas shopping season the parking meters downtown were covered with plastic bags that bore tags reading: 'Do not open until after Christmas. Free parking courtesy of the mayor and city council.' In other words, facing the prospect of an increased demand for already scarce space the city fathers reinstituted the system of the commons. (Cynically, we suspect that they gained more votes than they lost by this retrogressive act.)

In an approximate way, the logic of the commons has been understood for a long time, perhaps since the discovery of agriculture or the invention of private property in real estate. But it is understood mostly only in special cases which are not sufficiently generalized. Even at this late date, cattlemen leasing national land on the western ranges demonstrate no more than an ambivalent understanding, in constantly pressuring federal authorities to increase the head count to the point where overgrazing produces erosion and weed-dominance. Likewise, the oceans of the world continue to suffer from the survival of the philosophy of the commons. Maritime nations still respond automatically to the shibboleth of the 'freedom of the seas'. Professing to believe in the 'inexhaustible resources of the oceans', they bring species after species of fish and whales closer to extinction.

The National Parks present another instance of the working out of the tragedy of the commons. At present, they are open to all, without limit. The parks themselves are limited in extent—there is only one Yosemite Valley—whereas population seems to grow without limit. The values that visitors seek in the parks are steadily eroded. Plainly, we must soon cease to treat the parks as commons or they will be of no value to anyone.

What shall we do? We have several options. We might sell them off as private property. We might keep them as public property, but allocate the right to enter them. The allocation might be on the basis of wealth, by the use of an auction system. It might be on the basis of merit, as defined by some agreed-upon standards. It might be by lottery. Or it might be on a first-come, first-served basis, administered to long queues. These, I think, are all the reasonable possibilities. They are all objectionable. But we must choose—or acquiesce in the destruction of the commons that we call our National Parks.

W. D. Hamilton

■ Bill Hamilton, dear friend and Oxford colleague, died from a haemorrhage in 2000. He had just returned from an expedition to the Congo jungle. A tragic end to a brilliant life. He was an evolutionary theorist of immense distinction: several of his obituaries compared his importance in the late twentieth century to Darwin's in the late nineteenth. Yet he was a man of painful modesty and diffidence. Towards the end of his life he published a unique memoir consisting of autobiographical and reflective notes threaded through his collected scientific papers (he wrote no other books). By comparison with the major contributions that he made to almost all branches of evolutionary theory, 'Geometry for the Selfish Herd', from which the first extract is taken, is but a vignette. My second selection is from one of Hamilton's autobiographical notes, and it conveys the agonizing self-doubt, which—at least on the face of it—was so much a part of his complex character. ■

from 'GEOMETRY FOR THE SELFISH HERD'

A Model of Predation in One Dimension

Imagine a circular lily pond. Imagine that the pond shelters a colony of frogs and a water-snake. The snake preys on the frogs but only does so at a certain time of day—up to this time it sleeps on the bottom of the pond. Shortly before the snake is due to wake up all the frogs climb out onto the rim of the pond. This is because the snake prefers to catch frogs in the water. If it can't find any, however, it rears its head out of the water and surveys the disconsolate line sitting on the rim—it is supposed that fear of terrestial predators prevents the frogs from going back from the rim; the snake surveys this line and snatches *the nearest one.*

Now suppose that the frogs are given opportunity to move about on the rim before the snake appears, and suppose that initially they are dispersed in some rather random way. Knowing that the snake is about to appear, will all the frogs be content with their initial positions? No; each will have a better chance of not being nearest to the snake if he is situated in a narrow gap between two others. One can imagine that a frog that happens to have climbed out into a wide open space will want to improve his position. The part of the pond's perimeter on which the snake could appear and find a certain frog to be nearest to him may be termed that frog's 'domain of danger': its length is half that of the gap between the neighbours on either side. Figure 10a shows the best move for one particular frog and how his domain of danger is diminished by it.

But usually neighbours will be moving as well and one can imagine a confused toing-and-froing in which the desirable narrow gaps are as elusive as the croquet hoops in Alice's game in Wonderland. From the positions in Figure 10a, assuming the outside frogs to be in gaps larger than any others shown, the moves in Figure 10b may be expected.

What will be the result of this communal exercise? Devious and unfair as usual, natural justice does not, in general, equalize the risks of these selfish frogs by spacing them out. On the contrary, with any reasonable assumptions about the exact jumping behaviour, they quickly collect in heaps. Except in the case of three frogs who start spaced out in an acute-angled triangle I know of no rule of jumping that can prevent them aggregating. Some occupy protected central positions from the start;

Figure 10. Selfish prey movements through jumping given nearest-prey predation in a one-dimensional habitat. (a) A particular prey's *domain of danger* is shown by the solid bar; this includes all points nearer to the given prey than to any other. Independent of position within a gap between neighbour prey, the domain is equal to half of the gap. The arrowed movement illustrates a jump into a narrower gap beyond a neighbour, thereby diminishing a domain. (b) All prey are assumed moving on the principle of (a), with each jumper passing its neighbour's position by one-third of the gap length beyond. Increase of aggregation can be seen.

some are protected only initially in groups destined to dissolve; some, on the margins of groups, commute wildly from one heap to another and yet continue to bear most of the risk. Figure 11 shows the result of a computer simulation experiment in which 100 frogs are initially spaced randomly round the pool. In each 'round' of jumping a frog stays put only if the 'gap' it occupies is smaller than both neighbouring gaps; otherwise it jumps into the smaller of these gaps, passing the neighbour's position by one-third of the gap-length. Note that at the termination of the experiment only the largest group is growing rapidly.

The idea of this round pond and its circular rim is to study cover-seeking behaviour in an edgeless universe. No apology, therefore, need be made even for the rather ridiculous behaviour that tends to arise in the later stages of the model process, in which frogs supposedly fly right round the circular rim to 'jump into' a gap on the other

10° segments of pool margin (degrees)

Position number	0				90							180							270								360
1	2 3 3 3	6 1 3 3	4 1 3 5	1 3 4 5 2 9 4	5 6 3 5 4	1 2 2	4 3
2	5 2 2	8 3 2	6 1 1 7	2 4 7	11 2	7 5 2 5 5	3 2	4 4
3	6 1 2	8 3 1	8	9	4 9	11	9 4 1 7 5	2 2	4 4
4	6 1	9 3	9	8	4 11	10	10 3	8 6	1 2	4 5
5	7	9 2	9	8	5 12	8	12 1	9 6	3	4 5
6	7	9	9	8	5 14	7	13	9 5	5	3 6
7	8	7	9	8	5 16	5	15	9 3	6	3 6
8	9	5	9	9	4 18	3	17	9 1	6	5 5
9	10	4	8	10	3 20	1	19	8	6	7 4
10	12	3	7	11	2 22		20	6	6	9 2
11	13	2	6	12	1 22		22	4	7	9 2
12	14		6	13	22		24	3	7	9 2
13	14		5	13	22		26	2	8	8 2
14	15		3	13	22		28	1	9	7 2
15	16		1	13	22		30		10	6 2
16	17			12	22		32		9	6 2
17	17			11	22		33		8	7 2
18	18			9	22		35		6	8 2
19	19			7	22		37		5	9 1

Figure 11. Gregarious behaviour of 100 frogs is shown in terms of the numbers found successively within 10° segments on the margin of the pool. The initial scatter (position 1) is random. Frogs jump simultaneously giving the series of positions shown. They pass neighbours' positions by one-third of the width of the gap. For further explanation, see text.

side of the aggregation. The model gives the hint which I wish to develop: that even when one starts with an edgeless group of animals, randomly or evenly spaced, the *selfish avoidance of a predator can lead to aggregation.*

Aggregations and Predators

It may seem a far cry from such a phantasy to the realities of natural selection. Nevertheless, I think there can be little doubt that behaviour which is similar in biological intention to that of the hypothetical frogs is an important factor in the gregarious tendencies of a very wide variety of animals. Most of the herds and flocks with which one is familiar show a visible closing-in of the aggregation in the presence of their common predators. Starlings do this in the presence of a sparrowhawk; sheep in the presence of a dog, or, indeed, any frightening stimulus. Parallel observations are available for the vast flocks of the quelea and for deer. No doubt a thorough search of the literature would reveal many other examples. The phenomenon in fish must be familiar to anyone who has tried to catch minnows or sand eels with a net in British waters. Almost any sudden stimulus causes schooling fish to cluster more tightly, and fish have been described as packing, in the presence of predators, into balls so tight that they cannot swim and such that some on top are thrust above the surface of the water. A shark has been described as biting mouthfuls from a school of fish 'much in the manner of a person eating an apple'.

* * *

from NARROW ROADS OF GENELAND

Reading about them was not enough, they were too fantastic. Only in 1963 in Brazil did I first thoroughly believe in fig wasps, the tiny black insects that both breed in and pollinate all fig trees. Worldwide this means 600 species, not just the one cultivated species that is most familiar to Northern Europeans for its dried fruits in the brown, oblong packets. The 600 mainly inhabit tropical countries where many are giants of the forest. To this great assemblage the almost shrubby European fig stands an outlier in several ways. For example, besides its geographic placement as the world's most northerly fig, and its small stature, it is dioecious—that is, has male and female trees separate like holly or yew. Again, generally each fig species has one kind of wasp adapted to pollinate it, but the European can sometimes have none; for the most practical and novel pollination feature of the European fig is that although it has a wasp not all its varieties need it, or at least not for their crops. The best varieties still use the wasp to set fruit, but some others produce fruit with no viable seeds and therefore no need for pollination.

If, however, you eat only the best and are now worried by the idea that you swallow a small wasp with each fig, first let me say that the wasps doing the job for the trees are tiny—2 mm or less—and, second, that you will be out of line with much of your successful ancestry if upset at the idea of eating these few small flakes of chitinous bran and protein derived from an embalmed and long dead pollinator. All through the later Tertiary period fig-tree fruiting was certainly a cause for celebration for forest wildlife and our ancestors must have whooped and brachiated towards the fruiting crowns just as our great ape cousins and most monkeys do today. Arriving there, they consumed figs that had many more wasps—hundreds per fruit instead of just one or two—compared with our figs today. This protein addition to a largely vegetarian diet and the perceptual/spatial problems involved in reaching the bonanza must have been among the factors pushing ahead the enlargement of the anthropoid brain, and it is perhaps the extra wasp protein that makes figs such a favourite food not only for primates but for a great variety of birds and bats.

Rumours of the winged and targeted symbiotic pollination of fig trees had intrigued me before my 1963 visit to Brazil and seeing the reality was no disappointment. Yet for more than 10 years I had no time, either in libraries or during my visits to the tropics, to look into the points that remained puzzling. When I did come to take a more serious interest in 1975 and developed the data for the paper in this chapter, it was during my third visit to Brazil in 1975, and even then the study came about more by accident than by plan.

In 1974, I concentrated my courses at Imperial College into a short intensive period so that I could be free from teaching for most of 1975. My idea was unconnected with figs and was rather to pursue the 'rotten-wood' interest of the last chapter into the tropical realm: I would go to Brazil, look at tropical rotting-wood insects, and make comparisons with my observations in temperate Britain. To help with my expenses, especially because my wife Christine and our two small children were coming with me, I had arranged to teach a course in population genetics at a campus in southern Brazil where several people of Warwick Kerr's Rio Claro group were established and I was therefore known. Two Faculdades, of Medicine and also of Philosophy, Science and Letters, had been built together on a common campus on the outskirts of Ribeirão Preto. Ribeirão Preto is a small but wealthy coffee and sugar-cane city and centres a region of the best soils of likewise small and rich São Paulo State. (For the town 'small' means about 100 000 inhabitants; for the state it means about the size of England—this to be contrasted to, say, Amazonas State, which is the size of Europe.) The course I taught was my first wholly at a graduate level and my class was just six. To my surprise I enjoyed the teaching. I had the difficulty of speaking in Portuguese but, against this, how different were these thoughtful, respectful Brazilians from the hundred-headed swarm I faced in my customary nine o'clock lectures at IC. There I was lucky if I could see one or two of the heads caring one Johannsen's polygenic bean for my Hardy–Weinberg equilibrium or my definition of linkage: during the classes most students would be good-humouredly chatting or perhaps deep in their morning newspapers. And towards the end of the course I knew they would have another topic: they would be discussing who was to lead this year's delegation to the professor to complain about the irrelevance and incomprehensibility of the lectures I was giving. In those days apparently 'inapplicable' subjects such as population genetics and evolution were not the zoology that aspiring entomologists and parasitologists had come to Britain's prime technical institute to learn. Instead they wanted insect physiology, practical parasitology, and the like—in short, knowledge to make a living with. It is an immodest comparison but what a pleasure it was to me later to read how Isaac Newton as a lecturer at Cambridge in the seventeenth century seemed to have had similar experiences to mine. Newton sometimes lectured to 'ye bare walls', as his assistant put it—a class of zero rushed in to learn from the world's greatest-ever scientist! At least mine came to the theatre; but then, perhaps, there was some check that they did. Anyway, in contrast, here were Brazilian students really wanting to listen.

Per Bak

from HOW NATURE WORKS

■ There is a genre of fashionable mathematical ideas, going under names such as 'complexity theory', 'self-organized criticality', and 'chaos theory', which are more interesting than my slightly put-downy word 'fashionable' might suggest. The trouble starts when they get into the hands of fashionable pseudo-intellectuals who don't really understand them but think they sound trendy. It is therefore good to go to the scientific source, and for 'self-organized criticality' the source is the Danish physicist Per Bak. Here he describes the famous sand-pile model, but unfortunately I don't have space to let him develop its implications. One measure of this anthology's success will emerge later if it entices readers to seek out the original books, in this case *How Nature Works*. ■

How can the universe start with a few types of elementary particles at the big bang, and end up with life, history, economics, and literature? The question is screaming out to be answered but it is seldom even asked. Why did the big bang not form a simple gas of particles, or condense into one big crystal? We see complex phenomena around us so often that we take them for granted without looking for further explanation. In fact, until recently very little scientific effort was devoted to understanding why nature is complex.

I will argue that complex behavior in nature reflects the tendency of large systems with many components to evolve into a poised, 'critical' state, way out of balance, where minor disturbances may lead to events, called avalanches, of all sizes. Most of the changes take place through catastrophic events rather than by following a smooth gradual path. The evolution to this very delicate state occurs without design from any outside agent. The state is established solely because of the dynamical interactions among individual elements of the system: the critical state is *self-organized*. Self-organized criticality is so far the only known general mechanism to generate complexity.

To make this less abstract, consider the scenario of a child at the beach letting sand trickle down to form a pile (Figure 12). In the beginning, the pile is flat, and the individual grains remain close to where they land. Their motion can be understood in terms of their physical properties. As the process continues, the pile becomes steeper, and there will be little sand slides. As time goes on, the sand slides become bigger and bigger. Eventually, some of the sand slides may even span all or most of the pile. At that point, the system is far out of balance, and its behavior can no longer be understood in terms of the behavior of the individual grains. The avalanches form a dynamic of their own, which can be understood only from a holistic description of the properties of the entire pile rather than from a reductionist description of individual grains: the sandpile is a complex system.

The complex phenomena observed everywhere indicate that nature operates at the self-organized critical state. The behavior of the critical sandpile mimics several phenomena observed across many sciences, which are associated with complexity.

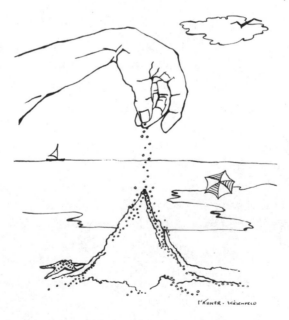

Figure 12. Sandpile. (Drawing by Ms. Elaine Wiesenfeld.)
[…]

The laws of physics can explain how an apple falls but not why Newton, a part of a complex world, was watching the apple. Nor does physics have much to say about the apple's origin. Ultimately, though, we believe that all the complex phenomena, including biological life, do indeed obey physical laws: we are simply unable to make the connection from atoms in which we know that the laws are correct, through the chemistry of complicated organic molecules, to the formation of cells, and to the arrangement of those cells into living organisms. There has never been any proof of a metaphysical process not following the laws of physics that would distinguish living matter from any other. One might wonder whether this state of affairs means that we cannot find general 'laws of nature' describing why the ordinary things that we actually observe around us are complex rather than simple.

The question of the origin of complexity from simple laws of physics—maybe the biggest puzzle of all—has only recently emerged as an active science. One reason is that high-speed computers, which are essential in this study, have not been generally available before. However, even now the science of complexity is shrouded in a good deal of skepticism—it is not clear how any general result can possibly be helpful, because each science works well within its own domain.

Because of our inability to directly calculate how complex phenomena at one level arise from the physical mechanisms working at a deeper level, scientists sometimes throw up their hands and refer to these phenomena as 'emergent'. They just pop out of nowhere. Geophysics emerges from astrophysics. Chemistry emerges from physics. Biology emerges from chemistry and geophysics, and so on. Each science develops its own jargon, and works with its own objects and concepts. Geophysicists talk about tectonic plate motion and earthquakes without reference to astrophysics, biologists describe the properties and evolution of species without reference to geophysics, economists describe human monetary transactions without reference to biology, and so on. There is nothing wrong with that! Because of the seeming intractability of emergent phenomena, no other modus operandi is possible. If no new phenomena emerged in large systems out of the dynamics of systems working at a lower level, then we would need no scientists but particle physicists, since there would be no other areas to cover. But then there would be no particle physicists. Quality, in some way, emerges from quantity.

Martin Gardner

THE FANTASTIC COMBINATIONS OF JOHN CONWAY'S NEW SOLITAIRE GAME 'LIFE'

■ The mathematician John Conway developed a game, unfortunately called Life, which must be one of the most unexpectedly seminal contributions ever made by a mathematician to the thinking processes of non-mathematicians. Who would have thought that so much could *emerge* (to use the technical term) from such simple rules? The lesson will not be lost on thoughtful people wondering how the complex and beautiful world that surrounds us could have emerged, all unguided, from primordially simple beginnings (the brilliantly inventive computer scientist Stephen Wolfram has written a gigantic book, *A New Kind of Science* on this kind of emergence). The exposition of Conway's game that I have chosen is not Conway's own but that of Martin Gardner, whose mathematical teases, games, puzzles, and tantalizers have delighted generations of readers of *Scientific American* and earned him a devoted coterie of admirers all around the world. Here, then, is Gardner's column from the October 1970 issue of *Scientific American*, complete with exercises for the reader. ■

Most of the work of John Horton Conway, a mathematician at Gonville and Caius College of the University of Cambridge, has been in pure mathematics. For instance, in 1967 he discovered a new group—some call it 'Conway's constellation'—that includes all but two of the then known sporadic groups. (They are called 'sporadic' because they fail to fit any classification scheme.) It is a breakthrough that has had exciting repercussions in both group theory and number theory. It ties in closely with an earlier discovery by John Leech of an extremely dense packing of unit spheres in a space of 24 dimensions where each sphere touches 196,560 others. As Conway has remarked, 'There is a lot of room up there.'

In addition to such serious work Conway also enjoys recreational mathematics. Although he is highly productive in this field, he seldom publishes his discoveries. One exception was his paper on 'Mrs Perkins' Quilt', a dissection problem discussed in 'Mathematical Games' for September 1966. My topic for July 1967, was sprouts, a topological pencil-and-paper game invented by Conway and M. S. Paterson. Conway has been mentioned here several other times.

This month we consider Conway's latest brainchild, a fantastic solitaire pastime he calls 'life'. Because of its analogies with the rise, fall and alterations of a society of living organisms, it belongs to a growing class of what are called 'simulation games'—games that resemble real-life processes. To play life you must have a fairly large checkerboard and a plentiful supply of flat counters of two colors. (Small checkers or poker chips do nicely.) An Oriental 'go' board can be used if you can find flat counters that are small enough to fit within its cells. (Go stones are unusable because they are not flat.) It is possible to work with pencil and graph paper but it is much easier, particularly for beginners, to use counters and a board.

The basic idea is to start with a simple configuration of counters (organisms), one to a cell, then observe how it changes as you apply Conway's 'genetic laws' for births, deaths and survivals. Conway chose his rules carefully, after a long period of experimentation, to meet three desiderata:

1. There should be no initial pattern for which there is a simple proof that the population can grow without limit.
2. There should be initial patterns that *apparently* do grow without limit.
3. There should be simple initial patterns that grow and change for a considerable period of time before coming to an end in three possible

ways: fading away completely (from overcrowding or from becoming too sparse), settling into a stable configuration that remains unchanged thereafter, or entering an oscillating phase in which they repeat an endless cycle of two or more periods.

In brief, the rules should be such as to make the behavior of the population unpredictable.

Conway's genetic laws are delightfully simple. First note that each cell of the checkerboard (assumed to be an infinite plane) has eight neighboring cells, four adjacent orthogonally, four adjacent diagonally. The rules are:

1. Survivals. Every counter with two or three neighboring counters survives for the next generation.
2. Deaths. Each counter with four or more neighbors dies (is removed) from overpopulation. Every counter with one neighbor or none dies from isolation.
3. Births. Each empty cell adjacent to exactly three neighbors—no more, no fewer—is a birth cell. A counter is placed on it at the next move.

It is important to understand that all births and deaths occur *simultaneously*. Together they constitute a single generation or, as we shall call it, a 'move' in the complete 'life history' of the initial configuration. Conway recommends the following procedure for making the moves:

1. Start with a pattern consisting of black counters.
2. Locate all counters that will die. Identify them by putting a black counter on top of each.
3. Locate all vacant cells where births will occur. Put a white counter on each birth cell.
4. After the pattern has been checked and double-checked to make sure no mistakes have been made, remove all the dead counters (piles of two) and replace all newborn white organisms with black counters.

You will now have the first generation in the life history of your initial pattern. The same procedure is repeated to produce subsequent generations. It should be clear why counters of two colors are needed. Because births and deaths occur simultaneously, newborn counters play no role in causing other deaths or births. It is essential, therefore, to be able to distinguish them from live counters of the previous generation while you check the pattern to be sure no errors have been made. Mistakes are very easy to make, particularly when first playing the game. After playing it for a while you will gradually make fewer mistakes, but even experienced players must exercise great care in checking every new generation before removing the dead counters and replacing newborn white counters with black.

You will find the population constantly undergoing unusual, sometimes beautiful and always unexpected change. In a few cases the society eventually dies out (all counters vanishing), although this may not happen until after a great many generations. Most starting patterns either reach stable figures—Conway calls them 'still lifes'—that cannot change or patterns that oscillate forever. Patterns with no initial symmetry tend to become symmetrical. Once this happens the symmetry cannot be lost, although it may increase in richness.

Conway conjectures that no pattern can grow without limit. Put another way, any configuration with a finite number of counters cannot grow beyond a finite upper limit to the number of counters on the field. This is probably the deepest and most difficult question posed by the game. Conway has offered a prize of $50 to the first person who can prove or disprove the conjecture before the end of the year. One way to disprove it would be to discover patterns that keep adding counters to the field: a 'gun' (a configuration that repeatedly shoots out moving objects such as the 'glider', to be explained below) or a 'puffer train' (a configuration that moves but leaves behind a trail of 'smoke'). I shall forward all proofs to Conway, who will act as the final arbiter of the contest.

Let us see what happens to a variety of simple patterns.

A single organism or any pair of counters, wherever placed, will obviously vanish on the first move.

A beginning pattern of three counters also dies immediately unless at least one counter has two neighbors. Figure 13 shows the five triplets that do not fade on the first move. (Their orientation is of course irrelevant.) The first three [a, b, c] vanish on the second move. In connection with c it is worth noting that a single diagonal chain of counters, however long,

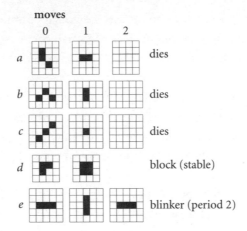

Figure 13. The fate of five triplets in 'life'.

loses its end counters on each move until the chain finally disappears. The speed a chess king moves in any direction is called by Conway (for reasons to be made clear later) the 'speed of light'. We say, therefore, that a diagonal chain decays at each end with the speed of light.

Pattern *d* becomes a stable 'block' (two-by-two square) on the second move. Pattern *e* is the simplest of what are called 'flip-flops' (oscillating figures of period 2). It alternates between horizontal and vertical rows of three. Conway calls it a 'blinker'.

Figure 14 shows the life histories of the five tetrominoes (four rook-wise-connected counters). The square [*a*] is, as we have seen, a still-life figure. Tetrominoes *b* and *c* reach a stable figure, called a 'beehive', on the second move. Beehives are frequently produced patterns. Tetromino *d* becomes a beehive on the third move. Tetromino *e* is the most interesting of the lot. After nine moves it becomes four isolated blinkers, a flip-flop called 'traffic lights'. It too is a common configuration. Figure 15 shows the 12 commonest forms of still life.

The reader may enjoy experimenting with the 12 pentominoes (all patterns of five rookwise-connected counters) to see what happens to each. He will find that six vanish before the fifth move, two quickly reach a stable pattern of seven counters and three in a short time become traffic lights. The only pentomino that does not end quickly (by vanishing, becoming stable or oscillating) is the R pentomino ('*a*' *in Figure 16*). Its

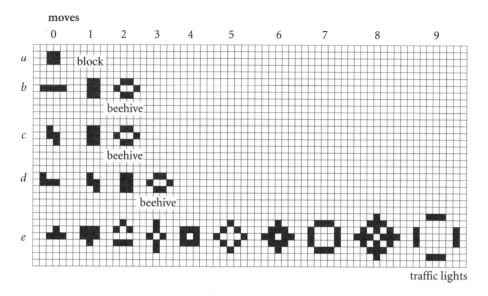

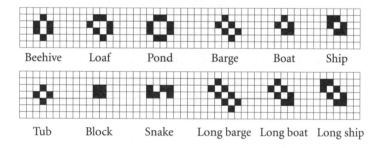

Figure 14. The life histories of the five tetrominoes.

Beehive   Loaf   Pond   Barge   Boat   Ship

Tub   Block   Snake   Long barge   Long boat   Long ship

Figure 15. The commonest stable forms.

fate is not yet known. Conway has tracked it for 460 moves. By then it has thrown off a number of gliders. Conway remarks: 'It has left a lot of miscellaneous junk stagnating around, and has only a few small active regions, so it is not at all obvious that it will continue indefinitely. After 48 moves it has become a figure of seven counters on the left and two symmetric regions on the right which, if undisturbed, would grow into a honey farm (four beehives) and traffic lights. However, the honey farm gets eaten into pretty quickly and the four blinkers forming the traffic lights disappear one by one into the rest of a rather blotchy population.'

For long-lived populations such as this one Conway sometimes uses a PDP 7 computer with a screen on which he can observe the changes. The program was written by M. J. T. Guy and S. R. Bourne. Without its help some discoveries about the game would have been difficult to make.

As easy exercises to be answered next month the reader is invited to discover the fate of the Latin cross ['*b*' in Figure 16], the swastika [*c*], the letter *H*[*d*], the beacon [*e*], the clock [*f*], the toad [*g*] and the pinwheel [*h*]. The last three figures were discovered by Simon Norton. If the center counter of the *H* is moved up one cell to make an arch (Conway calls it 'pi'), the change is unexpectedly drastic. The *H* quickly ends but pi has a long history. Not until after 173 moves has it settled down to five blinkers, six blocks and two ponds. Conway also has tracked the life histories of all the hexominoes, and all but seven of the heptominoes.

One of the most remarkable of Conway's discoveries is the five-counter glider shown in Figure 17. After two moves it has shifted slightly and been reflected in a diagonal line. Geometers call this a 'glide reflection'; hence the figure's name. After two more moves the glider has righted itself and moved one cell diagonally down and to the right from its initial position. We mentioned above that the speed of a chess king is called the speed of light. Conway chose the phrase because it is the highest speed at which any kind of movement can occur on the board. No pattern can replicate itself rapidly enough to move at such speed.

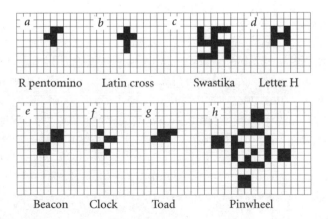

Figure 16. The R pentomino (*a*) and exercises for the reader.

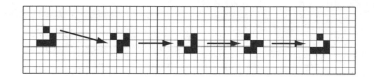

Figure 17. The 'glider'.

Conway has proved that the maximum speed diagonally is a fourth the speed of light. Since the glider replicates itself in the same orientation after four moves, and has travelled one cell diagonally, one says that it glides across the field at a fourth the speed of light.

Movement of a finite figure horizontally or vertically into empty space, Conway has also shown, cannot exceed half the speed of light. Can any reader find a relatively simple figure that travels at such a speed? Remember, the speed is obtained by dividing the number of moves required to replicate a figure by the number of cells it has shifted. If a figure replicates in four moves in the same orientation after traveling two unit squares horizontally or vertically, its speed will be half that of light. I shall report later on any discoveries by readers of any figures that crawl across the board in any direction at any speed, however low. Figures that move in this way are extremely hard to find. Conway knows of only four, including the glider, which he calls 'spaceships' (the glider is a 'featherweight spaceship'; the others have more counters). He has asked me to keep the three heavier spaceships secret as a challenge to readers. Readers are also urged to search for periodic figures other than the ones given here.

Figure 18 depicts three beautiful discoveries by Conway and his collaborators. The stable honey farm ['*a*' *in the illustration*] results after 14 moves from a horizontal row of seven counters. Since a five-by-five block in one move produces the fourth generation of this life history, it becomes a honey farm after 11 moves. The 'figure 8' [*b*], an oscillator found by Norton, both resembles an 8 and has a period of 8. The form *c*, called 'pulsar *CP* 48–56–72', is an oscillator with a life cycle of period 3. The state shown here has 48 counters, state two has 56 and state three has 72, after which the pulsar returns to 48 again. It is generated in 32 moves by a heptomino consisting of a horizontal row of five counters with one counter directly below each end counter of the row.

Conway has tracked the life histories of a row of *n* counters through *n* = 20. We have already disclosed what happens through *n* = 4. Five

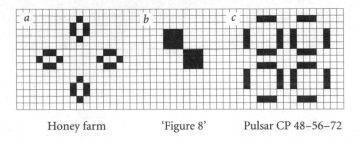

Honey farm   'Figure 8'   Pulsar CP 48–56–72

Figure 18. Three remarkable patterns, one stable and two oscillating.

counters result in traffic lights, six fade away, seven produce the honey farm, eight end with four blinkers and four blocks, nine produce two sets of traffic lights, and 10 lead to the 'pentadecathlon', with a life cycle of period 15. Eleven counters produce two blinkers, 12 end with two bee-hives, 13 with two blinkers, 14 and 15 vanish, 16 give 'big traffic lights' (eight blinkers), 17 end with four blocks, 18 and 19 fade away and 20 generate two blocks.

Rows consisting of sets of five counters, an empty cell separating adjacent sets, have also been tracked by Conway. The 5–5 row generates the pulsar *CP* 48–56–72 in 21 moves, 5–5–5 ends with four blocks, 5–5–5–5 ends with four honey farms and four blinkers, 5–5–5–5–5 terminates with a 'spectacular display of eight gliders and eight blinkers. Then the gliders crash in pairs to become eight blocks.' The form 5–5–5–5–5–5 ends with four blinkers, and 5–5–5–5–5–5–5, Conway remarks, 'is mar-velous to sit watching on the computer screen'. He has yet to track it to its ultimate destiny, however.

Lancelot Hogben

from MATHEMATICS FOR THE MILLION

■   To an earlier generation, a Martin Gardner-like role was played by Lancelot Hogben, actually a biologist but also the author of the celebrated

Mathematics for the Million. Hogben's approach was historical more than puzzle-setting. As I have already remarked of Haldane, Hogben was a man of the left whose politics sometimes obtruded into his writing in a way that seems dated to us today, and his historical treatment of each branch of mathematics emphasized its economic importance, innovation always being driven by need. As it happens, this bias is only slightly evident in the extract I have chosen, which is a felicitous treatment of Zeno's famous paradox of Achilles and the tortoise (see also Douglas Hofstadter, below), and how today we solve it with the use of the mathematical concept of limits and the convergence of infinite series. ■

In the course of the adventure upon which we are going to embark we shall constantly find that we have no difficulty in answering questions which tortured the minds of very clever mathematicians in ancient times. This is not because you and I are very clever people. It is because we inherit a social culture which has suffered the impact of material forces foreign to the intellectual life of the ancient world. The most brilliant intellect is a prisoner within its own social inheritance. An illustration will help to make this quite definite at the outset.

The Eleatic philosopher Zeno set all his contemporaries guessing by propounding a series of conundrums, of which the one most often quoted is the paradox of Achilles and the tortoise. Here is the problem about which the inventors of school geometry argued till they had speaker's throat and writer's cramp. Achilles runs a race with the tortoise. He runs ten times as fast as the tortoise. The tortoise has 100 yards' start. Now, says Zeno, Achilles runs 100 yards and reaches the place where the tortoise started. Meanwhile the tortoise has gone a tenth as far as Achilles, and is therefore 10 yards ahead of Achilles. Achilles runs this 10 yards. Meanwhile the tortoise has run a tenth as far as Achilles, and is therefore 1 yard in front of him. Achilles runs this 1 yard. Meanwhile the tortoise has run a tenth of a yard and is therefore a tenth of a yard in front of Achilles. Achilles runs this tenth of a yard. Meanwhile the tortoise goes a tenth of a tenth of a yard. He is now a hundredth of a yard in front of Achilles. When Achilles has caught up this hundredth of a yard, the tortoise is a thousandth of a yard in front. So, argued Zeno, Achilles is always getting nearer the tortoise, but can never quite catch him up.

You must not imagine that Zeno and all the wise men who argued the point failed to recognize that Achilles really did get past the tortoise. What troubled them was, where is the catch? You may have been asking the same question. The important point is that you did not ask it for the same reason which prompted them. What is worrying you is why they thought up funny little riddles of that sort. Indeed, what you are really concerned with is an *historical* problem. I am going to show you in a minute that the problem is not one which presents any *mathematical* difficulty to you. You know how to translate it into size language, because you inherit a social culture which is separated from theirs by the collapse of two great civilizations and by two great social revolutions. The difficulty of the ancients was not an historical difficulty. It was a mathematical difficulty. They had not evolved a size language into which this problem could be freely translated.

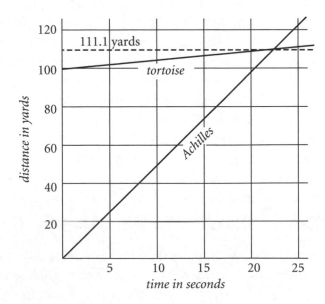

Figure 19. The Race of Achilles and the Tortoise. Greek geometry, which was timeless, could not make it obvious that Achilles would overtake the tortoise. The new geometry of the Newtonian century put time in the picture, thereby showing when and where the two came abreast.

The Greeks were not accustomed to speed limits and passenger-luggage allowances. They found any problem involving division very much more difficult than a problem involving multiplication. They had no way of doing division to any order of accuracy, because they relied for calculation on the mechanical aid of the counting frame or abacus. They could not do sums on paper. For all these and other reasons which we shall meet again and again, the Greek mathematician was unable to see something that we see without taking the trouble to worry about whether we see it or not. If we go on piling up bigger and bigger quantities, the pile goes on growing more rapidly without any end as long as we go on adding more. If we can go on adding larger and larger quantities indefinitely without coming to a stop, it seemed to Zeno's contemporaries that we ought to be able to go on adding smaller and still smaller quantities indefinitely without reaching a limit. They thought that in one case the pile goes on for ever, growing more rapidly, and in the other it goes on for ever, growing more slowly. There was nothing in their number language to suggest that when the engine slows beyond a certain point, it chokes off.

To see this clearly, let us first put down in numbers the distance which the tortoise traverses at different stages of the race after Achilles starts. As we have described it above, the tortoise moves 10 yards in stage 1, 1 yard in stage 2, one-tenth of a yard in stage 3, one-hundredth of a yard in stage 4, etc. Suppose we had a number language like the Greeks and Romans, or the Hebrews, who used letters of the alphabet. Using the one that is familiar to us because it is still used for clocks, graveyards, and law-courts, we might write the total of all the distances the tortoise ran before Achilles caught him up like this:

$$X + I + \frac{I}{X} + \frac{I}{C} + \frac{I}{M}, \text{ and so on}[1]$$

We have put 'and so on' because the ancient world got into great difficulties when it had to handle numbers more than a few thousands. Apart from the fact that we have left the tail of the series to your imagination (and do not forget that the tail is most of the animal if it goes on for ever), notice another disadvantage about this script. There is absolutely nothing to suggest to you how the distances at each stage of the race are connected with one another. Today we have a number vocabulary which makes this relation perfectly evident, when we write it down as:

$$10 + 1 + \frac{1}{10} + \frac{1}{100} + \frac{1}{1,000} + \frac{1}{10,000} + \frac{1}{100,000} + \frac{1}{1,000,000}, \text{ and so on}$$

In this case we put 'and so on' to save ourselves trouble, not because we have not the right number-words. These number-words were borrowed from the Hindus, who learnt to write number language after Zeno and Euclid had gone to their graves. A social revolution, the Protestant Reformation, gave us schools which made this number language the common property of mankind. A second social upheaval, the French Revolution, taught us to use a reformed spelling. Thanks to the Education Acts of the nineteenth century, this reformed spelling is part of the common fund of knowledge shared by almost every sane individual in the English-speaking world. Let us write the last total, using this reformed spelling, which we call decimal notation. That is to say:

$$10 + 1 + 0.1 + 0.01 + 0.001 + 0.0001 + 0.00001 + 0.000001, \text{ and so on}$$

We have only to use the reformed spelling to remind ourselves that this can be put in a more snappy form:

$$11.111111, \text{ etc.,}$$

or still better:

$$11.\dot{1}$$

We recognize the fraction 0.1 as a quantity that is less than 2/10 and more than 1/10. If we have not forgotten the arithmetic we learned at school, we may even remember that 0.$\dot{1}$ corresponds with the fraction 1/9. This means that, the longer we make the sum, 0.1 + 0.01 + 0.001, etc., the nearer it gets to 1/9, and it never grows bigger than 1/9. The total of all the yards the tortoise moves till there is no distance between himself and Achilles makes up just 1/9 yards, and no more. You will now begin to see what was meant by saying that the riddle presents no mathematical difficulty to you. You yourself have a number language constructed so that it can take into account a possibility which mathematicians describe by a very impressive name. They call it the convergence of an infinite series to a limiting value. Put in plain words, this only means that, if you go on piling up smaller and smaller quantities as long as you can, you *may* get a pile of which the size is not made measurably larger by adding any more.

The immense difficulty which the mathematicians of the ancient world experienced when they dealt with a process of division carried on indefinitely, or with what modern mathematicians call infinite series, limits, transcendental numbers, irrational quantities, and so forth, provides an example of a great social truth borne out by the whole history of human knowledge. Fruitful intellectual activity of the cleverest people draws its strength from the common knowledge which all of us share. Beyond a certain point clever people can never transcend the limitations of the social culture they inherit. When clever people pride themselves on their own isolation, we may well wonder whether they are very clever after all. Our studies in mathematics are going to show us that whenever the culture of a people loses contact with the common life of mankind and becomes exclusively the plaything of a leisure class, it is becoming a priestcraft. It is destined to end, as does all priestcraft, in superstition. To be proud of intellectual isolation from the common life of mankind and to be disdainful of the great social task of education is as stupid as it is wicked. It is the end of progress in knowledge. No society, least of all so intricate and mechanized a society as ours, is safe in the hands of a few clever people.

1. The Romans did not actually have the convenient method of representing proper fractions used above for illustrative purposes.

Ian Stewart

from 'THE MIRACULOUS JAR'

█ The infinite is much abused in figures of speech ('Mozart is an infinitely better composer than...' 'Bradman was an infinitely better batsman than...'—no he wasn't, he was a certain amount better). Infinity means something much more precise than that. Ian Stewart is a modern mathematician who carries on the tradition of Lancelot Hogben and Martin Gardner as purveyor of the wonders of the mathematical imagination. In 'The Miraculous Jar' he looks at the daunting idea of infinity in the careful way that mathematicians do, and explains how some great mathematicians have sought to tame it. █

Pick up any mathematics book or journal and you probably won't get more than a page or two into it before you bump headlong against infinity. For example, selecting a few books from my shelf at random, the second *line* of the introduction to Harold Davenport's *The Higher Arithmetic* refers to 'the natural numbers 1, 2, 3,...' and the sequence of dots is intended to imply that the natural numbers are infinite in extent. Page 2 of *Numerical Analysis* by Lee Johnson and Dean Riess quotes the infinite series for the exponential function. B. A. F. Wehrfritz's *Infinite Linear Groups*—need I say more? Page 1 of *Nonlinear Dynamics and Turbulence*, edited by G. I. Barenblatt, G. Iooss, and D. D. Joseph, refers to 'the Navier–Stokes equations or a finite-dimensional Galerkin approximation', correctly leading us to infer that mathematicians treat the full Navier–Stokes equations as an infinite-dimensional object. Page 434 of *Winning Ways* by Elwyn Berlekamp, John Conway, and Richard Guy talks of a game whose position is '∞, 0, ±1, ±4', where ∞ is the standard symbol for 'infinity'. You may feel that page 434 is stretching the point a bit, but it's the sixth page in volume 2, and I didn't check volume 1.

Infinity is, according to Philip Davis and Reuben Hersh, the 'Miraculous Jar of Mathematics'. It is miraculous because its contents are inexhaustible. Remove one object from an infinite jar and you have, not one fewer, but *exactly the same number* left. It is paradoxes like this that forced our forefathers to be wary of arguments involving appeals to the infinite. But the lure of infinity is too great. It is such a marvellous place to lose awkward things in. The number of mathematical proofs that succeed by pushing everything difficult out to infinity, and watching it vanish altogether, is itself almost infinite. But what do we really mean by infinity? Is it wild nonsense, or can it be tamed? Are the infinities of mathematics real, or are they clever fakes, the wolf of the infinite in finite sheep's clothing?

Hilbert's Hotel

If you are laying a table and each place-setting has one knife and one fork, then you know that there are just as many knives as forks. This is true whether you lay out an intimate candlelit dinner for two or a Chinese banquet for 2,000, and you don't need to know how many places are set to be sure the numbers agree. This observation is the cornerstone of the number concept. Two sets of objects are said to be in *one-to-one correspondence* if to each object in one there corresponds a unique object in the other, and vice versa. Sets that can be placed in one-to-one correspondence contain the same number of objects.

When the sets are infinite, however, paradoxes arise. For example, Hilbert described an imaginary hotel with infinitely many rooms, numbered 1, 2, 3,…One evening, when the hotel is completely full, a solitary guest arrives seeking lodging. The resourceful hotel manager moves each guest up a room, so that the inhabitant of room 1 moves to room 2, room 2 to 3, and so on. With all guests relocated, room 1 becomes free for the new arrival! Next day an Infinity Tours coach arrives, containing infinitely many new guests. This time the manager moves the inhabitant of room 1 to room 2, room 2 to 4, room 3 to 6,…, room n to $2n$. This frees all odd-numbered rooms, so coach passenger number 1 can go into room 1, number 2 to room 3, number 3 to room 5, and, in general, number n to room $2n-1$. Even if infinitely many infinite coachloads of tourists arrive, everybody can be accommodated.

Similar paradoxes have been noted throughout history. Proclus, who wrote commentaries on Euclid in about AD 450, noted that the diameter of a circle divides it into two halves, so there must be twice as many halves as diameters. Philosophers in the Middle Ages realized that two concentric circles can be matched one-to-one by making points on the same radius correspond; so a small circle has just as many points as a large one. In Galileo's *Mathematical Discourses and Demonstrations* the sagacious Salviati raises the same problem: 'If I ask how many are the Numbers Square, you can answer me truly, that they be as many as are their proper roots; since every Square hath its Root, and every Root its Square, nor hath any Square more than one sole Root, or any Root more than one sole Square.' To this the seldom-satisfied Simplicius replies: 'What is to be resolved on this occasion?' And Salviati cops out with: 'I see no other decision that it may admit, but to say, that all Numbers are infinite; Squares are infinite; and that neither is the multitude of Squares less than all Numbers, nor this greater than that: and in conclusion, that the Attributes of Equality, Majority, and Minority have no place in Infinities, but only in terminate quantities.'

Infinity in Disguise

Galileo's answer to the paradoxes is that infinity behaves differently from anything else, and is best avoided. But sometimes it's very hard to avoid it. The problem of infinity arose more insistently in the early development of calculus, with the occurrence of infinite series. For example, what is

$$1 + \frac{1}{2} + \frac{1}{4} + \frac{1}{8} + \frac{1}{16} + \dots ?$$

It's easy to see that as the number of terms increases, the sum gets closer and closer to 2. So it's attractive to say that the whole infinite sum is *exactly* 2. Newton made infinite series the foundation of his methods for differentiating and integrating functions. So the problem of making sense of them must be faced. And infinite series are themselves paradoxical. For example, what does the series

$$1 - 1 + 1 - 1 + 1 - 1 + \dots$$

add up to? Written like this

$$(1-1) + (1-1) + (1-1) + \dots$$

the sum is clearly 0. On the other hand,

$$1 - (1-1) - (1-1) - \dots$$

is clearly 1. So $0 = 1$, and the whole of mathematics collapses in a contradiction.

Calculus was much too important for its practitioners to be put off by minor snags and philosophical problems like this. Eventually the matter was settled by reducing statements about infinite sums to more convoluted ones about finite sums. Instead of talking about an infinite sum $a_0 + a_1 + a_2 + \dots$ having value a, we say that the *finite* sum $a_0 + a_1 + \dots + a_n$ can be made to differ from a by less than any assigned error ε, provided n is taken larger than some N (depending on ε). Only if such an a exists does the series *converge*, that is, is the sum considered to make sense. In the same way the statement 'there are infinitely many integers' can be replaced by the finite version 'given any integer, there exists a larger one'. As Gauss put it in 1831: 'I protest against the use of an infinite quantity as an actual entity; this is never allowed in mathematics. The infinite is only a manner of speaking, in which one properly speaks of limits to which certain ratios can come as near as desired, while others are permitted to increase without bound.' Today, in any university course on analysis, students are taught to handle the infinite in this way. A typical problem might be: 'prove that $(n^2 + n)/n^2$ tends to 1 as n tends to infinity'. Woe betide the student who answers '$(\infty^2 + \infty)/\infty^2 = \infty/\infty = 1$'. But also woe betide him who writes '$(n^2 + n)/n^2 = 1 + 1/n$, now let $n = \infty$ to get $1 + 1/\infty = 1 + 0 = 1$', although this is arguably correct. (Before you're allowed to write sloppy things like that you must prove your mathematical machismo by going through the tortuous circumlocutions needed to make it unobjectionable. Once you've learned, the hard way, not to be sloppy, nobody minds if you are!)

This point of view goes back to Aristotle, and is described as *potential infinity*. We do not assert that an actual infinite exists, but we rephrase our assertion in a form that permits quantities to be as large as is necessary at the time. No longer do we see the miraculous jar as containing a true infinity of objects; we just observe that however many we take out, there's always another one in there. Put that way, it sounds like a pretty dubious distinction; but on a philosophical level it avoids the sticky question: 'How much stuff is there in that jar?'

Sum Crisis!

However, there were still bold souls who continued to play about with the idea of 'actual' infinity; to think of an infinite set not as a process 1, 2, 3,... which could in principle be continued beyond any chosen point, but as a completed, infinite whole. One of the first was Bernard Bolzano, who wrote a book called *Paradoxes of the Infinite* in 1851. But Bolzano's main interest was in providing solid foundations for calculus, and he decided that actually infinite sets aren't really needed there.

In the late nineteenth century there was something of a crisis in mathematics. Not fancy philosophical paradoxes about infinity, but a solid down to earth crisis that affected the day to day technique of working mathematicians, in the theory of Fourier series. A Fourier series looks something like this:

$$f(x) = \cos x + \frac{1}{2}\cos 2x + \frac{1}{3}\cos 3x + \ldots$$

and was developed by Joseph Fourier in his work on heat flow. The question is, when does such a series have a sum? Different mathematicians were obtaining contradictory answers. The whole thing was a dreadful mess, because too many workers had substituted plausible 'physical' arguments for good logical mathematics. It needed sorting out, urgently. Basically the answer is that a Fourier series makes good sense provided the set of values x, at which the function f behaves badly, is not itself too nasty. Mathematicians were forced to look at the fine structure of sets of points on the real line. In 1874 this problem led Georg Cantor to develop a theory of *actually* infinite sets, a topic that he developed over the succeeding years. His brilliantly original ideas attracted attention and some admiration, but his more conservatively minded contemporaries made little attempt to conceal their distaste. Cantor did two things. He founded Set Theory (without which today's mathematicians would find themselves tongue-tied, so basic a language has it become), and he discovered in so doing that some infinities are bigger than others.

Cantor's Paradise

Cantor started by making a virtue out of what everyone else had regarded as a vice. He *defined* a set to be infinite if it can be put in one-to-one correspondence with a proper part (subset) of itself. Two sets are equivalent or have the same *cardinal* if they can be put in one-to-one correspondence with each other. The smallest infinite set is that comprising the natural numbers $\{0, 1, 2, 3, \ldots\}$. Its cardinal is denoted by the symbol $\aleph_0$ (aleph-zero) and this is the smallest infinite number. It has all sorts of weird properties, such as

$$\aleph_0 + 1 = \aleph_0,\ \aleph_0 + \aleph_0 = \aleph_0,\ \aleph_0^2 = \aleph_0$$

but nevertheless it leads to a consistent version of arithmetic for infinite numbers. (What do you expect infinity to do when you double it, anyway?) Any set with cardinal $\aleph_0$ is said to be *countable*. Examples include the sets of negative integers, all integers, even numbers, odd numbers, squares, cubes, primes, and—more surprisingly—rationals. We are used to the idea that there are far more rationals than integers, because the integers have large gaps between them whereas the rationals are densely distributed. But that intuition is misleading because it forgets that one-to-one correspondences don't have to respect the order in which points occur. A rational p/q is defined by a pair (p, q) of integers, so the number of rationals is $\aleph_0^2$. But this is just $\aleph_0$ as we've seen.

After a certain amount of this sort of thing, one starts to wonder whether *every* infinite set is countable. Maybe Salviati was right, and $\aleph_0$ is just a fancy symbol for ∞. Cantor showed this isn't true. The set of real numbers is uncountable. There is an infinity bigger than the infinity of natural numbers! The proof is highly original. Roughly, the idea is to assume that the reals are countable, and argue for a contradiction. List them out, as decimal expansions. Form a new decimal whose first digit after the decimal point is different from that of the first on the list; whose second digit differs from that of the second on the list; and in general whose nth digit differs from that of the nth on the list. Then this new number cannot be anywhere in the list, which is absurd since the list was assumed to be complete. This is Cantor's 'diagonal argument', and it has cropped up ever since in all sorts of important problems. Building on this, Cantor was able to give a dramatic proof that transcendental numbers must exist. Recall that a number is transcendental if it does not satisfy any polynomial equation with rational coefficients. Examples include $\pi = 3.14159\ldots$ and the base of natural logarithms $e = 2.71828\ldots$, although it took mathematicians many years to prove that suspicion. In 1873 Charles Hermite proved that e is transcendental. The degree of difficulty can be judged from a letter that he wrote: 'I do not dare to attempt to show the transcendence of π. If others undertake it, no one will be happier than I about their success, but believe me, my dear friend, this cannot fail to cost them some effort.' In 1882 Ferdinand Lindemann made the effort, and succeeded in adapting Hermite's approach to deal with π.

Cantor showed that you do not need these enormously difficult theorems to demonstrate that trancendental numbers *exist*, by proving in a very simple manner that the set of algebraic numbers is countable. Since the full set of reals is uncountable, there must exist numbers that are not algebraic. End of proof (which is basically a triviality); collapse of audience in incredulity. In fact Cantor's argument shows more: it shows that there must be uncountably many transcendentals! There are *more* transcendental numbers than algebraic ones; and you can prove it without ever exhibiting a single example of either. It must have seemed like magic, not mathematics.

Even Cantor had his moments of disbelief. When, after three years of trying to demonstrate the opposite, he proved that n-dimensional space has exactly the same number of points as 1-dimensional space, he wrote: 'I see it but I do not believe it.' Others felt a little more strongly, for example Paul du Bois-Reymond: 'It appears repugnant to common sense.' There were also some paradoxes whose resolution did not just require imaginative development of a new but consistent intuition. For example, Cantor showed that, given any infinite cardinal, there is a larger one. There are infinitely many different infinities. But now consider the cardinal of the set of all cardinals. This must be larger than any cardinal whatsoever, including itself! This problem was eventually resolved by restricting the concept of 'set', but I wouldn't say people are totally happy about that answer even today.

Mathematicians were divided on the importance of Cantor's ideas. Leopold Kronecker attacked them publicly and vociferously for a decade; at one point Cantor had a nervous breakdown. But Kronecker had a very restrictive philosophy of mathematics—'God made the integers, all else is the work of Man'—and he was no more likely to approve of Cantor's theories than the Republican Party is likely to turn the Mid-West over to collective farming. Poincaré said that later generations would regard them as 'a disease from which one has recovered'. Hermann Weyl opined that Cantor's infinity of infinities was 'a fog on a fog'. On the other hand Adolf Hurwitz and Hadamard discovered important applications of Set Theory to analysis, and talked about them at prestigious international conferences. Hilbert, the leading mathematician of his age, said in 1926: 'No one shall expel us from the paradise which Cantor created', and praised his ideas as 'the most astonishing product of mathematical thought'. As with other strikingly original ideas, only those who were prepared to make an effort to understand and *use* them in their own work came to appreciate them. The commentators on the sidelines, smugly negative, let their sense of self-importance override their imagination and taste. Today the fruits of Cantor's labours form the basis of the whole of mathematics.

Claude E. Shannon and Warren Weaver

from THE MATHEMATICAL THEORY OF COMMUNICATION

Claude Shannon was another mathematician who delighted in the playful *jeux d'esprits* of his noble subject, but he is best known as the father of information theory. As befits his employment in the Bell Telephone labs, he wanted to construct a kind of economics of communication. The information content of a message is related to its surprise value. If the recipient already knows most of what is being said, the information content is slight. 'It rained today in Death Valley' has higher information content than 'It rained today in Oxford'. It is a little surprising, although not really when you think further about it, that the difference lies solely in the prior expectations of the recipient. Intuitive ideas such as this led Shannon to a precise metric of information content. His unit is the 'bit', defined as the quantity of information needed to halve the recipient's prior uncertainty. For example, if you already know that a baby has been born, it could be a boy or a girl. Two (approximately equiprobable alternatives are possible so the information involved is one bit. In the case of suits of playing cards, the information involved is two bits (the power to which 2 has to be raised to equal the number of possibilities, 4 in this case). For choices involving a day of the week, the associated information is 2.8 bits (you have to raise 2 to the power 2.8 to yield 7—which is to say that the base two logarithm of 7 is 2.8). In practice, the calculation of information content is a little more complicated, and in Shannon's hands it developed into a whole mathematical theory of information. Fascinatingly, the natural and sensible (logarithmic) formula for information content that he came up with is mathematically the same as the formula developed by Ludwig Boltzmann and still used by physicists for entropy. Shannon's book on information was written jointly with his colleague Warren Weaver. Shannon wrote the difficult bits and Weaver wrote a more popular account. The extract that follows is from Weaver's portion.

A Communication System and Its Problems

The communication system considered may be symbolically represented as follows:

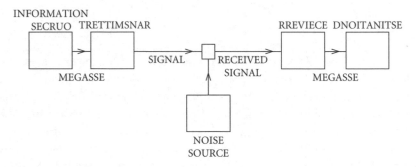

Figure 20.

The *information source* selects a desired *message* out of a set of possible messages (this is a particularly important remark, which requires considerable explanation later). The selected message may consist of written or spoken words, or of pictures, music, etc.

The *transmitter* changes this *message* into the *signal* which is actually sent over the *communication channel* from the transmitter to the *receiver*. In the case of telephony, the channel is a wire, the signal a varying electrical current on this wire; the transmitter is the set of devices (telephone transmitter, etc.) which change the sound pressure of the voice into the varying electrical current. In telegraphy, the transmitter codes written words into sequences of interrupted currents of varying lengths (dots, dashes, spaces). In oral speech, the information source is the brain, the transmitter is the voice mechanism producing the varying sound pressure (the signal) which is transmitted through the air (the channel). In radio, the channel is simply space (or the aether, if any one still prefers that antiquated and misleading word), and the signal is the electromagnetic wave which is transmitted.

The *receiver* is a sort of inverse transmitter, changing the transmitted signal back into a message, and handing this message on to the destination. When I talk to you, my brain is the information source, yours the destination; my vocal system is the transmitter, and your ear and the associated eighth nerve is the receiver.

In the process of being transmitted, it is unfortunately characteristic that certain things are added to the signal which were not intended by the information source. These unwanted additions may be distortions of sound (in telephony, for example) or static (in radio), or distortions in shape or shading of picture (television), or errors in transmission (telegraphy or facsimile), etc. All of these changes in the transmitted signal are called *noise*.

The kind of questions which one seeks to ask concerning such a communication system are:

a. How does one measure *amount of information?*
b. How does one measure the *capacity* of a communication channel?
c. The action of the transmitter in changing the message into the signal often involves a *coding process*. What are the characteristics of an efficient coding process? And when the coding is as efficient as possible, at what rate can the channel convey information?
d. What are the general characteristics of *noise?* How does noise affect the accuracy of the message finally received at the destination? How can one minimize the undesirable effects of noise, and to what extent can they be eliminated?
e. If the signal being transmitted is *continuous* (as in oral speech or music) rather than being formed of *discrete* symbols (as in written speech, telegraphy, etc.), how does this fact affect the problem?

Information

The word *information*, in this theory, is used in a special sense that must not be confused with its ordinary usage. In particular, *information* must not be confused with meaning.

In fact, two messages, one of which is heavily loaded with meaning and the other of which is pure nonsense, can be exactly equivalent, from the present viewpoint, as regards information. It is this, undoubtedly, that Shannon means when he says that 'the semantic aspects of communication are irrelevant to the engineering aspects'. But this does not mean that the engineering aspects are necessarily irrelevant to the semantic aspects.

To be sure, this word information in communication theory relates not so much to what you *do* say, as to what you *could* say.

That is, information is a measure of one's freedom of choice when one selects a message. If one is confronted with a very elementary situation where he has to choose one of two alternative messages, then it is arbitrarily said that the information, associated with this situation, is unity. Note that it is misleading (although often convenient) to say that one or the other message conveys unit information. The concept of information applies not to the individual messages (as the concept of meaning would), but rather to the situation as a whole, the unit information indicating that in this situation one has an amount of freedom of choice, in selecting a message, which it is convenient to regard as a standard or unit amount.

The two messages between which one must choose, in such a selection, can be anything one likes. One might be the text of the King James Version of the Bible, and the other might be 'Yes'. The transmitter might code these two messages so that 'zero' is the signal for the first, and 'one' the signal for the second; or so that a closed circuit (current flowing) is the signal for the first, and an open circuit (no current flowing) the signal for the second. Thus the two positions, closed and open, of a simple relay, might correspond to the two messages.

To be somewhat more definite, the amount of information is defined, in the simplest cases, to be measured by the logarithm of the number of available choices. It being convenient to use logarithms[1] to the base 2, rather than common or Briggs' logarithm to the base 10, the information, when there are only two choices, is proportional to the logarithm of 2 to the base 2. But this is unity; so that a two-choice situation is characterized by information of unity, as has already been stated above. This unit of information is called a 'bit', this word, first suggested by John W. Tukey, being a condensation of 'binary digit'. When numbers are expressed in the binary system there are only two digits, namely 0 and 1; just as ten digits, 0 to 9 inclusive, are used in the decimal number system which employs 10 as a base. Zero and one may be taken symbolically to represent any two choices, as noted above; so that 'binary digit' or 'bit' is natural to associate with the two-choice situation which has unit information.

If one has available say 16 alternative messages among which he is equally free to choose, then since $16 = 2^4$ so that $\log_2 16 = 4$, one says that this situation is characterized by 4 bits of information.

It doubtless seems queer, when one first meets it, that information is defined as the *logarithm* of the number of choices. But in the unfolding of the theory, it becomes more and more obvious that logarithmic measures are in fact the natural ones. At the moment, only one indication of this will be given. It was mentioned above that one simple on-or-off relay, with its two positions labelled, say, 0 and 1 respectively, can handle a unit information situation, in which there are but two message choices. If one relay can handle unit information, how much can be handled by say three relays? It seems very reasonable to want to say that three relays could handle three times as much information as one. And this indeed is the way it works out if one uses the logarithmic definition of information. For three relays are capable of responding to 2^3 or 8 choices, which symbolically might be written as 000, 001, 011, 100, 110, 101, 111, in the first of which all three relays are open, and in the last of which all three relays are closed. And the logarithm to the base 2 of 2^3 is 3, so that the logarithmic measure assigns three units of information to this situation, just as one would wish. Similarly, doubling the available time squares the number of possible messages, and doubles the logarithm; and hence doubles the information if it is measured logarithmically.

The remarks thus far relate to artificially simple situations where the information source is free to choose only between several definite messages—like a man picking out one of a set of standard birthday greeting telegrams. A more natural and more important situation is that in which the information source makes a sequence of choices from some set of elementary symbols, the selected sequence then forming the message. Thus a man may pick out one word after another, these individually selected words then adding up to form the message.

At this point an important consideration which has been in the background, so far, comes to the front for major attention. Namely, the role which probability plays in the generation of the message. For as the successive symbols are chosen, these choices are, at least from the point of view of the communication system, governed by probabilities; and in fact by probabilities which are not independent, but which, at any stage of the process, depend upon the preceding choices. Thus, if we are concerned with English speech, and if the last symbol chosen is 'the', then the probability that the next word be an article, or a verb form other than a verbal, is very small. This probabilistic influence stretches over more than two words, in fact. After the three words 'in the event' the probability for 'that' as the next word is fairly high, and for 'elephant' as the next word is very low.

That there are probabilities which exert a certain degree of control over the English language also becomes obvious if one thinks, for example, of the fact that in our language the dictionary contains no words whatsoever in which the initial letter j is followed by b, c, d, f, g, j, k, l, q, r, t, v, w, x, or z; so that the probability is actually zero that an initial j be followed by any of these letters. Similarly, anyone would agree that the probability is low for such a sequence of words as 'Constantinople fishing nasty pink'. Incidentally, it is low, but not zero; for it is perfectly possible to think of a passage in which one sentence closes with 'Constantinople fishing', and the next begins with 'Nasty pink'. And we might observe in passing that the unlikely four-word sequence under discussion *has* occurred in a single good English sentence, namely the one above.

A system which produces a sequence of symbols (which may, of course, be letters or musical notes, say, rather than words) according to certain probabilities is called a *stochastic process*, and the special case of a stochastic process in which the probabilities depend on the previous events, is called a *Markoff process* or a Markoff chain. Of the Markoff processes which might conceivably generate messages, there is a special class which is of primary importance for communication theory, these being what are called *ergodic processes*. The analytical details here are complicated and the reasoning so deep and involved that it has taken some of the best efforts of the best mathematicians to create the associated theory; but the rough nature of an ergodic process is easy to understand. It is one which produces a sequence of symbols which would be a poll-taker's dream, because any reasonably large sample tends to be representative of the sequence as a whole. Suppose that two persons choose samples in different ways, and study what trends their statistical properties would show as the samples become larger. If the situation is ergodic, then those two persons, however they may have chosen their samples, agree in their estimates of the properties of the whole. Ergodic systems, in other words, exhibit a particularly safe and comforting sort of statistical regularity.

Now let us return to the idea of *information*. When we have an information source which is producing a message by successively selecting discrete symbols (letters, words, musical notes, spots of a certain size, etc.), the probability of choice of the various symbols at one stage of the process being dependent on the previous choices (i.e., a Markoff process), what about the information associated with this procedure?

The quantity which uniquely meets the natural requirements that one sets up for 'information' turns out to be exactly that which is known in thermodynamics as *entropy*. It is expressed in terms of the various probabilities involved—those of getting to certain stages in the process of forming messages, and the probabilities that, when in those stages, certain symbols be chosen next. The formula, moreover, involves the *logarithm* of probabilities, so that it is a natural generalization of the logarithmic measure spoken of above in connection with simple cases.

To those who have studied the physical sciences, it is most significant that an entropy-like expression appears in the theory as a measure of information. Introduced by Clausius nearly one hundred years ago, closely associated with the name of Boltzmann, and given deep meaning by Gibbs in his classic work on statistical mechanics, entropy has become so basic and pervasive a concept that Eddington remarks 'The law that entropy always increases—the second law of thermodynamics—holds, I think, the supreme position among the laws of Nature.'

In the physical sciences, the entropy associated with a situation is a measure of the degree of randomness, or of 'shuffledness' if you will, in the situation; and the tendency of physical systems to become less and less organized, to become more and more perfectly shuffled, is so basic that Eddington argues that it is primarily this tendency which gives time its arrow—which would reveal to us, for example, whether a movie of the physical world is being run forward or backward.

Thus when one meets the concept of entropy in communication theory, he has a right to be rather excited—a right to suspect that one has hold of something that may turn out to be basic and important. That information be measured by entropy is, after all, natural when we remember that information, in communication theory, is associated with the amount of freedom of choice we have in constructing messages. Thus for a communication source one can say, just as he would also say it of a thermodynamic ensemble, 'This situation is highly organized, it is not characterized by a large degree of randomness or of choice—that is to say, the information (or the entropy) is low.' We will return to this point later, for unless I am quite mistaken, it is an important aspect of the more general significance of this theory.

Having calculated the entropy (or the information, or the freedom of choice) of a certain information source, one can compare this to the maximum value this entropy could have, subject only to the condition that the source continue to employ the same symbols. The ratio of the actual to the maximum entropy is called the *relative entropy* of the source. If the relative entropy of a certain source is, say 0.8, this roughly means that this source is, in its choice of symbols to form a message, about 80 per cent as free as it could possibly be with these same symbols. One minus the relative entropy is called the *redundancy*. This is the fraction of the structure of the message which is determined not by the free choice of the sender, but rather by the accepted statistical rules governing the use of the symbols in question. It is sensibly called redundancy, for this fraction of the message is in fact redundant in something close to the ordinary sense; that is to say, this fraction of the message is unnecessary (and hence repetitive or redundant) in the sense that if it were missing the message would still be essentially complete, or at least could be completed.

It is most interesting to note that the redundancy of English is just about 50 per cent,[2] so that about half of the letters or words we choose in writing or speaking are under our free choice, and about half (although we are not ordinarily aware of it) are really controlled by the statistical structure of the language.

Apart from more serious implications, which again we will postpone to our final discussion, it is interesting to note that a language must have at least 50 per cent of real freedom (or relative entropy) in the choice of letters if one is to be able to construct satisfactory crossword puzzles. If it has complete freedom, then every array of letters is a crossword puzzle. If it has only 20 per cent of freedom, then it would be impossible to construct crossword puzzles in such complexity and number as would make the game popular. Shannon has estimated that if the English language had only about 30 per cent redundancy, then it would be possible to construct three-dimensional crossword puzzles.

Before closing this section on information, it should be noted that the real reason that Level A analysis deals with a concept of information which characterizes the whole statistical nature of the information source, and is not concerned with the individual messages (and not at all directly concerned with the meaning of the individual messages) is that from the point of view of engineering, a communication system must face the problem of handling any message that the source can produce. If it is not possible or practicable to design a system which can handle everything perfectly, then the system should be designed to handle well the jobs it is most likely to be asked to do, and should resign itself to be less efficient for the rare task. This sort of consideration leads at once to the necessity of characterizing the statistical nature of the whole ensemble of messages which a given kind of source can and will produce. And *information*, as used in communication theory, does just this.

1. When $m^x = y$, then x is said to be the logarithm of y to the base m.
2. The 50 per cent estimate accounts only for statistical structure out to about eight letters, so that the ultimate value is presumably a little higher.

Alan Turing

from 'COMPUTING MACHINERY AND INTELLIGENCE'

I don't think I was exaggerating when I wrote of the English mathematician Alan Turing as follows:

As the pivotal intellect in the breaking of the German Enigma codes, Turing arguably made a greater contribution to defeating the Nazis than Eisenhower or Churchill. Thanks to Turing and his 'Ultra' colleagues at Bletchley Park, Allied generals in the field were consistently, over long periods of the war, privy to detailed German plans before the German generals had time to implement them. After the war, when Turing's role was no longer top secret, he should have been knighted and feted as a saviour of his nation. Instead, this gentle, stammering, eccentric genius was destroyed, for a 'crime', committed in private, which harmed nobody.

Turing ate an apple that he had injected with cyanide, having been arrested for homosexual activities in private (that's what Britain was like as late as 1954). Quite apart from his wartime service, Turing is the only other plausible candidate, along with John von Neumann, for the title of father of the computer. And in the field of philosophy, his 'imitation game', which is the subject of these extracts from his famous 1950 paper, 'Computing Machinery and Intelligence', is the starting-point for most discussions of the possibilities for artificial intelligence. Indeed, it is a good role model for one of philosophy's favourite techniques, the thought experiment. ■

The Imitation Game

I propose to consider the question, 'Can machines think?' This should begin with definitions of the meaning of the terms 'machine' and 'think'. The definitions might be framed so as to reflect so far as possible the normal use of the words, but this attitude is dangerous. If the meaning of the words 'machine' and 'think' are to be found by examining how they are commonly used it is difficult to escape the conclusion that the meaning and the answer to the question, 'Can machines think?' is to be sought in a statistical survey such as a Gallup poll. But this is absurd. Instead of attempting such a definition I shall replace the question by another, which is closely related to it and is expressed in relatively unambiguous words.

The new form of the problem can be described in terms of a game which we call the 'imitation game'. It is played with three people, a man (A), a woman (B), and an interrogator (C) who may be of either sex. The interrogator stays in a room apart from the other two. The object of the game for the interrogator is to determine which of the other two is the man and which is the woman. He knows them by labels X and Y, and at the end of the game he says either 'X is A and Y is B' or 'X is B and Y is A.' The interrogator is allowed to put questions to A and B thus:

C: Will X please tell me the length of his or her hair? Now suppose X is actually A, then A must answer. It is A's object in the game to try and cause C to make the wrong identification. His answer might therefore be

'My hair is shingled, and the longest strands are about nine inches long.'

In order that tones of voice may not help the interrogator the answers should be written, or better still, typewritten. The ideal arrangement is to have a teleprinter communicating between the two rooms. Alternatively the question and answers can be repeated by an intermediary. The object of the game for the third player (B) is to help the interrogator. The best strategy for her is probably to give truthful answers. She can add such things as 'I am the woman, don't listen to him!' to her answers, but it will avail nothing as the man can make similar remarks.

We now ask the question, 'What will happen when a machine takes the part of A in this game?' Will the interrogator decide wrongly as often when the game is played like this as he does when the game is played between a man and a woman? These questions replace our original, 'Can machines think?'

Critique of the New Problem

As well as asking, 'What is the answer to this new form of the question', one may ask, 'Is this new question a worthy one to investigate?' This latter question we investigate without further ado, thereby cutting short an infinite regress.

The new problem has the advantage of drawing a fairly sharp line between the physical and the intellectual capacities of a man. No engineer or chemist claims to be able to produce a material which is indistinguishable from the human skin. It is possible that at some time this might be done, but even supposing this invention available we should feel there was little point in trying to make a 'thinking machine' more human by dressing it up in such artificial flesh. The form in which we have set the problem reflects this fact in the condition which prevents the interrogator from seeing or touching the other competitors, or hearing their voices. Some other advantages of the proposed criterion may be shown up by specimen questions and answers. Thus:

Q: Please write me a sonnet on the subject of the Forth Bridge.

A: Count me out on this one. I never could write poetry.

Q: Add 34957 to 70764.

A: (Pause about 30 seconds and then give as answer) 105621.

Q: Do you play chess?

A: Yes.

Q: I have K at my K1, and no other pieces. You have only K at K6 and R at R1. It is your move. What do you play?

A: (After a pause of 15 seconds) R-R8 mate.

The question and answer method seems to be suitable for introducing almost any one of the fields of human endeavour that we wish to include. We do not wish to penalize the machine for its inability to shine in beauty competitions, nor to penalize a man for losing in a race against an aeroplane. The conditions of our game make these disabilities irrelevant. The 'witnesses' can brag, if they consider it advisable, as much as they please about their charms, strength or heroism, but the interrogator cannot demand practical demonstrations.

The game may perhaps be criticised on the ground that the odds are weighted too heavily against the machine. If the man were to try and pretend to be the machine he would clearly make a very poor showing. He would be given away at once by slowness and inaccuracy in arithmetic. May not machines carry out something which ought to be described as thinking but which is very different from what a man does? This objection is a very strong one, but at least we can say that if, nevertheless, a machine can be constructed to play the imitation game satisfactorily, we need not be troubled by this objection.

It might be urged that when playing the 'imitation game' the best strategy for the machine may possibly be something other than imitation of the behaviour of a man. This may be, but I think it is unlikely that there is any great effect of this kind. In any case there is no intention to investigate here the theory of the game, and it will be assumed that the best strategy is to try to provide answers that would naturally be given by a man.

[...]

In the process of trying to imitate an adult human mind we are bound to think a good deal about the process which has brought it to the state that it is in. We may notice three components,

(a) The initial state of the mind, say at birth,
(b) The education to which it has been subjected,
(c) Other experience, not to be described as education, to which it has been subjected.

Instead of trying to produce a programme to simulate the adult mind, why not rather try to produce one which simulates the child's? If this were then subjected to an appropriate course of education one would obtain the adult brain. Presumably the child-brain is some-thing like a note-book as one buys it from the stationers. Rather little mechanism, and lots of blank sheets. (Mechanism and writing are from our point of view almost synonymous.) Our hope is that there is so little mechanism in the child-brain that something like it can be easily programmed. The amount of work in the education we can assume, as a first approximation, to be much the same as for the human child.

We have thus divided our problem into two parts. The child-programme and the education process. These two remain very closely connected. We cannot expect to find a good child-machine at the first attempt. One must experiment with teaching one such machine and see how well it learns. One can then try another and see if it is better or worse. There is an obvious connection between this process and evolution, by the identifications

Structure of the child-machine = Hereditary material
Changes of the child-machine = Mutations
Natural selection = Judgment of the experimenter

One may hope, however, that this process will be more expeditious than evolution. The survival of the fittest is a slow method for measur-ing advantages. The experimenter, by the exercise of intelligence, should be able to speed it up. Equally important is the fact that he is not restricted to random mutations. If he can trace a cause for some weakness he can probably think of the kind of mutation which will improve it.

It will not be possible to apply exactly the same teaching process to the machine as to a normal child. It will not, for instance, be provided with legs, so that it could not be asked to go out and fill the coal scuttle. Possibly it might not have eyes. But however well these deficiencies might be overcome by clever engineering, one could not send the creature to school without the other children making excessive fun of it. It must be given some tuition. We need not be too concerned about the legs, eyes, etc. The example of Miss Helen Keller shows that education can take place provided that communication in both directions between teacher and pupil can take place by some means or other.

We normally associate punishments and rewards with the teaching process. Some simple child-machines can be constructed or programmed on this sort of principle. The machine has to be so constructed that events which shortly preceded the occurrence of a punishment-signal are unlikely to be repeated, whereas a reward-signal increased the probability of repetition of the events which led up to it. These definitions do not presuppose any feelings on the part of the machine. I have done some experiments with one such child-machine, and succeeded in teaching it a few things, but the teaching method was too unorthodox for the experiment to be considered really successful.

The use of punishments and rewards can at best be a part of the teaching process. Roughly speaking, if the teacher has no other means of communicating to the pupil, the amount of information which can reach him does not exceed the total number of rewards and punishments applied. By the time a child has learnt to repeat 'Casabianca' he would probably feel very sore indeed, if the text could only be discovered by a 'Twenty Questions' technique, every 'NO' taking the form of a blow. It is necessary therefore to have some other 'unemotional' channels of communication. If these are available it is possible to teach a machine by punishments and rewards to obey orders given in some language, e.g. a symbolic language. These orders are to be transmitted through the 'unemotional' channels. The use of this language will diminish greatly the number of punishments and rewards required.

Opinions may vary as to the complexity which is suitable in the child-machine. One might try to make it as simple as possible consistently with the general principles. Alternatively one might have a complete system of logical inference 'built in'.[1] In the latter case the store would be largely occupied with definitions and propositions. The propositions would have various kinds of status, e.g. well-established facts, conjectures, mathematically proved theorems, statements given by an authority, expressions having the logical form of proposition but not belief-value. Certain propositions may be described as 'imperatives'. The machine should be so constructed that as soon as an imperative is classed as 'well-established' the appropriate action automatically takes place. To illustrate this, suppose the teacher says to the machine, 'Do your home-work now'. This may cause 'Teacher says "Do your homework now"' to be included amongst the well-established facts. Another such fact might be, 'Everything that teacher says is true'. Combining these may eventually lead to the imperative, 'Do your homework now', being included amongst the well-established facts, and this, by the construction of the machine, will mean that the homework actually gets started, but the effect is very satisfactory. The processes of inference used by the machine need not be such as would satisfy the most exacting logicians. There might for instance be no hierarchy of types. But this need not mean that type fallacies will occur, any more than we are bound to fall over unfenced cliffs. Suitable imperatives (expressed *within* the systems, not forming part of the rules *of* the system) such as 'Do not use a class unless it is a subclass of one which has been mentioned by teacher' can have a similar effect to 'Do not go too near the edge'.

The imperatives that can be obeyed by a machine that has no limbs are bound to be of a rather intellectual character, as in the example (doing homework) given above. Important amongst such imperatives will be ones which regulate the order in which the rules of the logical system concerned are to be applied. For at each stage when one is using a logical system, there is a very large number of alternative steps, any of which one is permitted to apply, so far as obedience to the rules of the logical system is concerned. These choices make the difference between a brilliant and a footling reasoner, not the difference between a sound and a fallacious one. Propositions leading to imperatives of this kind might be 'When Socrates is mentioned, use the syllogism in Barbara' or 'If one method has been proved to be quicker than another, do not use the slower method'. Some of these may be 'given by authority', but others may be produced by the machine itself, *e.g.* by scientific induction.

The idea of a learning machine may appear paradoxical to some readers. How can the rules of operation of the machine change? They should describe completely how the machine will react whatever its history might be, whatever changes it might undergo. The rules are thus quite time-invariant. This is quite true. The explanation of the paradox is that the rules which get changed in the learning process are of a rather less pretentious kind, claiming only an ephemeral validity. The reader may draw a parallel with the Constitution of the United States.

An important feature of a learning machine is that its teacher will often be very largely ignorant of quite what is going on inside, although he may still be able to some extent to predict his pupil's behaviour. This should apply most strongly to the later education of a machine arising from a child-machine of well-tried design (or programme). This is in clear contrast with normal procedure when using a machine to do computations: one's object is then to have a clear mental picture of the state of the machine at each moment in the computation. This object can only be achieved with a struggle. The view that 'the machine can only do what we know how to order it to do' appears strange in face of this. Most of the programmes which we can put into the machine will result in its doing something that we cannot make sense of at all, or which we regard as completely random behaviour. Intelligent behaviour presumably consists in a departure from the completely disciplined behaviour involved in computation, but a rather slight one, which does not give rise to random behaviour, or to pointless repetitive loops. Another important result of preparing our machine for its part in the imitation game by a process of teaching and learning is that 'human fallibility' is likely to be omitted in a rather natural way, i.e. without special 'coaching'…Processes that are learnt do not produce a hundred per cent certainty of result; if they did they could not be unlearnt.

It is probably wise to include a random element in a learning machine. A random element is rather useful when we are searching for a solution of some problem. Suppose for instance we wanted to find a number between 50 and 200 which was equal to the square of the sum of its digits, we might start at 51 then try 52 and go on until we got a number that worked. Alternatively we might choose numbers at random until we got a good one. This method has the advantage that it is unnecessary to keep track of the values that have been tried, but the disadvantage that one may try the same one twice, but this is not very important if there are several solutions. The systematic method has the disadvantage that there may be an enormous block without any solutions in the region which has to be investigated first. Now the learning process may be regarded as a search for a form of behaviour which will satisfy the teacher (or some other criterion). Since there is probably a very large number of satisfactory solutions the random method seems to be better than the systematic. It should be noticed that it is used in the analogous process of evolution. But there the systematic method is not possible. How could one keep track of the different genetical combinations that had been tried, so as to avoid trying them again?

We may hope that machines will eventually compete with men in all purely intellectual fields. But which are the best ones to start with? Even this is a difficult decision. Many people think that a very abstract activity, like the playing of chess, would be best. It can also be maintained that it is best to provide the machine with the best sense organs that money can buy, and then teach it to understand and speak English. This process could follow the normal teaching of a child. Things would be pointed out and named, etc. Again I do not know what the right answer is, but I think both approaches should be tried.

We can only see a short distance ahead, but we can see plenty there that needs to be done.

1. Or rather 'programmed in' for our child-machine will be programmed in a digital computer. But the logical system will not have to be learnt.

Albert Einstein

from 'WHAT IS THE THEORY OF RELATIVITY?'

▇ Albert Einstein made great use of thought (*Gedanken*) experiments, not just to explain his ideas to others but in developing the ideas in his own head. It seems only right to have another piece from Einstein, for this section of the book. I thought about including an extract from his popular exposition of relativity, but English was not his native language, and there are more accessible treatments available, two of which we will encounter shortly. Instead, here is a more general essay by him, in which he muses on different classes of theory, with special reference to his own. ▇

We can distinguish various kinds of theories in physics. Most of them are constructive. They attempt to build up a picture of the more complex phenomena out of the materials of a relatively simple formal scheme from which they start out. Thus the kinetic theory of gases seeks to reduce mechanical, thermal, and diffusional processes to movements of molecules—i.e. to build them up out of the hypothesis of molecular motion. When we say that we have succeeded in understanding a group of natural processes, we invariably mean that a constructive theory has been found which covers the processes in question.

Along with this most important class of theories there exists a second, which I will call 'principle-theories'. These employ the analytic, not the synthetic, method. The elements which form their basis and starting-point are not hypothetically constructed but empirically discovered ones, general characteristics of natural processes, principles that give rise to mathematically formulated criteria which the separate processes or the theoretical representations of them have to satisfy. Thus the science of thermodynamics seeks by analytical means to deduce necessary conditions, which separate events have to satisfy, from the universally experienced fact that perpetual motion is impossible.

The advantages of the constructive theory are completeness, adaptability, and clearness, those of the principle theory are logical perfection and security of the foundations.

The theory of relativity belongs to the latter class. In order to grasp its nature, one needs first of all to become acquainted with the principles on which it is based. Before I go into these, however, I must observe that the theory of relativity resembles a building consisting of two separate stories, the special theory and the general theory. The special theory, on which the general theory rests, applies to all physical phenomena with the exception of gravitation; the general theory provides the law of gravitation and its relations to the other forces of nature.

It has, of course, been known since the days of the ancient Greeks that in order to describe the movement of a body, a second body is needed to which the movement of the first is referred. The movement of a vehicle is considered in reference to the earth's surface, that of a planet to the totality of the visible fixed stars. In physics the body to which events are spatially referred is called the coordinate system. The laws of the mechanics of Galileo and Newton, for instance, can only be formulated with the aid of a coordinate system.

The state of motion of the coordinate system may not, however, be arbitrarily chosen, if the laws of mechanics are to be valid (it must be free from rotation and acceleration). A coordinate system which is admitted in mechanics is called an 'inertial system'. The state of motion of an inertial system is according to mechanics not one that is determined uniquely by nature. On the contrary, the following definition holds good: a coordinate system that is moved uniformly and in a straight line relative to an inertial system is likewise an inertial system. By the 'special principle of relativity' is meant the generalization of this definition to include any natural event whatever: thus, every universal law of nature which is valid in relation to a coordinate system C, must also be valid, as it stands, in relation to a coordinate system C', which is in uniform translatory motion relatively to C.

The second principle, on which the special theory of relativity rests, is the 'principle of the constant velocity of light in vacuo'. This principle asserts that light in vacuo always has a definite velocity of propagation (independent of the state of motion of the observer or of the source of the light). The confidence which physicists place in this principle springs from the successes achieved by the electrodynamics of Maxwell and Lorentz.

Both the above-mentioned principles are powerfully supported by experience, but appear not to be logically reconcilable. The special theory of relativity finally succeeded in reconciling them logically by a modification of kinematics—i.e. of the doctrine of the laws relating to space and time (from the point of view of physics). It became clear that to speak of the simultaneity of two events had no meaning except in relation to a given coordinate system, and that the shape of measuring devices and the speed at which clocks move depend on their state of motion with respect to the coordinate system.

But the old physics, including the laws of motion of Galileo and Newton, did not fit in with the suggested relativist kinematics. From the latter, general mathematical conditions issued, to which natural laws had to conform, if the above-mentioned two principles were really to apply. To these, physics had to be adapted. In particular, scientists arrived at a new law of motion for (rapidly moving) mass points, which was admirably confirmed in the case of electrically charged particles. The most important upshot of the special theory of relativity concerned the inert masses of corporeal systems. It turned out that the inertia of a system necessarily depends on its energy-content, and this led straight to the notion that inert mass is simply latent energy. The principle of the conservation of mass lost its independence and became fused with that of the conservation of energy.

The special theory of relativity, which was simply a systematic development of the electrodynamics of Maxwell and Lorentz, pointed beyond itself, however. Should the independence of physical laws of the state of motion of the coordinate system be restricted to the uniform translatory motion of coordinate systems in respect to each other? What has nature to do with our coordinate systems and their state of motion? If it is necessary for the purpose of describing nature, to make use of a coordinate system arbitrarily introduced by us, then the choice of its state of motion ought to be subject to no restriction; the laws ought to be entirely independent of this choice (general principle of relativity).

The establishment of this general principle of relativity is made easier by a fact of experience that has long been known, namely, that the weight and the inertia of a body are controlled by the same constant (equality of inertial and gravitational mass). Imagine a coordinate system which is rotating uniformly with respect to an inertial system in the Newtonian manner. The centrifugal forces which manifest themselves in relation to this system must, according to Newton's teaching, be regarded as effects of inertia. But these centrifugal forces are, exactly like the forces of gravity, proportional to the masses of the bodies. Ought it not to be possible in this case to regard the coordinate system as stationary and the centrifugal forces as gravitational forces? This seems the obvious view, but classical mechanics forbid it.

This hasty consideration suggests that a general theory of relativity must supply the laws of gravitation, and the consistent following up of the idea has justified our hopes.

But the path was thornier than one might suppose, because it demanded the abandonment of Euclidean geometry. This is to say, the laws according to which solid bodies may be arranged in space do not completely accord with the spatial laws attributed to bodies by Euclidean geometry. This is what we mean when we talk of the 'curvature of space'. The fundamental concepts of the 'straight line', the 'plane', etc., thereby lose their precise significance in physics.

In the general theory of relativity the doctrine of space and time, or kinematics, no longer figures as a fundamental independent of the rest of physics. The geometrical behavior of bodies and the motion of clocks rather depend on gravitational fields, which in their turn are produced by matter.

George Gamow

from MR TOMPKINS

■ I have already introduced George Gamow in connection with a delightful comic poem about the demise of the Steady State Theory of the universe. The poem came from one of Gamow's stories featuring *Mr Tompkins*: this playful scientist chose to explain difficult scientific ideas through the medium of fiction. Others who have done something similar are Russell

Stannard, whose 'Uncle Albert' we shall meet a few pages on, and, in the nineteenth century, Edwin Abbott with his *Flatland*. What follows is the episode where Mr Tompkins learns what it would be like to live in a universe that was curved round on itself. We will find out more about this idea of Einstein's in the next extract. ■

The Pulsating Universe

After dinner on their first evening in the Beach Hotel with the old professor talking about cosmology, and his daughter chatting about art, Mr Tompkins finally got to his room, collapsed on to the bed, and pulled the blanket over his head. Botticelli and Bondi, Salvador Dali and Fred Hoyle, Lemaître and La Fontaine got all mixed up in his tired brain, and finally he fell into a deep sleep....

Sometime in the middle of the night he woke up with a strange feeling that instead of lying on a comfortable spring mattress he was lying on something hard. He opened his eyes and found himself prostrated on what he first thought to be a big rock on the seashore. Later he discovered that it was actually a very big rock, about 30 feet in diameter, suspended in space without any visible support. The rock was covered with some green moss, and in a few places little bushes were growing from cracks in the stone. The space around the rock was illuminated by some glimmering light and was very dusty. In fact, there was more dust in the air than he had ever seen, even in the films representing dust storms in the middle west. He tied his handkerchief round his nose and felt, after this, considerably relieved. But there were more dangerous things than the dust in the surrounding space. Very often stones of the size of his head and larger were swirling through the space near his rock, occasionally hitting it with a strange dull sound of impact. He noticed also one or two rocks of approximately the same size as his own, floating through space at some distance away. All this time, inspecting his surroundings, he was clinging hard to some protruding edges of his rock in constant fear of falling off and being lost in the dusty depths below. Soon, however, he became bolder, and made an attempt to crawl to the edge of his rock and to see whether there was really nothing underneath, supporting it. As he was crawling in this way, he noticed, to his great surprise, that he did not fall off, but that his weight was constantly pressing him to the surface of the rock, although he had covered already more than a quarter of its circumference. Looking from behind a ridge of loose stones on the spot just underneath the place where he originally found himself, he discovered nothing to support the rock in space. To his great surprise, however, the glimmering light revealed the tall figure of his friend the old professor standing apparently with his head down and making some notes in his pocket-book.

Now Mr Tompkins began slowly to understand. He remembered that he was taught in his schooldays that the earth is a big round rock moving freely in space around the sun. He also remembered the picture of two antipodes standing on the opposite sides of the earth. Yes, his rock was just a very small stellar body attracting everything to its surface, and he and the old professor were the only population of this little planet. This consoled him a little; there was at least no danger of falling off!

'Good morning,' said Mr Tompkins, to divert the old man's attention from his calculations.

The professor raised his eyes from his note-book. 'There are no mornings here,' he said, 'there is no sun and not a single luminous star in this universe. It is lucky that the bodies here show some chemical process on their surface, otherwise I should not be able to observe the expansion of this space', and he returned again to his note-book.

Mr Tompkins felt quite unhappy; to meet the only living person in the whole universe, and to find him so unsociable! Unexpectedly, one of the little meteorites came to his help; with a crashing sound the stone hit the book in the hands of the professor and threw it, travelling fast through space, away from their little planet. 'Now you will never see it again,' said Mr Tompkins, as the book got smaller and smaller, flying through space.

'On the contrary,' replied the professor. 'You see, the space in which we now are is not infinite in its extension. Oh yes, yes, I know that you have been taught in school that space is infinite, and that two parallel lines never meet. This, however, is not true either for the space in which the rest of humanity lives, or for the space in which we are now. The first one is of course very large indeed; the scientists estimated its present

Figure 21. There are no mornings here.

dimensions to be about 10,000,000,000,000,000,000,000 miles, which, for an ordinary mind, is fairly infinite. If I had lost my book there, it would take an incredibly long time to come back. Here, however, the situation is rather different. Just before the note-book was torn out of my hands, I had figured out that this space is only about five miles in diameter, though it is rapidly expanding. I expect the book back in not more than half an hour.'

'But,' ventured Mr Tompkins, 'do you mean that your book is going to behave like the boomerang of an Australian native, and, by moving along a curved trajectory, fall down at your feet?'

'Nothing of the sort,' answered the professor. 'If you want to understand what really happens, think about an ancient Greek who did not know that the earth was a sphere. Suppose he has given somebody instructions to go always straight northwards. Imagine his astonishment when his runner finally returns to him from the south. Our ancient Greek did not have a notion about travelling round the world (round the earth, I mean in this case), and he would be sure that his runner had lost his way and had taken a curved route which brought him back. In reality his man was going all the time along the straightest line one can draw on the surface of the earth, but he travelled round the world and thus came back from the opposite direction. The same thing is going to happen to my book, unless it is hit on its way by some other stone and thus deflected from the straight track. Here, take these binoculars, and see if you can still see it.'

Mr Tompkins put the binoculars to his eyes, and, through the dust which somewhat obscured the whole picture, he managed to see the professor's note-book travelling through space far far away. He was somewhat surprised by the pink colouring of all the objects, including the book, at that distance.

'But,' he exclaimed after a while, 'your book is returning, I see it growing larger.'

'No,' said the professor, 'it is still going away. The fact that you see it growing in size, as if it were coming back, is due to a peculiar focusing effect of the closed spherical space on the rays of light. Let us return to our ancient Greek. If the rays of light could be kept going all the time along the curved surface of the earth, let us say by refraction of the atmosphere, he would be able, using powerful binoculars, to see his runner all the time during the journey. If you look on the globe, you will see that the straightest lines on its surface, the meridians, first diverge from one pole, but, after passing the equator, begin to converge towards the opposite pole. If the rays of light travelled along the meridians, you, located, for example, at one pole, would see the person going away from you growing smaller and smaller only until he crossed the equator. After this point you would see him growing larger and it would seem to you that he was returning, going, however, backwards. After he had reached the opposite pole, you would see him as large as if he were standing right by your side. You would not be able, however, to touch him, just as you cannot touch the image in a spherical mirror. On this basis of two-dimensional analogy, you can imagine what happens to the light rays in the strangely curved three-dimensional space. Here, I think the image of the book is quite close now.' In fact, dropping the binoculars, Mr Tompkins could see that the book was only a few yards away. It looked, however, very strange indeed! The contours were not sharp, but rather washed out, the formulae written by the professor on its pages could be hardly recognized, and the whole book looked like a photo-graph taken out of focus and underdeveloped.

'You see now,' said the professor, 'that this is only the image of the book, badly distorted by light travelling across one half of the universe. If you want to be quite sure of it, just notice how the stones behind the book can be seen through its pages.'

Mr Tompkins tried to reach the book, but his hand passed through the image without any resistance.

'The book itself,' said the professor, 'is now very close to the opposite pole of the universe, and what you see here are just two images of it. The second image is just behind you and when both images coincide, the real book will be exactly at the opposite pole.' Mr Tompkins didn't hear; he was too deeply absorbed in his thoughts, trying to remember how the images of objects are formed in elementary optics by concave mirrors and lenses. When he finally gave it up, the two images were again reced-ing in opposite directions.

'But what makes the space curved and produce all these funny effects?' he asked the professor.

'The presence of ponderable matter,' was the answer. 'When Newton discovered the law of gravity, he thought that gravity was just an ordi-nary force, the same type of force as, for example, is produced by an elastic string stretched between two bodies. There always remains, however, the mysterious fact that all bodies, independent of their weight and size, have the same acceleration and move the same way under the action of gravity, provided you eliminate the friction of air and that sort of thing, of course. It was Einstein who first made it clear that the primary action of ponderable matter is to produce the cur-vature of space and that the trajectories of all bodies moving in the field of gravity are curved just because space itself is curved. But I think it is too hard for you to understand, without knowing sufficient mathematics.'

'It is,' said Mr Tompkins. 'But tell me, if there were no matter, would we have the kind of geometry I was taught at school, and would parallel lines never meet?'

'They would not,' answered the professor, 'but neither would there be any material creature to check it.'

'Well, perhaps Euclid never existed, and therefore could construct the geometry of absolutely empty space?'

But the professor apparently did not like to enter into this metaphysical discussion.

In the meantime the image of the book went off again far away in the original direction, and started coming back for the second time. Now it was still more damaged than before, and could hardly be recognized at all, which, according to the professor, was due to the fact that the light rays had travelled this time round the whole universe.

'If you turn your head once more,' he said to Mr Tompkins, 'you will see my book finally coming back after completing its journey round the world.' He stretched his hand, caught the book, and pushed it into his pocket.

Paul Davies

from THE GOLDILOCKS ENIGMA

▪ We stay with the difficult idea of the curvature of space, in the hands of the theoretical physicist Paul Davies. He is one of the most prolific writers of science for the general reader, and there are few among the deep problems of physics and cosmology that he has not touched upon. This extract is from his recent book, *The Goldilocks Enigma* (in Britain, but *The Cosmic Jackpot* in America. At the risk of becoming a bore with this buzzing bee in my bonnet, shouldn't we actually start boycotting publishers who confuse everybody with this kind of gratuitous renaming of books as they cross the Atlantic?) *The Goldilocks Enigma* is a lovely title, derived from the fact that the bed, chair, and porridge that Goldilocks enjoyed were all 'just right'. This extract is Paul Davies's description of curved space and of Einstein's idea that the universe itself is a hypersphere, which is finite but unbounded, such that no one galaxy is closer to the 'edge' than any other, for there is no edge, any more than our spherical Earth has an edge. WMAP stands for Wilkinson

Microwave Anisotropy Probe, which is a satellite sent up to map the cosmic background radiation left over from the Big Bang. As Davies explains, results from WMAP indicate that Einstein may have been wrong about the shape of the universe, though not about General Relativity itself. ■

Warped Space

So far I've been concentrating on astronomical discoveries. But cosmology would not be a true science if it lacked a framework of physical theory within which these discoveries can be understood. The theoretical basis of modern cosmology was established almost a century ago by Einstein in the form of his general theory of relativity. The theory was published in 1915, in the dark days of the First World War, but this did not stop astronomers and physicists from both sides of the conflict taking a keen interest in what it had to say about cosmology. The general theory of relativity, or general relativity as it is normally abbreviated, was designed to replace Newton's seventeenth-century theory of gravitation. In cosmology, gravitation is the dominating force, overwhelming all others because of the vast mass of the cosmos, so it is to gravitational theory that cosmologists turn to understand the expanding universe.

Einstein's brilliance was to spot that although gravitation manifests itself as a force, it may also be understood in a completely different way, in terms of 'warped geometry'. Let me explain what this means. The rules of geometry we learn at school date from the time of ancient Greece: the subject is often referred to as Euclidean geometry, after Euclid, who wrote it all down. There are many well-known theorems that can be proved from Euclid's axioms, for example the famous one named after Pythagoras. Another well-known theorem is that the angles of any triangle add up to two right angles (180°). These properties of lines, circles, triangles, and so on are watertight, but they come with one important proviso: they apply to *flat* surfaces. The theorems work correctly on blackboards and sheets of paper on school desktops, but they do not work on curved or warped surfaces such as globes. Pilots and navigators are well aware of this, and they have to use different geometrical rules to cope with the Earth's curvature. For example, on the Earth's surface a triangle can contain *three* right angles (see Figure 22).

If two-dimensional surfaces can be either flat (Euclidean geometry) or warped (non-Euclidean geometry), could *three*-dimensional space also have either 'flat' (Euclidean) geometry or warped geometry? Before Einstein, almost everyone assumed that space had 'flat' or Euclidean geometry, straightforwardly extended from the rules we learn for two dimensions. But there is no logical reason why that must be so. Some nineteenth-century mathematicians toyed with the idea that the geometry of three-dimensional space could be a generalization of curved surface geometry. They worked out the geometrical rules for this 'warped space', but at the time it was treated simply as a mathematical game. All that changed with general relativity. Einstein proposed that a gravitational

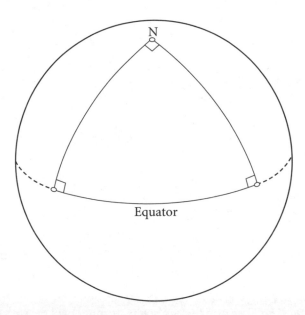

Figure 22. On a spherical surface the rules of geometry differ from those on a flat sheet. For example, a triangle may contain three right angles—such as this triangle on Earth's surface, with its apex at the North Pole and its base along the equator. The two-dimensional spherical surface represented here is an analogue of warped three-dimensional space.

field can warp three-dimensional space, necessitating the use of non-Euclidean geometry to describe it.

What, then, is curved space? One way to imagine it is to think of a triangle drawn around the sun (see Figure 23). Importantly, this must be a *flat* triangle (i.e. it lies in a plane). Now measure the angles and add them up. If Euclid's geometry applies to this situation, the result will be 180°. Einstein, however, claimed that the answer should be slightly greater than 180°, even though the triangle is flat, because the sun's gravitational field warps the three-dimensional geometry of the space around it. This experiment can actually be done (more or less) by bouncing radar waves off Venus and Mercury and doing triangulation. It turns out that Einstein is right—space really is curved rather than flat. (There is an important terminological nicety here: when cosmologists talk about 'flat' space, they

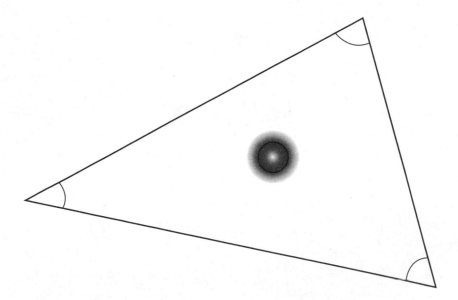

Figure 23. If a *flat* triangle is drawn round the sun, the angles add up to a bit more than 180° because the sun's gravitational field distorts the geometry of space in its vicinity. An equivalent way to think about this phenomenon is that the sides of the triangle are the straightest lines possible in the curved geometry. If light beams were directed along the sides of the triangle, it would seem to the receiver on the far side of the sun that the beams had been slightly bent by the sun's gravity.

don't mean space flattened to a pancake, they mean three-dimensional space with Euclidean geometry.) Sometimes the curved geometry near the sun is described by saying that the sun's gravity bends light rays passing near it, in which case the triangle would have distorted angles because the sides are wonky. This is true: it is an equivalent way to think about curved space, with the important point that the wonky sides are actually the straightest possible lines that can be drawn in the warped geometry, so it isn't just a matter of straightening the bent light beams out and recovering Euclid's results. The space is irreducibly curved, and no amount of manipulation will make it conform to Euclidean rules.

The warping of space around the sun, although detectable, is nevertheless tiny. Its existence was confirmed by the English astronomer Arthur Eddington, who measured the bending of light by observing the slight displacements in the positions of stars in the same part of the sky as the sun during the 1919 total eclipse. Eddington's star beams were bent by the amount that general relativity predicted, and this dramatic confirmation elevated Einstein to celebrity status. The spacewarp is small because the sun's gravitational field is weak by astronomical standards. Today we know of other objects in space with much larger gravitational fields that bend light more noticeably. One striking example occurs when a galaxy interposes itself between Earth and a more distant light source. Under these conditions, the galaxy bends the light around it on all sides, rather like a lens, causing the image of the distant source to be smeared out in an arc. In some cases, the image forms a complete ring, known appropriately enough as an Einstein ring. The most extreme bending of light—or warping of space—occurs around a black hole. In this case the spacewarp is so strong that it actually traps light completely, preventing it from escaping.

I have simplified the foregoing account in one important respect. In his earlier, so-called special theory of relativity, published in 1905, Einstein demonstrated that space is linked to time in a manner that makes it natural to consider the whole package—*spacetime*—together. Space has three dimensions and time has one, making four dimensions in all. Hermann Minkowski, one of Einstein's mathematics teachers, worked out how to modify the rules of Euclidean geometry to describe four-dimensional spacetime. When Einstein went on to generalize his theory of relativity in 1915 to include gravitation, he proposed that it is spacetime that is warped, and not merely space. Distorted spacetime geometry may imply warped space, warped time, or both. In the foregoing discussion of the spacewarp around the sun, I have ignored the time aspect. That is important too, and the sun's (tiny) timewarp has also been measured. In fact, Earth's even smaller timewarp is measurable; it manifests itself by the fact that clocks tick very slightly faster at higher altitudes—on a mountain top, say—than at sea level.

Einstein's Finite but Unbounded Universe

The observable universe contains about 10^{50} tonnes of visible matter in the form of stars, gas and dust, all of which combines to create a powerful gravitational field. Because gravitation warps the geometry of space, an interesting question immediately arises: what is the overall shape of the universe? By this, I don't mean how the galaxies are distributed in space. What I am referring to is *the shape of space itself*, considered on the grand scale of the cosmos. This was the problem Einstein set out to address in 1917, two years after he first presented his general theory of relativity. By applying the idea of warped space to cosmology he was able to construct a mathematical model of the entire universe. Although the model turned out to be rather wide of the mark, it introduced into cosmology several important features.

As I have explained, the sun creates a small distortion of space in its vicinity. Other stars create similar localized distortions. The question I am now considering is how all these distortions combine together. Will the curvature be cumulative, so that when we come to consider clusters of galaxies the spacewarp will be getting seriously big, or will the distortions tend to cancel each other out? In Einstein's mathematical model of the universe the curvature accumulates so that, averaged out over billions of light years, the shape of space resembles a three-dimensional version of the surface of a sphere, which is referred to as a hypersphere. Don't worry if you can't envisage a hypersphere. The important point is that it makes good mathematical sense, and its properties are easy to calculate by generalizing the geometry of familiar two-dimensional spherical surfaces.

An important property concerns the volume of space. In Einstein's hyperspherical universe space is finite (just as the Earth's spherical surface is finite). This means that space (in Einstein's model) does not extend for ever—thus contradicting what my father taught me. Another important property of Einstein's universe is that it is uniform (on average). The same is true, of course, of the surface of a sphere. There are no distinguishing features that single out any particular spot on a spherical surface as special; there is no centre or boundary. (The Earth has a centre, of course, but the *surface* of the Earth has no centre.) So Einstein's universe would look the same from any galaxy, precisely as astronomers observe. It is therefore finite yet unbounded—unbounded in the sense that there is no edge or barrier to prevent an object travelling from one place to any other place in the universe. Yet there are a limited number of places to go, in the same sense that there are a limited number of places to visit on the Earth's surface. And just as one can circumnavigate the Earth by always aiming straight ahead— returning home from the opposite direction—so one could in principle go round the Einstein universe, by aiming in a straight line, never deviating and returning from the opposite direction to that in which you had set out. Indeed, with a powerful enough telescope, you could look right around the Einstein universe and see the back of your own head!

[...]

One of the difficulties people have in conceptualizing a hypersphere is with the troublesome issue of 'what lies in the middle'. They think of a spherical two-dimensional surface, such as a round balloon, and say, 'well, the balloon has air inside it'. The issue of what the Einstein universe 'encloses' is a bit of a red herring, however. We humans, and the universe we perceive (at least those parts of it we have so far perceived), are restricted to three dimensions of space, so the issue of what, if anything, lies 'inside' Einstein's three-dimensional hyperspherical space is moot. If it helps, you can envisage this 'interior' region as a fourth dimension of space (empty, or filled with green cheese for that matter), but because we are trapped in the hyperspherical three-dimensional 'surface' it doesn't make a jot of difference to us whether the interior region is there or not, or what it contains. Much the same goes for the exterior region, the analogue of the space outside the balloon.

To ram this point home, since it proves so hard for people to grasp, try to put yourself in the position of a pancake-like creature restricted to life on the surface of a round balloon. The pancake might conjecture about what lies inside the balloon (air, empty space, green cheese…), but whatever is there doesn't affect the pancake's actual experiences because it cannot access the space inside the balloon, or receive any information from it. Furthermore, it is not even necessary for there to be *anything at all* (even empty space) inside a spherical surface for an inhabitant of the surface to deduce its sphericity. That is to say, the pancake doesn't need a God's eye-view of the balloon to conclude that its world is spherical—closed and finite, yet without boundary. The pancake can deduce this entirely by observations it can make from the confines of the spherical surface: the sphericity is *intrinsic* to the surface, and does not depend on it being embedded in an enveloping three-dimensional space. How can the pancake tell? Well, for example, by drawing triangles and measuring whether the angles add up to more than 180°. Or the pancake could circumnavigate its world. In the same vein, humans could deduce that we are living in a closed, finite, hyper-spherical Einstein space without reference to any higher-dimensional embedding or enveloping space, merely by doing geometry *within* the space. So the existence or otherwise of an 'interior' or 'exterior' region of the Einstein universe, not to mention what it consists of, is quite simply irrelevant. But if you would like to imagine inaccessible empty space there for ease of visualization, then go ahead. It makes no difference.

What Shape is the Universe?

All this is well and good, but was Einstein actually right about the universe being shaped like a hypersphere? Here, WMAP has been of immense help. Obviously, if the universe were seriously lopsided, it would show up in the pattern of microwaves from the sky. The fact that this radiation is so uni-form already indicates that the universe, out as far as we can see, is at least fairly regular in shape. But what shape is it? Resorting again to a two-dimensional analogy, we can immediately identify two perfectly regular shapes: an infinite flat sheet and a perfect sphere. But there is a third shape, a sort of inverse of the sphere. Remember that on a sphere a triangle has angles adding up to more than 180°. Technically, the sphere is defined to be curved *positively*. What about a uniform surface on which the angles add up to *less* than 180°? This is a space with *negative* curvature. Such a surface exists, and it looks a bit like a saddle, but infinitely extended (see Figure 24). All three surfaces—with zero, positive, and negative curvature—can be generalized to three dimensions. Since the 1920s, when cosmologists first realized that there were three different shapes for uniform space, they have wanted to know which one our universe most closely resembles.

A direct assault on the problem has been tried many times. Because the geometry of the three different spaces is different, astronomers ought to be able to tell simply by looking. Measuring the angles of a triangle over cosmic distances isn't feasible, but there are other possibilities. Returning again to two dimensions, imagine drawing a series of concentric circles on a flat sheet. The area enclosed by each circle rises in proportion to the square of the radius: double the radius and the area is four times as great. But on the surface of a sphere this relationship goes wrong: the area increases with radius *less* rapidly. That's easy to see, because if you tried to flatten a cap, you would have to cut wedges out of it, so it would fail to cover a disk of equivalent radius on a flat sheet. Similarly, the area on the saddle shape increases *more* rapidly than the square of the radius. Converting all this into three dimensions means

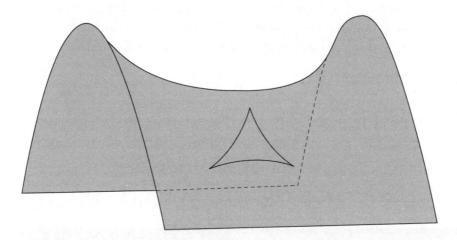

Figure 24. It is possible that, on the cosmological scale, space is uniform but curved outwards instead of inwards. The two-dimensional surface shown here is the analogue of such a negatively curved, three-dimensional space. It is infinite and homogeneous. The negatively curved geometry manifests itself in the distortion of a triangle, whose angles add up to less than 180°.

that the volume of a region of space will rise as the cube of the radius if the universe is flat (three-dimensionally flat, remember, not pancake-flat). If the universe is a hypersphere, as Einstein suggested, the volume will increase less rapidly with radius, and if it is a 'hypersaddle' it will increase more rapidly. The volume of a region of space can be assessed by counting the number of galaxies it contains.

Some astronomers tried to establish the geometry of the universe in this direct way, but their results were inconclusive on account of the difficulty of measuring precise distances to far-flung galaxies, and other technical complications. However, the answer can be inferred from the WMAP data, by measuring the sizes of the temperature fluctuations.... Before WMAP was launched, theorists had already worked out how big the physical sizes of the strongest fluctuations should be. Converting that into apparent angular size in the sky depends on the geometry of space: if the universe is positively curved it would make the angles appear larger, while negative curvature would make them smaller. If the universe is geometrically flat (i.e. has Euclidean geometry), the angular size of the strongest hot and cold fluctuations should be about 1° across. The results that flowed back from the satellite were definitive. The fluctuations were very close to 1° in size, a result confirmed by ground-based and balloon-based experiments. Cosmologists then declared that, to within observational accuracy of about 2 per cent, space is flat.

Russell Stannard

from THE TIME AND SPACE OF UNCLE ALBERT

▓ I have already mentioned Russell Stannard's 'Uncle Albert' series of children's books. The wise and kind Uncle Albert (we are meant to think 'Einstein' of course, although he is never named) sends his little niece Gedanken off on thought experiments—thought *experiences*, rather, since

this is fiction—from which she returns, having learned an important principle of modern physics. In this extract, Gedanken chases a light beam and
comes back with some understanding of Special Relativity. I cannot claim
really to understand relativity to this day, but even (or perhaps especially)
at my advanced age I find that Uncle Albert brings me closest to it. ▪

––––––––

'...It's my job to look after you and see that everything runs smoothly.
All you have to do is...well...just enjoy it. Right. Are you ready?'

Gedanken looked happier and nodded.

'OK. Strap yourself in.'

She fastened her seat belt.

'Now, all you have to worry about is that red button in front of
you. When you want to start up the motor, you just press the button.
Keep your finger on it for as long as you want the rocket to fire. All
right?'

'Yes,' she nodded, getting more and more excited.

'OK, when you're ready, you can blast off.'

Gedanken took a deep breath, reached out and pressed the button.
Immediately, from the rear of the spacecraft, there came a throaty roar
of engine noise. She felt herself flattened against the back of the seat.

'How thrilling!' she thought.

After a while the computer called out, 'You can let go of the button
when you like.'

She did so, and the engine noise died away. No longer was she pressed
against her seat. She had a floating sensation and felt she would go drift-
ing off if the seat belt hadn't continued to hold her in position. It was
very pleasant once one got used to it.

'That's our speed, is it?' asked Gedanken, pointing to a digital read-
out just above the button in front of her. It said SPEED RELATIVE TO
EARTH and was reading '0.500 times the speed of light'.

'That's right,' said the computer. 'We're now travelling at half the
speed of light.'

'Then why aren't we slowing down?'

'Why should we?'

'Because the engine's off. We ought to be slowing down.'

'This is not a bicycle,' the computer said in a superior tone of voice. 'This is a spacecraft. Once a spacecraft is up to speed it doesn't need pushing any more. Out here, there is no air or anything to slow us down. So, we just keep cruising along. We only need the rocket motor when we want to *change* speed—when we want to go faster or slower.'

'How does a motor make you go slower?'

'By going into reverse, of course.'

Above the digital read-out was a large window facing out of the front of the spacecraft. Gedanken looked at the stars laid out before her. With each moment she was feeling more relaxed and at home. She was really beginning to enjoy it all.

'Right,' interrupted the computer, 'if you're ready, we had better get on with our mission. Uncle Albert has programmed some instructions into me somewhere. I'll have a look in my memory bank.'

'We're to catch up with a light beam,' Gedanken volunteered eagerly.

There was a pause, and the computer resumed, 'Ah yes. Quite right. Catch a light beam, he says. How peculiar. Suppose he knows what he's doing. All right, can you see one? There ought to be plenty about with all those stars out there.'

Gedanken peered out of the window. Suddenly she pointed excitedly, 'There! Is that one?'

Directly outside, a shimmering glow sped by. It seemed to Gedanken to have a face—an impish face…and…yes, it was giggling! She definitely heard a faint high-pitched giggle, like an over-excited schoolgirl, and a teasing voice that called: 'Go on, catch me if you can.' As it got further away the fuzzy light patch got smaller and fainter.

'Yes,' cried the computer, 'that's one. After it before it gets away!'

'Easy,' shouted Gedanken. 'We're already doing half its speed.' And with that she pressed the button, the engines roared into life once more, and they were giving chase.

After a few minutes the computer called out, 'OK. That should do it. We ought to have caught up with it by now.'

Gedanken released the button and looked for the light beam. Her face fell.

'Oh. It's further away than ever,' she said.

'What? Further away...That's impossible.'

'But it is.'

'Can't be. What speed are we doing?' asked the computer.

'Er...0.900. Nine-tenths the speed of light—I think.'

'Is that all?'

'Yes,' she confirmed.

'How odd. According to my calculations we ought easily to be doing the speed of light.'

Gedanken heard a far-off voice giggling and laughing: 'You'll have to do better than that.'

'Give it another go,' said the computer. 'I'll put you on full power this time.'

Gedanken pressed the button again. The engine noise was deafening—far louder than before. The spacecraft shuddered and shook in a most alarming manner. It was as though it were about to shatter into tiny pieces.

After what seemed an age, the computer instructed her to let go the button, and she thankfully did so. She peered out of the window. At first she could see nothing. Then she spotted the light beam.

'Oh no. It's miles away now—and it's still going away from us.'

'Impossible. What's our speed now?'

'0.999 times the speed of light.'

'Most irregular. The answer I've got is quite different.' Then, with a note of disgust in its voice, it added, 'I must have been misprogrammed. I've heard of this sort of thing. Never thought it would happen to me, though. How embarrassing. I'll have to do a check on myself.'

'...can't wait...can't wait...must be on my way...'

Gedanken heard the light beam's voice dying away in the distance. The fuzzy patch of light faded—and finally disappeared.

Brian Greene

from THE ELEGANT UNIVERSE

■ You (I suspect) and (certainly) I, are not equipped to understand intuitively what it means to say that we live in a universe of more than three dimensions. This, I believe, is because the decisions that our ancestors' brains were called upon to make to assist their genetic survival could not be improved by perceiving or understanding more than three dimensions. If our ancestors had had to survive in cosmic megaspaces, or in the microspaces of the quantum world, we'd see things differently. But, positioned as we are in the middle of Martin Rees's logarithmic scale, our brains evolved a system of perception that could not cope with more than three dimensions. Perhaps Paul Davies, Mr Tompkins's old professor, and Gedanken's Uncle Albert have softened us up to the point where we are ready to take delivery of what Brian Greene now has to tell us. There may be extra dimensions, not only on the grand, Einsteinian megascale of the cosmos, but also tiny coiled up dimensions, tiny on the seriously tiny scale of the quantum, at every point inside the very fabric of space. This idea, in turn, is necessary to soften us up for the even stranger ideas of String Theory which, controversially, could be paving the way to the long hoped-for 'Theory of Everything'. Brian Greene is one of the main architects of string theory, and it is fitting that he should be the one to explain to us the idea of tiny coiled-up dimensions. ■

The suggestion that our universe might have more than three spatial dimensions may well sound fatuous, bizarre, or mystical. In reality, though, it is concrete and thoroughly plausible. To see this, it's easiest to shift our sights temporarily from the whole universe and think about a more familiar object, such as a long, thin garden hose.

Imagine that a few hundred feet of garden hose is stretched across a canyon, and you view it from, say, a quarter of a mile away. From this distance, you will easily perceive the long, unfurled, horizontal extent of the hose, but unless you have uncanny eyesight, the *thickness* of the hose will be difficult to discern. From your distant vantage point, you would think that if an ant were constrained to live on the hose, it would have only *one* dimension in which to walk: the left-right dimension along the hose's length. If someone asked you to specify where the ant was at a given moment, you would need to give only *one* piece of data: the distance of the ant from the left (or the right) end of the hose. The upshot is that from a quarter of a mile away, a long piece of garden hose appears to be a one-dimensional object.

In reality, we know that the hose *does* have thickness. You might have trouble resolving this from a quarter mile, but by using a pair of binoculars you can zoom in on the hose and observe its girth directly. From this magnified perspective, you see that a little ant living on the hose actually has *two* independent directions in which it can walk: along the left-right dimension spanning the length of the hose as already identified, *and* along the 'clockwise-counterclockwise dimension' around the circular part of the hose. You now realize that to specify where the tiny ant is at any given instant, you must actually give *two* pieces of data: where the ant is along the length of the hose, and where the ant is along its circular girth. This reflects the fact the surface of the garden hose is two-dimensional.

Nonetheless, there is a clear difference between these two dimensions. The direction along the length of the hose is long, extended, and easily visible. The direction circling around the thickness of the hose is short, 'curled up', and harder to see. To become aware of the circular dimension, you have to examine the hose with significantly greater precision.

This example underscores a subtle and important feature of spatial dimensions: they come in two varieties. They can be large, extended, and therefore directly manifest, or they can be small, curled up, and much more difficult to detect. Of course, in this example you did not have to exert a great deal of effort to reveal the 'curled-up' dimension encircling the thickness of the hose. You merely had to use a pair of binoculars. However, if you had a very thin garden hose—as thin as a hair or a capillary—detecting its curled-up dimension would be more difficult.

In a paper he sent to Einstein in 1919, [Theodor] Kaluza made an astounding suggestion. He proposed that the spatial fabric of the universe might possess more than the three dimensions of common experience. The motivation for this radical thesis, as we will discuss shortly, was Kaluza's realization that it provided an elegant and compelling framework for weaving together Einstein's general relativity and Maxwell's electromagnetic theory into a single, unified conceptual framework. But, more immediately, how can this proposal be squared with the apparent fact that we *see* precisely three spatial dimensions?

The answer, implicit in Kaluza's work and subsequently made explicit and refined by the Swedish mathematician Oskar Klein in 1926, is that *the spatial fabric of our universe may have both extended and curled-up dimensions*. That is, just like the horizontal extent of the garden hose, our universe has dimensions that are large, extended, and easily visible—the three spatial dimensions of common experience. But like the circular girth of a garden hose, the universe may also have additional spatial dimensions that are tightly curled up into a tiny space—a space so tiny that it has so far eluded detection by even our most refined experimental equipment.

To gain a clearer image of this remarkable proposal, let's reconsider the garden hose for a moment. Imagine that the hose is painted with closely spaced black circles along its girth. From far away, as before, the garden hose looks like a thin, one-dimensional line. But if you zoom in with binoculars, you can detect the curled-up dimension, even more easily after our paint job, and you see the image illustrated in Figure 25. This figure emphasizes that the surface of the garden hose is two-dimensional, with one large, extended dimension and one small, circular dimension. Kaluza and Klein proposed that our spatial universe is similar, but that it has three large, extended spatial dimensions and one small, circular dimension—for a total of four spatial dimensions. It is difficult to draw something with that many dimensions, so for visualization purposes we must settle for an illustration incorporating two large dimensions and one small, circular dimension. We illustrate this in Figure 26, in which we magnify the fabric of space in much the same way that we zoomed in on the surface of the garden hose.

The lowest image in the figure shows the apparent structure of space—the ordinary world around us—on familiar distance scales such as meters. These distances are represented by the largest set of grid lines.

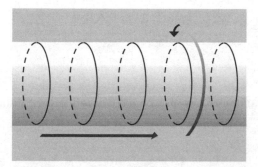

Figure 25. The surface of the garden hose is two-dimensional: one dimension (its horizontal extent), emphasized by the straight arrow, is long and extended; the other dimension (its circular girth), emphasized by the circular arrow, is short and curled up.

In the subsequent images, we zoom in on the fabric of space by focusing our attention on ever smaller regions, which we sequentially magnify in order to make them easily visible. At first as we examine the fabric of space on shorter distance scales, not much happens; it appears to retain the same basic form as it has on larger scales, as we see in the first three levels of magnification. However, as we continue on our journey toward the most microscopic examination of space—the fourth level of magnification in Figure 26—a new, curled-up, circular dimension becomes apparent, much like the circular loops of thread making up the pile of a tightly woven piece of carpet. Kaluza and Klein suggested that the extra circular dimension exists at *every* point in the extended dimensions, just as the circular girth of the garden hose exists at every point along its unfurled, horizontal extent. (For visual clarity, we have drawn only an illustrative sample of the circular dimension at regularly spaced points in the extended dimensions.) We show a close-up of the Kaluza–Klein vision of the microscopic structure of the spatial fabric in Figure 27.

The similarity with the garden hose is manifest, although there are some important differences. The universe has three large, extended space dimensions (only two of which we have actually drawn), compared with the garden hose's one, and, more important, we are now describing the spatial fabric of the *universe* itself, not just an object, like

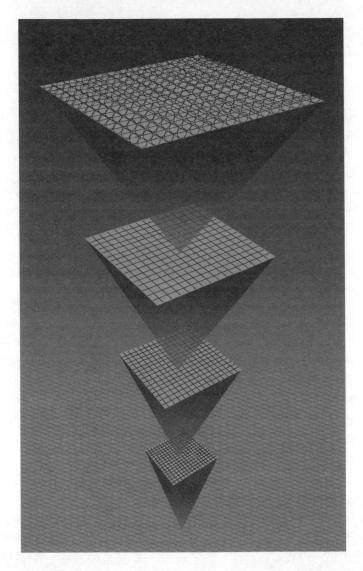

Figure 26. Each subsequent level represents a huge magnification of the spatial fabric displayed in the previous level. Our universe may have extra dimensions—as we see by the fourth level of magnification—so long as they are curled up into a space small enough to have as yet evaded direct detection.

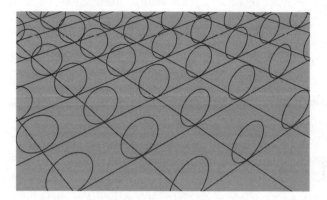

Figure 27. The grid lines represent the extended dimensions of common experience, whereas the circles are a new, tiny, curled-up dimension. Like the circular loops of thread making up the pile of a carpet, the circles exist at every point in the familiar extended dimensions—but for visual clarity we draw them as spread out on intersecting grid lines.

the garden hose, that exists *within* the universe. But the basic idea is the same: like the circular girth of the garden hose, if the additional curled-up, circular dimension of the universe is extremely small, it is much harder to detect than the manifest, large, extended dimensions. In fact, if its size is small enough, it will be beyond detection by even our most powerful magnifying instruments. And, of utmost importance, the circular dimension is *not* merely a circular bump within the familiar extended dimensions as the illustration might lead you to believe. Rather, the circular dimension is a *new* dimension, one that exists at every point in the familiar extended dimensions just as each of the up-down, left-right, and back-forth dimensions exists at every point as well. It is a new and independent direction in which an ant, if it were small enough, could move. To specify the spatial location of such a microscopic ant, we would need to say where it is in the three familiar extended dimensions (represented by the grid) and *also* where it is in the circular dimension. We would need *four* pieces of spatial information; if we add in time, we get a total of five pieces of space-time information—one more than we normally would expect.

And so, rather surprisingly, we see that although we are aware of only three extended spatial dimensions, Kaluza's and Klein's reasoning shows that this does not preclude the existence of additional curled-up dimensions, at least if they are very small. The universe may very well have more dimensions than meet the eye.

How small is 'small'? Cutting-edge equipment can detect structures as small as a billionth of a billionth of a meter. So long as an extra dimension is curled up to a size less than this tiny distance, it is too small for us to detect. In 1926 Klein combined Kaluza's initial suggestion with some ideas from the emerging field of quantum mechanics. His calculations indicated that the additional circular dimension might be as small as the Planck length[10^{-35}m], far shorter than experimental accessibility. Since then, physicists have called the possibility of extra tiny space dimensions *Kaluza–Klein theory*.

Stephen Hawking

from A BRIEF HISTORY OF TIME

■ No collection of science writing would be complete without something from Stephen Hawking, and not only because of the prodigious sales figures of *A Brief History of Time*. I am one of those who made it all the way through to the end (we few, we happy few, we band of brothers). There is no way around it, there are no easy short cuts, modern physics is a hard struggle, but it is worth the effort. What Hawking has to tell us is one of the greatest stories ever. It is a privilege to live in a century where such an epic can be told, and a privilege to hear it from one of its most distinguished discoverers. ■

In order to understand what you would see if you were watching a star collapse to form a black hole, one has to remember that in the theory of relativity there is no absolute time. Each observer has his own measure of time. The time for someone on a star will be different from that for someone at a distance, because of the gravitational field of the star. Suppose an intrepid astronaut on the surface of the collapsing star, collapsing inward with it, sent a signal every second, according to his watch, to his spaceship orbiting about the star. At some time on his watch, say 11.00, the star would shrink below the critical radius at which the gravitational field becomes so strong nothing can escape, and his signals would no longer reach the spaceship. As 11.00 approached, his companions watching from the spaceship would find the intervals between successive signals from the astronaut getting longer and longer, but this effect would be very small before 10.59.59. They would have to wait only very slightly more than a second between the astronaut's 10.59.58 signal and the one that he sent when his watch read 10.59.59, but they would have to wait for ever for the 11.00 signal. The light waves emitted from the surface of the star between 10.59.59 and 11.00, by the astronaut's watch, would be spread out over an infinite period of time, as seen from the spaceship. The time interval between the arrival of successive waves at the spaceship would get longer and longer, so the light from the star would appear redder and redder and fainter and fainter. Eventually, the star would be so dim that it could no longer be seen from the spaceship: all that would be left would be a black hole in space. The star would, however, continue to exert the same gravitational force on the spaceship, which would continue to orbit the black hole.

This scenario is not entirely realistic, however, because of the following problem. Gravity gets weaker the farther you are from the star, so the gravitational force on our intrepid astronaut's feet would always be greater than the force on his head. This difference in the forces would stretch our astronaut out like spaghetti or tear him apart before the star had contracted to the critical radius at which the event horizon formed! However, we believe that there are much larger objects in the universe, like the central regions of galaxies, that can also undergo gravitational collapse to produce black holes; an astronaut on one of these would not be torn apart before the black hole formed. He would not, in fact, feel anything special as he reached the critical radius, and could pass the point of no return without noticing it. However, within just a few hours, as the region continued to collapse, the difference in the gravitational forces on his head and his feet would become so strong that again it would tear him apart.

The work that Roger Penrose and I did between 1965 and 1970 showed that, according to general relativity, there must be a singularity of infinite density and space-time curvature within a black hole. This is rather like the big bang at the beginning of time, only it would be an end of time for the collapsing body and the astronaut. At this singularity the laws of science and our ability to predict the future would break down. However, any observer who remained outside the black hole would not be affected by this failure of predictability, because neither light nor any other signal could reach him from the singularity. This remarkable fact led Roger Penrose to propose the cosmic censorship hypothesis, which might be paraphrased as 'God abhors a naked singularity'. In other words, the singularities produced by gravitational collapse occur only in places, like black holes, where they are decently hidden from outside view by an event horizon. Strictly, this is what is known as the weak cosmic censorship hypothesis: it protects observers who remain outside the black hole from the consequences of the breakdown of predictability that occurs at the singularity, but it does nothing at all for the poor unfortunate astronaut who falls into the hole.

There are some solutions of the equations of general relativity in which it is possible for our astronaut to see a naked singularity: he may be able to avoid hitting the singularity and instead fall through a 'wormhole' and come out in another region of the universe. This would offer great possibilities for travel in space and time, but unfortunately it seems that these solutions may all be highly unstable; the least disturbance, such as the presence of an astronaut, may change them so that the astronaut could not see the singularity until he hit it and his time came to an end. In other words, the singularity would always lie in his future and never in his past. The strong version of the cosmic censorship hypothesis states that in a realistic solution, the singularities would always lie either entirely in the future (like the singularities of gravitational collapse) or entirely in the past (like the big bang). It is greatly to be hoped that some version of the censorship hypothesis holds because close to naked singularities it may be possible to travel into the past. While this would be fine for writers of science fiction, it would mean that no-one's life would ever be safe: someone might go into the past and kill your father or mother before you were conceived!

The event horizon, the boundary of the region of space-time from which it is not possible to escape, acts rather like a one-way membrane around the black hole: objects, such as unwary astronauts, can fall through the event horizon into the black hole, but nothing can ever get out of the black hole through the event horizon. (Remember that the event horizon is the path in space-time of light that is trying to escape from the black hole, and nothing can travel faster than light.) One could well say of the event horizon what the poet Dante said of the entrance to Hell: 'All hope abandon, ye who enter here.' Anything or anyone who falls through the event horizon will soon reach the region of infinite density and the end of time.

[...]

When we combine quantum mechanics with general relativity, there seems to be a new possibility that did not arise before: that space and time together might form a finite, four-dimensional space without singularities or boundaries, like the surface of the earth but with more dimensions. It seems that this idea could explain many of the observed features of the universe, such as its large-scale uniformity and also the smaller-scale departures from homogeneity, like galaxies, stars, and even human beings. It could even account for the arrow of time that we observe. But if the universe is completely self-contained, with no singularities or boundaries, and completely described by a unified theory, that has profound implications for the role of God as Creator.

Einstein once asked the question: 'How much choice did God have in constructing the universe?' If the no boundary proposal is correct, he had no freedom at all to choose initial conditions. He would, of course, still have had the freedom to choose the laws that the universe obeyed. This, however, may not really have been all that much of a choice; there may well be only one, or a small number, of complete unified theories, such as the heterotic string theory, that are self-consistent and allow the existence of structures as complicated as human beings who can investigate the laws of the universe and ask about the nature of God.

Even if there is only one possible unified theory, it is just a set of rules and equations. What is it that breathes fire into the equations and makes a universe for them to describe? The usual approach of science of constructing a mathematical model cannot answer the questions of why there should be a universe for the model to describe. Why does the universe go to all the bother of existing? Is the unified theory so compelling that it brings about its own existence? Or does it need a creator, and, if so, does he have any other effect on the universe? And who created him?

Up to now, most scientists have been too occupied with the development of new theories that describe *what* the universe is to ask the question *why*. On the other hand, the people whose business it is to ask *why*, the philosophers, have not been able to keep up with the advance of scientific theories. In the eighteenth century, philosophers considered the whole of human knowledge, including science, to be their field and discussed questions such as: did the universe have a beginning? However, in the nineteenth and twentieth centuries, science became too technical and mathematical for the philosophers, or anyone else except a few specialists. Philosophers reduced the scope of their enquiries so much that Wittgenstein, the most famous philosopher of this century, said, 'The sole remaining task for philosophy is the analysis of language.' What a comedown from the great tradition of philosophy from Aristotle to Kant!

However, if we do discover a complete theory, it should in time be understandable in broad principle by everyone, not just a few scientists. Then we shall all, philosophers, scientists and just ordinary people, be able to take part in the discussion of the question of why it is that we and the universe exist. If we find the answer to that, it would be the ultimate triumph of human reason—for then we would know the mind of God.

PART IV

WHAT SCIENTISTS DELIGHT IN

S. Chandrasekhar

from TRUTH AND BEAUTY

▦ Subrahmanyan Chandrasekhar was a distinguished astrophysicist, originally from India but later an American citizen, who won the Nobel Prize for his work on the development (often wrongly called evolution) of stars. I have long found poetic inspiration in the personal reflection which ends the following extract from his lecture on creativity, contrasting it favourably with the famous last lines of Keats's Ode on a Grecian Urn,

> 'Beauty is truth, truth beauty,'—that is all
> Ye know and all ye need to know. ▦

[I] am frankly puzzled by the difference that appears to exist in the patterns of creativity among the practitioners in the arts and the practitioners in the sciences: for, in the arts as in the sciences, the quest is after the same elusive quality: beauty. But what is beauty?

In a deeply moving essay on 'The Meaning of Beauty in the Exact Sciences', Heisenberg gives a definition of beauty which I find most apposite. The definition, which Heisenberg says goes back to antiquity, is that 'beauty is the proper conformity of the parts to one another and to the whole'. On reflection, it does appear that this definition touches the essence of what we may describe as 'beautiful': it applies equally to *King Lear*, the *Missa Solemnis*, and the *Principia*.

There is ample evidence that in science, beauty is often the source of delight. One can find many expressions of such delight scattered through the scientific literature. Let me quote a few examples.

Kepler:

Mathematics is the archetype of the beautiful.

David Hilbert (in his memorial address for Hermann Minkowski):

Our Science, which we loved above everything, had brought us together. It appeared to us as a flowering garden. In this garden there were well-worn paths where one might look around at leisure and enjoy one-self without effort, especially at the side of a congenial companion. But we also liked to seek out hidden trails and discovered many an unexpected view which was pleasing to our eyes; and when the one pointed it out to the other, and we admired it together, our joy was complete.

Hermann Weyl (as quoted by Freeman Dyson):

My work always tried to unite the true with the beautiful; but when I had to choose one or the other, I usually chose the beautiful.

Heisenberg (in a discussion with Einstein):

If nature leads us to mathematical forms of great simplicity and beauty—by forms I am referring to coherent systems of hypothesis, axioms, etc.—to forms that no one has previously encountered, we cannot help thinking that they are 'true', that they reveal a genuine feature of nature...You must have felt this too: the almost frightening simplicity and wholeness of the relationships which nature suddenly spreads out before us and for which none of us was in the least prepared.

All these quotations express thoughts that may appear vague or too general. Let me try to be concrete and specific.

The discovery by Pythagoras, that vibrating strings, under equal tension, sound together harmoniously if their lengths are in simple numerical ratios, established for the first time a profound connection between the intelligible and the beautiful. I think we may agree with Heisenberg that this is 'one of the truly momentous discoveries in the history of mankind'.

Kepler was certainly under the influence of the Pythagorean concept of beauty when he compared the revolution of the planets about the sun with a vibrating string and spoke of the harmonious concord of the different planetary orbits as the music of the spheres. It is known that Kepler was profoundly grateful that it had been reserved for him to discover, through his laws of planetary motion, a connection of the highest beauty.

A more recent example of the reaction of a great scientist, to this aspect of beauty at the moment of revelation of a great truth, is provided by Heisenberg's description of the state of his feeling when he found the key that opened the door to all the subsequent developments in the quantum theory.

Towards the end of May 1925, Heisenberg, ill with hay fever, went to Heligoland to be away from flowers and fields. There by the sea, he made rapid progress in resolving the difficulties in the quantum theory as it was at that time. He writes:

> Within a few days more, it had become clear to me what precisely had to take the place of the Bohr-Sommerfeld quantum conditions in an atomic physics working with none but observable magnitudes. It also became obvious that with this additional assumption, I had introduced a crucial restriction into the theory. Then I noticed that there was no guarantee that...the principle of the conservation of energy would apply...Hence I concentrated on demonstrating that the conservation law held; and one evening I reached the point where I was ready to determine the individual terms in the energy table [Energy Matrix]...When the first terms seemed to accord with the energy principle, I became rather excited, and I began to make countless arithmetical errors. As a result, it was almost three o'clock in the morning before the final result of my computations lay before me. The energy principle had held for all the terms, and I could no longer doubt the mathematical consistency and coherence of the kind of quantum mechanics to which my calculations pointed. At first, I was deeply alarmed. I had the feeling that, through the surface of atomic phenomena, I was looking at a strangely beautiful interior, and felt almost giddy at the thought that I now had to probe this wealth of mathematical structure nature had so generously spread out before me. I was far too excited to sleep, and so, as a new day dawned, I made for the southern tip of the island, where I had been longing to climb a rock jutting out into the sea. I now did so without too much trouble, and waited for the sun to rise.

May I allow myself at this point a personal reflection? In my entire scientific life, extending over forty-five years, the most shattering experience has been the realization that an exact solution of Einstein's equations of general relativity, discovered by the New Zealand mathematician, Roy Kerr, provides the *absolutely exact representation* of untold numbers of massive black holes that populate the universe. This 'shuddering before the beautiful', this incredible fact that a discovery motivated by a search after the beautiful in mathematics should find its exact replica in Nature, persuades me to say that beauty is that to which the human mind responds at its deepest and most profound. Indeed, everything I have tried to say in this connection has been stated more succinctly in the Latin mottos:

Simplex sigillum veri—The simple is the seal of the true.

and

Pulchritudo splendor veritatis—Beauty is the splendour of truth.

G. H. Hardy

from A MATHEMATICIAN'S APOLOGY

A waggish undergraduate contemporary of mine at Oxford, reading mathematics, told me the story of a certain pure mathematician whose ambition was to devise a theorem that was completely and utterly useless. 'But always some tiresome physicist would come along and find a use for it.' G. H. Hardy seems to have had the same ambition—he may even have been the object of the story. We have already seen C. P. Snow's memorable portrait of this eccentric mathematician, and the wonderful story of how he discovered and nurtured the Indian mathematical prodigy Ramanujan, who sadly died young. Now here is Hardy himself in later life, doing his best to convey the pure beauty of the purest mathematics. There is a poignancy to that 'in later life', for mathematicians are often said to be over the hill at some alarmingly early age, and Hardy himself, one of the greatest mathematicians of his generation, was wistfully aware of it:

> I had better say something here about the question of age, since it is particularly important for mathematicians. No mathematician should ever allow himself to forget that mathematics, more than any other art or science, is a young man's game...We may consider, for example, the career of a man who was certainly one of the world's three greatest mathematicians. Newton gave up mathematics at fifty, and had lost his enthusiasm long before; he had recognized no doubt by the time that he was forty that his great creative days were over. His greatest ideas of all, fluxions and the law of gravitation, came to him about 1666, when he was twenty-four...Galois died at twenty-one, Abel at twenty-seven, Ramanujan at thirty-three,

Riemann at forty. There have been men who have done great work a good deal later; Gauss's great memoir on differential geometry was published when he was fifty (though he had had the fundamental ideas ten years before). I do not know an instance of a major mathematical advance initiated by a man past fifty. If a man of mature age loses interest in and abandons mathematics, the loss is not likely to be very serious either for mathematics or for himself.

That was from *A Mathematician's Apology,* which he published at the age of 67. What follows is an extract from the same book. ■

It will be clear by now that, if we are to have any chance of making progress, I must produce examples of 'real' mathematical theorems, theorems which every mathematician will admit to be first-rate. And here I am very heavily handicapped by the restrictions under which I am writing. On the one hand my examples must be very simple, and intelligible to a reader who has no specialized mathematical knowledge; no elaborate preliminary explanations must be needed; and a reader must be able to follow the proofs as well as the enunciations. These conditions exclude, for instance, many of the most beautiful theorems of the theory of numbers, such as Fermat's 'two square' theorem or the law of quadratic reciprocity. And on the other hand my examples should be drawn from 'pukka' mathematics, the mathematics of the working professional mathematician; and this condition excludes a good deal which it would be comparatively easy to make intelligible but which trespasses on logic and mathematical philosophy.

I can hardly do better than go back to the Greeks. I will state and prove two of the famous theorems of Greek mathematics. They are 'simple' theorems, simple both in idea and in execution, but there is no doubt at all about their being theorems of the highest class. Each is as fresh and significant as when it was discovered—two thousand years have not written a wrinkle on either of them. Finally, both the statements and the proofs can be mastered in an hour by any intelligent reader, however slender his mathematical equipment.

1. The first is Euclid's[1] proof of the existence of an infinity of prime numbers.

The *prime numbers* or *primes* are the numbers

(A) $2, 3, 5, 7, 11, 13, 17, 19, 23, 29, \ldots$

which cannot be resolved into smaller factors.[2] Thus 37 and 317 are prime. The primes are the material out of which all numbers are built up by multiplication: thus $666 = 2.3.3.37$. Every number which is not prime itself is divisible by at least one prime (usually, of course, by several). We have to prove that there are infinitely many primes, i.e. that the series (A) never comes to an end.

Let us suppose that it does, and that

$$2, 3, 5, \cdots, P$$

is the complete series (so that P is the largest prime); and let us, on this hypothesis, consider the number Q defined by the formula

$$Q = (2.3.5.\cdots.P) + 1.$$

It is plain that Q is not divisible by any of $2, 3, 5, \cdots, P$; for it leaves the remainder 1 when divided by any one of these numbers. But, if not itself prime, it is divisible by *some* prime, and therefore there is a prime (which may be Q itself) greater than any of them. This contradicts our hypothesis, that there is no prime greater than P; and therefore this hypothesis is false.

The proof is by *reductio ad absurdum*, and *reductio ad absurdum*, which Euclid loved so much, is one of a mathematician's finest weapons.[3] It is a far finer gambit than any chess gambit: a chess player may offer the sacrifice of a pawn or even a piece, but a mathematician offers *the game*.

2. My second example is Pythagoras's[4] proof of the 'irrationality' of $\sqrt{2}$.

A 'rational number' is a fraction, $\frac{a}{b}$ where a and b are integers; we may suppose that a and b have no common factor, since if they had we could remove it. To say that '$\sqrt{2}$ is irrational' is merely another way of saying that 2 cannot be expressed in the form $(\frac{a}{b})^2$; and this is the same thing as saying that the equation

(B) $a^2 = 2b^2$

cannot be satisfied by integral values of a and b which have no common factor. This is a theorem of pure arithmetic, which does not demand any knowledge of 'irrational numbers' or depend on any theory about their nature.

We argue again by *reductio ad absurdum*; we suppose that (B) is true, a and b being integers without any common factor. It follows from (B) that a^2 is even (since $2b^2$ is divisible by 2), and therefore that a is even (since the square of an odd number is odd). If a is even then

(C) $$a = 2c$$

for some integral value of c; and therefore

$$2b^2 = a^2 = (2c)^2 = 4c^2$$

or

(D) $$b^2 = 2c^2.$$

Hence b^2 is even, and therefore (for the same reason as before) b is even. That is to say, a and b are both even, and so have the common factor 2. This contradicts our hypothesis, and therefore the hypothesis is false.

It follows from Pythagoras's theorem that the diagonal of a square is incommensurable with the side (that their ratio is not a rational number, that there is no unit of which both are integral multiples). For if we take the side as our unit of length, and the length of the diagonal is d, then, by a very familiar theorem also ascribed to Pythagoras,[5]

$$d^2 = 1^2 + 1^2 = 2,$$

so that d cannot be a rational number.

I could quote any number of fine theorems from the theory of numbers whose *meaning* anyone can understand. For example, there is what is called 'the fundamental theorem of arithmetic', that any integer can be resolved, *in one way only*, into a product of primes. Thus $666 = 2.3.3.37$, and there is no other decomposition; it is impossible that $666 = 2.11.29$ or that $13.89 = 17.73$ (and we can see so without working out the products). This theorem is, as its name implies, the foundation of higher arithmetic; but the proof, although not 'difficult', requires a certain amount of preface and might be found tedious by an unmathematical reader.

Another famous and beautiful theorem is Fermat's 'two square' theorem. The primes may (if we ignore the special prime 2) be arranged in two classes; the primes

$$5, 13, 17, 29, 37, 41, \ldots$$

which leave remainder 1 when divided by 4, and the primes

$$3, 7, 11, 19, 23, 31, \ldots$$

which leave remainder 3. All the primes of the first class, and none of the second, can be expressed as the sum of two integral squares: thus

$$5 = 1^2 + 2^2, 13 = 2^2 + 3^2,$$
$$17 = 1^2 + 4^2, 29 = 2^2 + 5^2;$$

but 3, 7, 11, and 19 are not expressible in this way (as the reader may check by trial). This is Fermat's theorem, which is ranked, very justly, as one of the finest of arithmetic. Unfortunately there is no proof within the comprehension of anybody but a fairly expert mathematician.

There are also beautiful theorems in the 'theory of aggregates' (*Mengenlehre*), such as Cantor's theorem of the 'non-enumerability' of the continuum. Here there is just the opposite difficulty. The proof is easy enough, when once the language has been mastered, but considerable explanation is necessary before the *meaning* of the theorem becomes clear. So I will not try to give more examples. Those which I have given are test cases, and a reader who cannot appreciate them is unlikely to appreciate anything in mathematics.

I said that a mathematician was a maker of patterns of ideas, and that beauty and seriousness were the criteria by which his patterns should be judged. I can hardly believe that anyone who has understood the two theorems will dispute that they pass these tests.

1. *Elements* IX 20. The real origin of many theorems in the *Elements* is obscure, but there seems to be no particular reason for supposing that this one is not Euclid's own.
2. There are technical reasons for not counting 1 as a prime.
3. The proof can be arranged so as to avoid a *reductio*, and logicians of some schools would prefer that it should be.
4. The proof traditionally ascribed to Pythagoras, and certainly a product of his school. The theorem occurs, in a much more general form, in Euclid (*Elements* X 9).
5. Euclid, *Elements* I 47.

Steven Weinberg

from DREAMS OF A FINAL THEORY

■ Nobel prizewinners sometimes disappoint. Obviously good at research, nevertheless you can't help feeling they ought to sound a bit more intelligent, or wise, witty or well-read. There are Nobelists who use the platform of the honour to promote bonkers ideas on psychic paranormalism. But there are no such anxieties about Steven Weinberg. It would be putting it mildly to say that he triumphantly lives up to what it says on the Nobel tin: a true intellectual as well as a brilliant theoretical physicist. His most famous book is probably *The First Three Minutes*. The following is from a later book, *Dreams of a Final Theory*. Beauty is important to physicists, and here we have a top physicist, at the height of his powers, using beautiful language to tell us why. ■

In 1974 Paul Dirac came to Harvard to speak about his historic work as one of the founders of modern quantum electrodynamics. Toward the end of his talk he addressed himself to our graduate students and advised them to be concerned only with the beauty of their equations, not with what the equations mean. It was not good advice for students, but the search for beauty in physics was a theme that ran throughout Dirac's work and indeed through much of the history of physics.

Some of the talk about the importance of beauty in science has been little more than gushing. I do not propose to use this chapter just to say more nice things about beauty. Rather, I want to focus more closely on the nature of beauty in physical theories, on why our sense of beauty is sometimes a useful guide and sometimes not, and on how the usefulness of our sense of beauty is a sign of our progress toward a final theory.

A physicist who says that a theory is beautiful does not mean quite the same thing that would be meant in saying that a particular painting or a piece of music or poetry is beautiful. It is not merely a personal expression of aesthetic pleasure; it is much closer to what a horse trainer means when she looks at a racehorse and says that it is a beautiful horse.

The horse trainer is of course expressing a personal opinion, but it is an opinion about an objective fact: that, on the basis of judgements that the trainer could not easily put into words, this is the kind of horse that wins races.

Of course, different horse trainers may judge a horse differently. That is what makes horse racing. But the horse trainer's aesthetic sense is a means to an objective end—the end of selecting horses that win races. The physicist's sense of beauty is also supposed to serve a purpose—it is supposed to help the physicist select ideas that help us to explain nature. Physicists, just as horse trainers, may be right or wrong in their judgements, but they are not merely enjoying themselves. They often *are* enjoying themselves, but that is not the whole purpose of their aesthetic judgements.

This comparison raises more questions than it answers. First, what *is* a beautiful theory? What are the characteristics of physical theories that give us a sense of beauty? A more difficult question: why does the physicist's sense of beauty work, when it does work? The stories [previously] told illustrated the rather spooky fact that something as personal and subjective as our sense of beauty helps us not only to invent physical theories but even to judge the validity of theories. Why are we blessed with such aesthetic insight? The effort to answer this question raises another question that is even more difficult, although perhaps it sounds trivial: what is it that the physicist wants to accomplish?

What is a beautiful theory? The curator of a large American art museum once became indignant at my use of the word 'beauty' in connection with physics. He said that in his line of work professionals have stopped using this word because they realize how impossible it is to define. Long ago the physicist and mathematician Henri Poincaré admitted that 'it may be very hard to define mathematical beauty, but that is just as true of beauty of all kinds'.

I will not try to define beauty, any more than I would try to define love or fear. You do not define these things; you know them when you feel them. Later, after the fact, you may sometimes be able to say a little to describe them, as I will try to do here.

By the beauty of a physical theory, I certainly do not mean merely the mechanical beauty of its symbols on the printed page. The metaphysical poet Thomas Traherne took pains that his poems should make pretty patterns on the page, but this is no part of the business of physics. I should also distinguish the sort of beauty I am talking about here from the quality that mathematicians and physicists sometimes call elegance. An elegant proof or calculation is one that achieves a powerful result with a minimum of irrelevant complication. It is not important for the beauty of a theory that its equations should have elegant solutions. The equations of general relativity are notoriously difficult to solve except in the simplest situations, but this does not detract from the beauty of the theory itself. As the physicist Leo Szilard (who invented the neutron chain reaction) used to say, 'Elegance is for tailors'.

Simplicity is part of what I mean by beauty, but it is a simplicity of ideas, not simplicity of a mechanical sort that can be measured by counting equations or symbols. Both Einstein's and Newton's theories of gravitation involve equations that tell us the gravitational forces produced by any given amount of matter. In Newton's theory there are three of these equations (corresponding to the three dimensions of space)—in Einstein's theory there are fourteen. In itself, this cannot be counted as an aesthetic advantage of Newton's theory over Einstein's. And in fact it is Einstein's theory that is more beautiful, in part because of the simplicity of his central idea about the equivalence of gravitation and inertia. That is a judgement on which scientists have generally agreed, and as we have seen it was largely responsible for the early acceptance of Einstein's theory.

There is another quality besides simplicity that can make a physical theory beautiful—it is the sense of inevitability that the theory may give us. In listening to a piece of music or hearing a sonnet one sometimes feels an intense aesthetic pleasure at the sense that nothing in the work could be changed, that there is not one note or one word that you would want to have different. In Raphael's *Holy Family* the placement of every figure on the canvas is perfect. This may not be of all paintings in the world your favorite, but, as you look at that painting, there is nothing that you would want Raphael to have done differently. The same is partly true (it is never more than partly true) of general relativity. Once you know the general physical principles adopted by Einstein, you understand that there is no other significantly different theory of gravitation to which Einstein could have been led. As Einstein said of general relativity, 'The chief attraction of the theory lies in its logical completeness. If a single one of the conclusions drawn from it proves wrong, it must be given up; to modify it without destroying the whole structure seems to be impossible.'

This is less true of Newton's theory of gravitation. Newton could have supposed that the gravitational force decreases with the inverse cube of distance rather than the inverse square if that is what the astronomical data had demanded, but Einstein could not have incorporated an inverse-cube law in his theory without scrapping its conceptual basis. Thus Einstein's fourteen equations have an inevitability and hence beauty that Newton's three equations lack. I think that this is what Einstein meant when he referred to the side of the equations that involve the gravitational field in his general theory of relativity as beautiful, as if made of marble, in contrast with the other side of the equations, referring to matter, which he said were still ugly, as if made of mere wood. The way that the gravitational field enters Einstein's equations is almost inevitable, but nothing in general relativity explained why matter takes the form it does.

The same sense of inevitability can be found (again, only in part) in our modern standard model of the strong and electroweak forces that act on elementary particles. There is one common feature that gives both general relativity and the standard model most of their sense of inevitability and simplicity: they obey *principles of symmetry*.

A symmetry principle is simply a statement that something looks the same from certain different points of view. Of all such symmetries, the simplest is the approximate bilateral symmetry of the human face. Because there is little difference between the two sides of your face, it looks the same whether viewed directly or when left and right are reversed, as when you look in a mirror. It is almost a cliché of filmmaking to let the audience realize suddenly that the actor's face they have been watching has been seen in a mirror; the surprise would be spoiled if people had two eyes on the same side of the face like flounders, and always on the same side.

Some things have more extensive symmetries than the human face. A cube looks the same when viewed from six different directions, all at right angles to each other, as well as when left and right are reversed. Perfect crystals look the same not only when viewed from various different directions but also when we shift our positions within the crystal by certain amounts in various directions. A sphere looks the same from any direction. Empty space looks the same from all directions and all positions.

Symmetries like these have amused and intrigued artists and scientists for millennia but did not really play a central role in science. We know many things about salt, and the fact that it is a cubic crystal and therefore looks the same from six different points of view does not rank among the most important. Certainly bilateral symmetry is not the most interesting thing about a human face. The symmetries that are really important in nature are not the symmetries of *things*, but the symmetries of *laws*.

A symmetry of the laws of nature is a statement that when we make certain changes in the point of view from which we observe natural phenomena, the laws of nature we discover do not change. Such symmetries are often called principles of *invariance*. For instance, the laws of nature that we discover take the same form however our laboratories are oriented; it makes no difference whether we measure directions relative to north or northeast or upward or any other direction. This was not so obvious to ancient or medieval natural philosophers; in everyday life there certainly seems to be a difference between up and down and horizontal directions. Only with the birth of modern science in the seventeenth century did it become clear that down seems different from up or north only because below us there happens to be a large mass, the earth, and not (as Aristotle thought) because the natural place of heavy or light things is downward or upward. Note that this symmetry does not say that up is the same as down; observers who measure distances upward or downward from the earth's surface report different descriptions of events such as the fall of an apple, but they discover the same laws, such as the law that apples are attracted by large masses like the earth.

The laws of nature also take the same form wherever our laboratories are located; it makes no difference to our results whether we do our experiments in Texas or Switzerland or on some planet on the other side of the galaxy. The laws of nature take the same form however we set our clocks; it makes no difference whether we date events from the Hegira or the birth of Christ or the beginning of the universe. This does not mean that nothing changes with time or that Texas is just the same as Switzerland, only that the laws discovered at different times and in different places are the same. If it were not for these symmetries the work of science would have to be redone in every new laboratory and in every passing moment.

Any symmetry principle is at the same time a principle of simplicity. If the laws of nature did distinguish among directions like up or down or north, then we would have to put something into our equations to keep track of the orientation of our laboratories, and they would be correspondingly less simple. Indeed, the very notation that is used by mathematicians and physicists to make our equations look as simple and compact as possible has built into it an assumption that all directions in space are equivalent.

Lee Smolin

from THE LIFE OF THE COSMOS

■ Another Nobel-prizewinning physicist (no disappointment, he) Murray Gell-Mann, once said of Lee Smolin: 'Smolin? Is he that young guy with the crazy ideas? He might not be wrong.' The ideas in question might seem crazy to a physicist but to a biologist they have the ring of warm familiarity. Smolin's solution to the Goldilocks Problem (why is the universe so favourable to life?) is gloriously Darwinian. Universes give birth to baby universes in black holes, and the daughter universes inherit the fundamental constants and laws of the parental physics. In the birth process, mutations occur, giving rise to a heterogeneous population of universes. Lineages of varying universes are subject to a kind of natural selection in favour of whatever traits assist in survival (some universes last longer than others, giving more time to reproduce) and reproduction (some universes are more likely to produce black holes than others). These traits happen to be the self-same ones as lead to universes friendly to biology (long life, whatever it takes to make stars and hence chemistry and hence biology). So the population of universes evolves to become Goldilocked into a biological future. The theory is laid out in the early part of *The Life of the Cosmos*, which I recommend. In a later chapter of the book, Smolin reflects on the same issues of beauty in physics and mathematics as were exercising the other authors in this section, and it is this that I reprint here. As it happens, it is an uncommonly beautiful and deep piece of writing, as Smolin balances rival ideas of why mathematics turns out to be so effective in describing the real world. Is it, as Smolin suspects, a consequence of statistical averaging over large numbers of small events? Or does the world we see reflect Platonic ideals of perfect mathematical form, as our next writer, Roger Penrose might prefer to think. ■

The Flower and the Dodecahedron

From Pythagoras to string theory, the desire to comprehend nature has been framed by the Platonic ideal that the world is a reflection of some perfect mathematical form. The power of this dream is undeniable, as we can see from the achievements it inspired, from Kepler's laws to Einstein's equations. Their example suggests that the goal of theoretical physics and cosmology should be the discovery of some beautiful mathematical structure that will underlie reality.

The proposals I have been discussing here, such as cosmological natural selection or the idea that processes of self-organization may account for the organization of the universe, go against this expectation. To explore these ideas means to give up, to some extent, the Platonic model of physical theory in favor of a conception in which the explanation for the world rests on the same kind of historical and statistical methodologies that underlie our understanding of biology. For this reason, if we are to take these kinds of ideas seriously we must examine the role that mathematics has come to play in our expectations of what a physical theory should be.

It is mathematics, more than anything else, that is responsible for the obscurity that surrounds the creative processes of theoretical physics. Perhaps the strangest moment in the life of a theoretical physicist is that in which one realizes, all of a sudden, that one's life is being spent in pursuit of a kind of mystical experience that few of one's fellow humans share. I'm sure that most scientists of all kinds are inspired by a kind of worship of nature, but what makes theoretical physicists peculiar is that our sense of connection with nature has nothing to do with any direct encounter with it. Unlike biologists or experimental physicists, what we confront in our daily work is not usually any concrete phenomena. Most of the time we wrestle not with reality but with mathematical representations of it.

Artists are aware that the highest beauty they can achieve comes not from reproducing nature, but from representing it. Theoretical physicists and mathematicians, more than other kinds of scientist, share this essentially aesthetic mode of working, for like artists we fashion constructions that, when they succeed, capture something about the real world, while at the same time remaining completely products of human imagination. But perhaps even artists do not get to share with us the expectation that our greatest creations may capture the deep and permanent reality behind mere transient experience.

This mysticism of the mathematical, the belief that at its deepest level reality may be captured by an equation or a geometrical construction, is the private religion of the theoretical physicist. Like other true mysticisms, it is something that cannot be communicated in words, but must be experienced. One must feel wordlessly the possibility that a piece of mathematics that one comprehends could also be the world.

I strongly suspect that this joy of seeing in one's mind a correspondence between a mathematical construction and something in nature has been experienced by most working physicists and mathematicians. The mathematics involved does not even have to be very complex; one can have this experience by comprehending a proof of the Pythagorean theorem and realizing at the same time that it must be true of every one of the right triangles that exist in the world. Or there can be a moment of clarity in which one really comprehends Newton's laws, and realizes simultaneously that what one has just grasped mentally is a logic that is realized in each of the countless things that move in the world. One feels at these moments a sense of joy and also—it must be said—of power, to have comprehended simultaneously a logical structure, constructed by the imagination, and an aspect of reality.

Because of this an education in physics or mathematics is a little like an induction into a mystical order. One may be fooled because there is no ceremony or liturgy, but this is just a sign that what we have here is a true mysticism. The wonder of the connection between mathematics and the world has sometimes been spoken about. For example, Eugene Wigner, who pioneered the use of the concept of symmetry in quantum theory, wrote about the 'unreasonable effectiveness of mathematics in physics'. But no one ever speaks of the experience of the realization of this connection. I strongly suspect, though, that it is an experience that everyone who becomes a theoretical physicist is struck by, early and often in their studies.

Of course, as one continues in one's studies, one shortly learns that neither Newton's laws nor Euclidean geometry actually do capture the world. But by that time one is hooked, captured by the possibility that a true image of the world could be held in the imagination. Even more, the ambition then rises in our young scientist that he or she may be the one who invents the formula that is the true mirror of the world. After all, given that there is a mathematical construction that is the complete description of reality, sooner or later someone is going to discover it. Why not you or me? And it is the ambition for this, the ultimate moment of comprehension and creativity, even more than the need for the admiration of one's peers, that keeps us fixed on what we write in our notebooks and draw on our blackboards.

Of course, what is both wonderful and terrifying is that there is absolutely no reason that nature at its deepest level must have anything to do with mathematics. Like mathematics itself, the faith in this shared mysticism of the mathematical scientist is an invention of human beings. No matter that one may make all sorts of arguments for it. We especially like to tell each other stories of the times when a beautiful piece of mathematics was first explored simply because it was beautiful, but later was found to represent a real phenomenon. This is certainly the story of non-Euclidean geometry, and it is the story of the triumph of the gauge principle, from its discovery in Maxwell's theory of electrodynamics to its fruition in general relativity and the standard model. But in spite of the obvious effectiveness of mathematics in physics, I have never heard a good a priori argument that the world must be organized according to mathematical principles.

Certainly, if one needs to believe that beyond the appearances of the world there lies a permanent and transcendent reality, there is no better choice than mathematics. No other conception of reality has led to so much success, in practical mastery of the world. And it is the only religion, so far as I know, that no one has ever killed for.

But if we are honest mathematicians, we must also admit that in many cases there is a simple, non-mathematical reason that an aspect of the world follows a mathematical law. Typically, this happens when a system is composed of an enormous number of independent parts, like a rubber band, the air in a room, or an electorate. The force on a rubber band increases proportionately to the distance stretched. But this reflects nothing deep, only that the force we feel is the sum of an enormous number of small forces between the atoms, each of which may react in a complicated, even unpredictable way, to the stretching. Similarly, there is no mystery or symmetry needed to explain why the air is spread uniformly in a room. Each atom moves randomly, it is just the statistics of enormous numbers. Perhaps the greatest nightmare of the Platonist is that, in the end, all of our laws will be like this, so that the root of all the beautiful regularities we have discovered will turn out to be more statistics, beyond which is only randomness or irrationality.

This is perhaps one reason why biology seems puzzling to some physicists. The possibility that the tremendous beauty of the living world might be, in the end, just a matter of randomness, statistics, and frozen accident stands as a genuine threat to the mystical conceit that reality can be captured in a single, beautiful equation. This is why it took me years to become comfortable with the possibility that the explanation for at least part of the laws of physics might be found in this same logic of randomness and frozen accident.

Roger Penrose

from THE EMPEROR'S NEW MIND

■ Sir Roger Penrose belongs to a family of enormous talent, bristling with world-class scientists, mathematicians and chess masters. He himself is probably the most distinguished of them all, one of the world's best-known mathematical physicists. His book, *The Emperor's New Mind* proposes a controversial, not to say eccentric, theory of consciousness (maybe eccentric is just what the hard problem of consciousness needs), but this is preceded by a crash course in physics, imparted in Penrose's original and thoughtful style. The extract that follows, from one of these less controversial but still stimulating chapters, ends with Gödel's Theorem, which is the subject of the next reading, from Douglas Hofstadter. ■

The question of mathematical truth is a very old one, dating back to the times of the early Greek philosophers and mathematicians—and, no doubt, earlier. However, some very great clarifications and startling *new* insights have been obtained just over the past hundred years, or so. It is these new developments that we shall try to understand. The issues are quite fundamental, and they touch upon the very question of whether our thinking processes can indeed be entirely algorithmic in nature. It is important for us that we come to terms with them.

In the late nineteenth century, mathematics had made great strides, partly because of the development of more and more powerful methods of mathematical proof. (David Hilbert and Georg Cantor, whom we have encountered before, and the great French mathematician Henri Poincaré, whom we shall encounter later, were three who were in the forefront of these developments.) Accordingly, mathematicians had been gaining confidence in the use of such powerful methods. Many of these methods involved the consideration of sets[1] with infinite numbers of members, and proofs were often successful for the very reason that it was possible to consider such sets as actual 'things'—completed existing wholes, with more than a mere potential existence. Many of these powerful ideas had sprung from Cantor's highly original concept of *infinite numbers*, which he had developed consistently using infinite sets.

However, this confidence was shattered when in 1902 the British logician and philosopher Bertrand Russell produced his now famous paradox (itself anticipated by Cantor, and a direct descendant of Cantor's 'diagonal slash' argument). To understand Russell's argument, we first need some feeling for what is involved in considering sets as completed wholes. We may imagine some set that is characterized in terms of a particular *property*. For example, the set of *red* things is characterized in terms of the property of *redness*: something belongs to that set if and only if it has redness. This allows us to turn things about, and talk about a property in terms of a single object, namely the entire set of things with that property. With this viewpoint, 'redness' *is* the set of all red things. (We may also conceive that some other sets are just 'there', their elements being characterized by no such simple property.)

This idea of defining concepts in terms of sets was central to the procedure, introduced in 1884 by the influential German logician Gottlob Frege, whereby *numbers* can be defined in terms of sets. For example, what do we mean by the actual number 3? We know what the property of 'threeness' is, but what is 3 itself? Now, 'threeness' is a property of *collections* of objects, i.e. it is a property of *sets*: a set has this particular property 'threeness' if and only if the set has precisely three members. The set of medal winners in a particular Olympic event has this property of 'threeness', for example. So does the set of tyres on a tricycle, or the set of leaves on a normal clover, or the set of solutions to the equation $x^3 - 6x^2 + 11x - 6 = 0$. What, then, is Frege's definition of the actual number 3? According to Frege, 3 must be a set *of* sets: the set of *all* sets with this property of 'threeness'. Thus a set has three members if and only if it belongs to Frege's set 3.

This may seem a little circular, but it is not, really. We can define *numbers* generally as totalities of equivalent sets, where 'equivalent' here means 'having elements that can be paired off one-to-one with each other' (i.e. in ordinary terms this would be 'having the same number of members'). The number 3 is then the particular one of these sets which has, as one of its members, a set containing, for example, just one apple, one orange, and one pear...There are also other definitions which can be given and which are rather more popular these days.

Now, what about the Russell paradox? It concerns a set R defined in the following way:

R is the set of all sets which are not members of themselves.

Thus, R is a certain collection of sets; and the criterion for a set X to belong to this collection is that the set X is itself not to be found amongst its *own* members.

Is it absurd to suppose that a set might actually be a member of itself? Not really. Consider, for example, the set I of *infinite* sets (sets with infinitely many members). There are certainly infinitely many *different* infinite sets, so I is itself infinite. Thus I indeed belongs to itself! How is it, then, that Russell's conception gives us a paradox? We ask: is Russell's very set R a member of itself or is it not? If it is *not* a member of itself then it should belong to R, since R consists precisely of those sets which are not members of themselves. Thus, R belongs to R after all—a contradiction. On the other hand, if R *is* a member of itself, then since 'itself' is actually R, it belongs to that set whose members are characterized by *not* being members of themselves, i.e. it is not a member of itself after all—again a contradiction![2]

This consideration was not a flippant one. Russell was merely using, in a rather extreme form, the same type of very general mathematical set-theoretic reasoning that the mathematicians were beginning to employ in their proofs. Clearly things had got out of hand, and it became appropriate to be much more precise about what kind of reasoning was to be allowed and what was not. It was obviously necessary that the allowed reasoning must be free from contradiction and that it should permit only true statements to be derived from statements previously known to be true. Russell himself, together with his colleague Alfred North Whitehead, set about developing a highly formalized mathematical system of axioms and rules of procedure, the aim being that it should be possible to translate all types of correct mathematical reasoning into their scheme. The rules were carefully selected so as to prevent the paradoxical types of reasoning that led to Russell's own paradox.

The specific scheme that Russell and Whitehead produced was a monumental piece of work. However, it was very cumbersome, and it turned out to be rather limited in the types of mathematical reasoning that it actually incorporated. The great mathematician David Hilbert, whom we first encountered in Chapter 2, embarked upon a much more workable and comprehensive scheme. *All* correct mathematical types of reasoning, for any particular mathematical area, were to be included. Moreover, Hilbert intended that it would be possible to *prove* that the scheme was free from contradiction. Then mathematics would be placed, once and for all, on an unassailably secure foundation.

However, the hopes of Hilbert and his followers were dashed when, in 1931, the brilliant 25-year-old Austrian mathematical logician Kurt Gödel produced a startling theorem which effectively destroyed the Hilbert programme. What Gödel showed was that any such precise ('formal') mathematical system of axioms and rules of procedure *whatever*, provided that it is broad enough to contain descriptions of simple arithmetical propositions (such as 'Fermat's last theorem') and provided that it is free from contradiction, must contain some statements which are neither provable nor disprovable by the means allowed within the system. The truth of such statements is thus 'undecidable' by the approved procedures. In fact, Gödel was able to show that the very statement of the consistency of the axiom system itself, when coded into the form of a suitable arithmetical proposition, must be one such 'undecidable' proposition.

1. A *set* just means a collection of things—physical objects or mathematical concepts—that can be treated as a whole. In mathematics, the elements (i.e. members) of a set are very often themselves sets, since sets can be collected together to form other sets. Thus one may consider sets of sets, or sets of sets of sets, etc.
2. There is an amusing way of expressing the Russell paradox in essentially commonplace terms. Imagine a library in which there are two catalogues, one of which lists precisely all the books in the library which somewhere refer to themselves and the other, precisely all the books which make no mention of themselves. In which catalogue is the second catalogue itself to be listed?

Douglas Hofstadter

from GÖDEL, ESCHER, BACH: AN ETERNAL GOLDEN BRAID

■ I have never met Douglas Hofstadter, but you only have to read his books to see that he is a one-off among scientists. He has written a book about language (with a title in punning French, which can't have gone down well with the marketing department). He taught himself Russian, apparently for the sole purpose of writing a verse translation of Eugene Onegin. He took over Martin Gardner's famous mathematical column in *Scientific American*. Several of these mathematical essays, collected in *Metamagical Themas*, exhibit his fascination with all things self-referential, and this is a dominant theme of two books on the nature of consciousness: his most recent work, *I am a Strange Loop* and his first book, the extraordinary bestseller *Gödel, Escher, Bach*. It is almost impossible to convey the flavour of this brain-teasing cocktail of art, music, mathematics, and dialogues. Here is one such dialogue, an exchange between Achilles and the Tortoise, which contains a deeply embedded (Hofstadter teases us on many layers) parable on Gödel's Theorem. ■

Contracrostipunctus

Achilles has come to visit his friend and jogging companion, the Tortoise, at his home.

Achilles: Heavens, you certainly have an admirable boomerang collection!

Tortoise: Oh, pshaw. No better than that of any other Tortoise. And now, would you like to step into the parlor?

Achilles: Fine. (*Walks to the corner of the room.*) I see you also have a large collection of records. What sort of music do you enjoy?

Tortoise: Sebastian Bach isn't so bad, in my opinion. But these days, I must say, I am developing more and more of an interest in a rather specialized sort of music.

Achilles: Tell me, what kind of music is that?

Tortoise: A type of music which you are most unlikely to have heard of. I call it 'music to break phonographs by'.

Achilles: Did you say 'to break phonographs by'? That is a curious concept. I can just see you, sledgehammer in hand, whacking one phonograph after another to pieces, to the strains of Beethoven's heroic masterpiece *Wellington's Victory*.

Tortoise: That's not quite what this music is about. However, you might find its true nature just as intriguing. Perhaps I should give you a brief description of it?

Achilles: Exactly what I was thinking.

Tortoise: Relatively few people are acquainted with it. It all began when my friend the Crab—have you met him, by the way?—paid me a visit.

Achilles: 'twould be a pleasure to make his acquaintance, I'm sure. Though I've heard so much about him, I've never met him.

Tortoise: Sooner or later I'll get the two of you together. You'd hit it off splendidly. Perhaps we could meet at random in the park one day…

Achilles: Capital suggestion! I'll be looking forward to it. But you were going to tell me about your weird 'music to smash phonographs by', weren't you?

Tortoise: Oh, yes. Well, you see, the Crab came over to visit one day. You must understand that he's always had a weakness for fancy gadgets, and at that time he was quite an aficionado for, of all things, record players. He had just bought his first record player, and being somewhat gullible, believed every word the salesman had told him about it—in particular, that it was capable of reproducing any and all sounds. In short, he was convinced that it was a Perfect phonograph.

Achilles: Naturally, I suppose you disagreed.

Tortoise: True, but he would hear nothing of my arguments. He staunchly maintained that any sound whatever was reproducible on his machine. Since I couldn't convince him of the contrary, I left it at that. But not long after that, I returned the visit, taking with me a record of a song which I had myself composed. The song was called 'I Cannot Be Played on Record Player 1'.

Achilles: Rather unusual. Was it a present for the Crab?

Tortoise: Absolutely. I suggested that we listen to it on his new phonograph, and he was very glad to oblige me. So he put it on. But unfortunately, after only a few notes, the record player began vibrating rather severely, and then with a loud 'pop', broke into a large number of fairly small pieces, scattered all about the room. The record was utterly destroyed also, needless to say.

Achilles: Calamitous blow for the poor fellow, I'd say. What was the matter with his record player?

Tortoise: Really, there was nothing the matter, nothing at all. It simply couldn't reproduce the sounds on the record which I had brought him, because they were sounds that would make it vibrate and break.

Achilles: Odd, isn't it? I mean, I thought it was a Perfect phonograph. That's what the salesman had told him, after all.

Tortoise: Surely, Achilles, you don't believe everything that salesmen tell you! Are you as naïve as the Crab was?

Achilles: The Crab was naïver by far! I know that salesmen are notorious prevaricators. I wasn't born yesterday!

Tortoise: In that case, maybe you can imagine that this particular salesman had somewhat exaggerated the quality of the Crab's piece of equipment…perhaps it was indeed less than Perfect, and could not reproduce every possible sound.

Achilles: Perhaps that is an explanation. But there's no explanation for the amazing coincidence that your record had those very sounds on it…

Tortoise: Unless they got put there deliberately. You see, before returning the Crab's visit, I went to the store where the Crab had bought his machine, and inquired as to the make. Having ascertained that, I sent off to the manufacturers for a description of its design. After receiving that by return mail, I analyzed the entire construction of the phonograph and discovered a certain set of sounds which, if they were produced anywhere in the vicinity, would set the device to shaking and eventually to falling apart.

Achilles: Nasty fellow! You needn't spell out for me the last details: that you recorded those sounds yourself, and offered the dastardly item as a gift…

Tortoise: Clever devil! You jumped ahead of the story! But that wasn't the end of the adventure, by any means, for the Crab did not believe that his record player was at fault. He was quite stubborn. So he went out and bought a new record player, this one even more expensive, and this time the salesman promised to give him double his money back in case the Crab found a sound which it could not reproduce exactly. So the Crab told me excitedly about his new model, and I promised to come over and see it.

Achilles: Tell me if I'm wrong—I bet that before you did so, you once again wrote the manufacturer, and composed and recorded a new song called 'I Cannot Be Played on Record Player 2', based on the construction of the new model.

Tortoise: Utterly brilliant deduction, Achilles. You've quite got the spirit.

Achilles: So what happened this time?

Tortoise: As you might expect, precisely the same thing. The phonograph fell into innumerable pieces, and the record was shattered.

Achilles: Consequently, the Crab finally became convinced that there can be no such thing as a Perfect record player.

Tortoise: Rather surprisingly, that's not quite what happened. He was sure that the next model up would fill the bill, and having twice the money, he—

Achilles: Oho—I have an idea! He could have easily outwitted you, by obtaining a LOW-fidelity phonograph—one that was not capable of reproducing the sounds which would destroy it. In that way, he would avoid your trick.

Tortoise: Surely, but that would defeat the original purpose—namely, to have a phonograph which could reproduce any sound whatsoever, even its own self-breaking sound, which is of course impossible.

Achilles: That's true. I see the dilemma now. If any record player—say Record Player X—is sufficiently high-fidelity, then when it attempts to play the song 'I Cannot Be Played on Record Player X', it will create just those vibrations which will cause it to break…So it fails to be Perfect. And yet, the only way to get around that trickery, namely for Record Player X to be of lower fidelity, even more directly ensures that it is not Perfect. It seems that every record player is vulnerable to one or the other of these frailties, and hence all record players are defective.

Tortoise: I don't see why you call them 'defective'. It is simply an inherent fact about record players that they can't do all that you might wish them to be able to do. But if there is a defect anywhere, it is not in THEM, but in your expectations of what they should be able to do! And the Crab was just full of such unrealistic expectations.

Achilles: Compassion for the Crab overwhelms me. High fidelity or low fidelity, he loses either way.

Tortoise: And so, our little game went on like this for a few more rounds, and eventually our friend tried to become very smart. He got wind of the principle upon which I was basing my own records, and decided to try to outfox me. He wrote to the phonograph makers, and described a device of his own invention, which they built to specification. He called it 'Record Player Omega'. It was considerably more sophisticated than an ordinary record player.

Achilles: Let me guess how: Did it have no moving parts? Or was it made of cotton? Or—

Tortoise: Let me tell you, instead. That will save some time. In the first place, Record Player Omega incorporated a television camera whose purpose it was to scan any record before playing it. This camera was hooked up to a small built-in computer, which would determine exactly the nature of the sounds, by looking at the groove-patterns.

Achilles: Yes, so far so good. But what could Record Player Omega do with this information?

Tortoise: By elaborate calculations, its little computer figured out what effects the sounds would have upon its phonograph. If it deduced that the sounds were such that they would cause the machine in its present configuration to break, then it did something very clever. Old Omega contained a device which could disassemble large parts of its phonograph subunit, and rebuild them in new ways, so that it could, in effect, change its own structure. If the sounds were 'dangerous', a new configuration was chosen, one to which the sounds would pose no threat, and this new configuration would then be built by the rebuilding subunit, under direction of the little computer. Only after this rebuilding operation would Record Player Omega attempt to play the record.

Achilles: Aha! That must have spelled the end of your tricks. I bet you were a little disappointed.

Tortoise: Curious that you should think so…I don't suppose that you know Gödel's Incompleteness Theorem backwards and forwards, do you?

Achilles: Know WHOSE Theorem backwards and forwards? I've never heard of anything that sounds like that. I'm sure it's fascinating, but I'd rather hear more about 'music to break records by'. It's an amusing little story. Actually, I guess I can fill in the end. Obviously, there was no point in going on, and so you sheepishly admitted defeat, and that was that. Isn't that exactly it?

Tortoise: What! It's almost midnight! I'm afraid it's my bedtime. I'd love to talk some more, but really I am growing quite sleepy.

Achilles: As am I. Well, I'll be on my way. (*As he reaches the door, he suddenly stops, and turns around.*) Oh, how silly of me! I almost forgot, I brought you a little present. Here. (*Hands the Tortoise a small, neatly wrapped package.*)

Tortoise: Really, you shouldn't have! Why, thank you very much indeed. I think I'll open it now. (*Eagerly tears open the package, and inside discovers a glass goblet.*) Oh, what an exquisite goblet! Did you know that I am quite an aficionado for, of all things, glass goblets?

Achilles: Didn't have the foggiest. What an agreeable coincidence!

Tortoise: Say, if you can keep a secret, I'll let you in on something: I'm trying to find a Perfect goblet: one having no defects of any sort in its shape. Wouldn't it be something if this goblet—let's call it 'G'—were the one? Tell me, where did you come across Goblet G?

Achilles: Sorry, but that's MY little secret. But you might like to know who its maker is.

Tortoise: Pray tell, who is it?

Achilles: Ever hear of the famous glassblower Johann Sebastian Bach? Well, he wasn't exactly famous for glassblowing—but he dabbled at the art as a hobby, though hardly a soul knows it—and this goblet is the last piece he blew.

Tortoise: Literally his last one? My gracious. If it truly was made by Bach, its value is inestimable. But how are you sure of its maker?

Achilles: Look at the inscription on the inside—do you see where the letters 'B', 'A', 'C', 'H' have been etched?

Tortoise: Sure enough! What an extraordinary thing. (*Gently sets Goblet G down on a shelf.*) By the way, did you know that each of the four letters in Bach's name is the name of a musical note?

Achilles: 'tisn't possible, is it? After all, musical notes only go from 'A' through 'G'.

Tortoise: Just so; in most countries, that's the case. But in Germany, Bach's own homeland, the convention has always been similar, except that what we call 'B', they call 'H', and what we call 'B-flat', they call 'B'. For instance, we talk about Bach's 'Mass in B Minor', whereas they talk about his 'H-moll Messe'. Is that clear?

Achilles:…hmm…I guess so. It's a little confusing: H is B, and B is B-flat. I suppose his name actually constitutes a melody, then.

Tortoise: Strange but true. In fact, he worked that melody subtly into one of his most elaborate musical pieces—namely, the final *Contrapunctus* in his *Art of the Fugue*. It was the last fugue Bach ever wrote. When I heard it for the first time, I had no idea how it would end. Suddenly, without warning, it broke off. And then…dead silence. I realized immediately that was where Bach died. It is an indescribably sad moment, and the effect it had on me was—shattering. In any case, B-A-C-H is the last theme of that fugue. It is hidden inside the piece. Bach didn't point it out explicitly, but if you know about it, you can find it without much trouble. Ah, me—there are so many clever ways of hiding things in music…

Achilles:…or in poems. Poets used to do very similar things, you know (though it's rather out of style these days). For instance, Lewis Carroll often hid words and names in the first letters (or characters) of the successive lines in poems he wrote. Poems which conceal messages that way are called 'acrostics'.

Tortoise: Bach, too, occasionally wrote acrostics, which isn't surprising. After all, counterpoint and acrostics, with their levels of hidden meaning, have quite a bit in common. Most acrostics, however, have only one hidden level—but there is no reason that one couldn't make a double-decker—an acrostic on top of an acrostic. Or one could make a 'contracrostic'—where the initial letters, taken in reverse order, form a message. Heavens! There's no end to the possibilities inherent in the form. Moreover, it's not limited to poets; anyone could write acrostics—even a dialogician.

Achilles: A dial-a-logician? That's a new one on me.

Tortoise: Correction: I said 'dialogician', by which I meant a writer of dialogues. Hmm…something just occurred to me. In the unlikely event that a dialogician should write a contrapuntal acrostic in homage to J. S. Bach, do you suppose it would be more proper for him to acrostically embed his own name—or that of Bach? Oh, well, why worry about such frivolous matters? Anybody who wanted to write such a piece could make up his own mind. Now getting back to Bach's melodic name, did you know that the melody B-A-C-H, if played upside down and backwards, is exactly the same as the original?

Achilles: How can anything be played upside down? Backwards, I can see—you get H-C-A-B—but upside down? You must be pulling my leg.

Tortoise: 'pon my word, you're quite a sceptic, aren't you? Well, I guess I'll have to give you a demonstration. Let me just go and fetch my fiddle—(*Walks into the next room, and returns in a jiffy with an ancient-looking violin.*)—and play it for you forwards and backwards and every which way. Let's see, now... (*Places his copy of the* Art of the Fugue *on his music stand and opens it to the last page.*) ... here's the last *Contrapunctus*, and here's the last theme...

The Tortoise begins to play: B-A-C-—but as he bows the final H, suddenly, without warning, a shattering sound rudely interrupts his performance. Both he and Achilles spin around, just in time to catch a glimpse of myriad fragments of glass tinkling to the floor from the shelf where Goblet G had stood, only moments before. And then... dead silence.

John Archibald Wheeler with Kenneth Ford

from GEONS, BLACK HOLES, AND QUANTUM FOAM

■   The American physicist John Archibald Wheeler provided one of my favourite quotations about the future of science.

> Surely someday, we can believe, we will grasp the central idea of it all as so simple, so beautiful, so compelling that we will all say to each other, Oh how could it have been otherwise! How could we all have been so blind for so long!

There are other versions of the quotation that are slightly different. Perhaps he said it more than once (it is good enough to bear repetition). In this extract from *Geons, Black Holes and Quantum Foam* (Wheeler coined the name 'Black Hole') he introduces his weird idea of 'It from Bit', where 'Bit' has the information theoretic sense coined by Shannon. I say 'weird', but it is no weirder than much else in modern physics. The weirdness reflects limitations in our evolved minds rather than in reality. The beautiful truth, when it comes, is bound to seem weird to most of us. We are fortunate to have unique thinkers among us like Wheeler, to lead us a little further into the 'Here be Dragons' badlands of modern physics than we might otherwise dare to enter.

Many students of chemistry and physics, entering upon their study of quantum mechanics, are told that quantum mechanics shows its essence in waves, or clouds, of probability. A system such as an atom is described by a wave function. This function satisfies the equation that Erwin Schrödinger published in 1926. The electron, in this description, is no longer a nugget of matter located at a point. It is pictured as a wave spread throughout the volume of the atom (or other region of space).

This picture is all right as far as it goes. It properly emphasizes the central role of probability in quantum mechanics. The wave function tells where the electron might be, not where it is. But, to my mind, the Schrödinger wave fails to capture the true essence of quantum mechanics. That essence, as the delayed-choice experiment shows, is *measurement*. A suitable experiment can, in fact, locate an electron at a particular place within the atom. A different experiment can tell how fast the electron is moving. The wave function is not central to what we actually know about an electron or an atom. It only tells us the likelihood that a particular experiment will yield a particular result. It is the experiment that provides actual information.

Measurement, the act of turning potentiality into actuality, is an act of choice, choice among possible outcomes. After the measurement, there are roads not taken. Before the measurement, all roads are possible—one can even say that all roads are being taken at once.

Thinking about quantum mechanics in this way, I have been led to think of analogies between the way a computer works and the way the universe works. The computer is built on yes–no logic. So, perhaps, is the universe. Did an electron pass through slit A or did it not? Did it cause counter B to click or counter C to click? These are the iron posts of observation.

Yet one enormous difference separates the computer and the universe—chance. In principle, the output of a computer is precisely determined by the input (remember the programmer's famous admonition: garbage in, garbage out). Chance plays no role. In the universe, by contrast, chance plays a dominant role. The laws of physics tell us only what *may* happen. Actual measurement tells us what *is* happening (or what *did* happen). Despite this difference, it is not unreasonable to imagine that information sits at the core of physics, just as it sits at the core of a computer.

Trying to wrap my brain around this idea of information theory as the basis of existence, I came up with the phrase 'it from bit'. The universe and all that it contains ('it') may arise from the myriad yes–no choices of measurement (the 'bits'). Niels Bohr wrestled for most of his life with the question of how acts of measurement (or 'registration') may affect reality. It is registration—whether by a person or a device or a piece of mica (anything that can preserve a record)—that changes potentiality into actuality. I build only a little on the structure of Bohr's thinking when I suggest that we may never understand this strange thing, the quantum, until we understand how information may underlie reality. Information may not be just what we *learn* about the world. It may be what *makes* the world.

An example of the idea of it from bit: when a photon is absorbed, and thereby 'measured'—until its absorption, it had no true reality—an unsplittable bit of information is added to what we know about the world, *and*, at the same time, that bit of information determines the structure of one small part of the world. It *creates* the reality of the time and place of that photon's interaction.

Another example: the surface area of the spherical horizon surrounding a black hole measures the black hole's entropy, and entropy is nothing more than the grand totality of lost information. For a black hole whose horizon spans even a few kilometers, the number of bits of lost information is large beyond any normal meaning of large, even beyond anything we call 'astronomical'. Nevertheless, it is not unimaginable.

We have an *it* (the area of the black hole's horizon) fixed by the number of *bits* of information shielded by that area.

Often quoted is the saying attributed to the architect Ludwig Mies van der Rohe, 'Less is more'. It is a good principle of design, even a good principle of physics research. In thinking about the world in the large, I have another phrase that I like, borrowed from my Princeton colleague Philip Anderson: 'More is different.' When you put enough elementary units together, you get something that is more than the sum of these units. A substance made of a great number of molecules, for instance, has properties such as pressure and temperature that no one molecule possesses. It may be a solid or a liquid or a gas, although no single molecule is solid or liquid or gas.

'More is different' may have something to do with 'it from bit'. The rich complexity of the universe as a whole does not in any way preclude an extremely simple element such as a bit of information from being what the universe is made of. When enough simple elements are stirred together, there is no limit to what can result.

David Deutsch

from THE FABRIC OF REALITY

■ David Deutsch is another one-off. A deep-thinking theoretical physicist, he is today's leading proponent of the (again weird, but possibly slightly less so than the competition) 'Many Worlds' interpretation of quantum theory, and pioneer of the futuristic idea of the quantum computer. His book, *The Fabric of Reality* is a remarkable amalgam of philosophy, quantum physics, evolutionary biology, and highly intelligent lateral thinking. In this brief extract from the book, Deutsch concludes a profound discussion of computation and virtual reality with a meditation on the virtual reality that is mathematics, and all that we think, imagine, and experience. ■

It is not customary to think of mathematics as being a form of virtual reality. We usually think of mathematics as being about abstract entities, such as numbers and sets, which do not affect the senses; and it might therefore seem that there can be no question of artificially rendering their effect on us. However, although mathematical entities do not affect the senses, the experience of doing mathematics is an external experience, no less than the experience of doing physics is. We make marks on pieces of paper and look at them, or we imagine looking at such marks—indeed, we cannot do mathematics without imagining abstract mathematical entities. But this means imagining an environment whose 'physics' embodies the complex and autonomous properties of those entities. For example, when we imagine the abstract concept of a line segment which has no thickness, we may imagine a line that is visible but imperceptibly wide. That much may, just about, be arranged in physical reality. But mathematically the line must continue to have no thickness when we view it under arbitrarily powerful magnification. That is not a property of any physical line, but it can easily be achieved in the virtual reality of our imagination.

Imagination is a straightforward form of virtual reality. What may not be so obvious is that our 'direct' experience of the world through our senses is virtual reality too. For our external experience is never direct; nor do we even experience the signals in our nerves directly—we would not know what to make of the streams of electrical crackles that they carry. What we experience directly is a virtual-reality rendering, conveniently generated for us by our unconscious minds from sensory data plus complex inborn and acquired theories (i.e. programs) about how to interpret them.

We realists take the view that reality is out there: objective, physical and independent of what we believe about it. But we never experience that reality directly. Every last scrap of our external experience is of virtual reality. And every last scrap of our knowledge—including our knowledge of the non-physical worlds of logic, mathematics and philosophy, and of imagination, fiction, art and fantasy—is encoded in the form of programs for the rendering of those worlds on our brain's own virtual-reality generator.

So it is not just science—reasoning about the physical world—that involves virtual reality. All reasoning, all thinking and all external experience are forms of virtual reality. These things are physical processes which so far have been observed in only one place in the universe, namely the vicinity of the planet Earth. [All] living processes involve virtual reality too, but human beings in particular have a special relationship with it. Biologically speaking, the virtual-reality rendering of their environment is the characteristic means by which human beings survive. In other words, it is the reason why human beings exist. The ecological niche that human beings occupy depends on virtual reality as directly and as absolutely as the ecological niche that koala bears occupy depends on eucalyptus leaves.

Primo Levi

from THE PERIODIC TABLE

Primo Levi is, of course, a major figure in world literature, which is why I broke my rule against translations for him. He also happened to be a chemist. *The Periodic Table* is a unique combination of autobiography and chemistry. Each chapter is headed by the name of an element (I have already made the comparison with the chromosomes by which Matt Ridley labelled his chapters, and with my own 'Tales'). Carbon, the chapter from which this extract is taken, also happens to be biographical in another sense. It is the biography of a particular carbon atom and it beautifully and succinctly conveys a rich array of scientific facts and wisdom. For example, we get a vivid idea of the superhuman timescale on which an atom plays out its existence; of the great underground reservoir of limestone (or chalk or coal) into which carbon atoms are locked up for aeons, punctuated by brief excursions into the atmosphere and into the living economy via plants and photosynthesis.

Our character lies for hundreds of millions of years, bound to three atoms of oxygen and one of calcium, in the form of limestone: it already has a very long cosmic history behind it, but we shall ignore it. For it time does not exist, or exists only in the form of sluggish variations in temperature, daily or seasonal, if, for the good fortune of this tale, its position is not too far from the earth's surface. Its existence, whose monotony cannot be thought of without horror, is a pitiless alternation of hots and colds, that is, of oscillations (always of equal frequency) a trifle more restricted and a trifle more ample: an imprisonment, for this potentially living personage, worthy of the Catholic Hell. To it, until this moment, the present tense is suited, which is that of description, rather than the past tense, which is that of narration—it is congealed in an eternal present, barely scratched by the moderate quivers of thermal agitation.

But, precisely for the good fortune of the narrator, whose story could otherwise have come to an end, the limestone rock ledge of which the atom forms a part lies on the surface. It lies within reach of man and his pickax (all honor to the pickax and its modern equivalents; they are still the most important intermediaries in the millennial dialogue between the elements and man): at any moment—which I, the narrator, decide out of pure caprice to be the year 1840—a blow of the pickax detached it and sent it on its way to the lime kiln, plunging it into the world of things that change. It was roasted until it separated from the calcium, which remained so to speak with its feet on the ground and went to meet a less brilliant destiny, which we shall not narrate. Still firmly clinging to two of its three former oxygen companions, it issued from the chimney and took the path of the air. Its story, which once was immobile, now turned tumultuous.

It was caught by the wind, flung down on the earth, lifted ten kilometers high. It was breathed in by a falcon, descending into its precipitous lungs, but did not penetrate its rich blood and was expelled. It dissolved three times in the water of the sea, once in the water of a cascading torrent, and again was expelled. It traveled with the wind for eight years: now high, now low, on the sea and among the clouds, over forests, deserts, and limitless expanses of ice; then it stumbled into capture and the organic adventure.

Carbon, in fact, is a singular element: it is the only element that can bind itself in long stable chains without a great expense of energy, and for life on earth (the only one we know so far) precisely long chains are required. Therefore carbon is the key element of living substance: but its promotion, its entry into the living world, is not easy and must follow an obligatory, intricate path, which has been clarified (and not yet definitively) only in recent years. If the elaboration of carbon were not a common daily occurrence, on the scale of billions of tons a week, wherever the green of a leaf appears, it would by full right deserve to be called a miracle.

The atom we are speaking of, accompanied by its two satellites which maintained it in a gaseous state, was therefore borne by the wind along a row of vines in the year 1848. It had the good fortune to brush against a leaf, penetrate it, and be nailed there by a ray of the sun. If my language here becomes imprecise and allusive, it is not only because of my ignorance: this decisive event, this instantaneous work *a tre*—of the carbon dioxide, the light, and the vegetal greenery—has not yet been described in definitive terms, and perhaps it will not be for a long time to come, so different is it from that other 'organic' chemistry which is the cumbersome, slow, and ponderous work of man: and yet this refined, minute, and quick-witted chemistry was 'invented' two or three billion years ago by our silent sisters, the plants, which do not experiment and do not discuss, and whose temperature is identical to that of the environment in which they live. If to comprehend is the same as forming an image, we will never form an image of a happening whose scale is a millionth of a millimeter, whose rhythm is a millionth of a second, and whose protagonists are in their essence invisible. Every verbal description must be inadequate, and one will be as good as the next, so let us settle for the following description.

Our atom of carbon enters the leaf, colliding with other innumerable (but here useless) molecules of nitrogen and oxygen. It adheres to a large and complicated molecule that activates it, and simultaneously receives the decisive message from the sky, in the flashing form of a packet of solar light: in an instant, like an insect caught by a spider, it is separated from its oxygen, combined with hydrogen and (one thinks) phosphorus, and finally inserted in a chain, whether long or short does not matter, but it is the chain of life. All this happens swiftly, in silence, at the temperature and pressure of the atmosphere, and gratis: dear colleagues, when we learn to do likewise we will be *sicut Deus*, and we will have also solved the problem of hunger in the world.

But there is more and worse, to our shame and that of our art. Carbon dioxide, that is, the aerial form of the carbon of which we have up till now spoken: this gas which constitutes the raw material of life, the permanent store upon which all that grows draws, and the ultimate destiny of all flesh, is not one of the principal components of air but rather a ridiculous remnant, an 'impurity,' thirty times less abundant than argon, which nobody even notices. The air contains 0.03 per cent; if Italy was air, the only Italians fit to build life would be, for example, the fifteen thousand inhabitants of Milazzo in the province of Messina. This, on the human scale, is ironic acrobatics, a juggler's trick, an incomprehensible display of omnipotence-arrogance, since from this ever renewed impurity of the air we come, we animals and we plants, and we the human species, with our four billion discordant opinions, our millenniums of history, our wars and shames, nobility and pride. In any event, our very presence on the planet becomes laughable in geometric terms: if all of humanity, about 250 million tons, were distributed in a layer of homogeneous thickness on all the emergent lands, the 'stature of man' would not be visible to the naked eye; the thickness one would obtain would be around sixteen thousandths of a millimeter.

Now our atom is inserted: it is part of a structure, in an architectural sense; it has become related and tied to five companions so identical with it that only the fiction of the story permits me to distinguish them. It is a beautiful ring-shaped structure, an almost regular hexagon, which however is subjected to complicated exchanges and balances with the water in which it is dissolved; because by now it is dissolved in water, indeed in the sap of the vine, and this, to remain dissolved, is both the obligation and the privilege of all substances that are destined (I was about to say 'wish') to change. And if then anyone really wanted to find out why a ring, and why a hexagon, and why soluble in water, well, he need not worry: these are among the not many questions to which our doctrine can reply with a persuasive discourse, accessible to everyone, but out of place here.

It has entered to form part of a molecule of glucose, just to speak plainly: a fate that is neither fish, flesh, nor fowl, which is intermediary, which prepares it for its first contact with the animal world but does not authorize it to take on a higher responsibility: that of becoming part of a proteic edifice. Hence it travels, at the slow pace of vegetal juices, from the leaf through the pedicel and by the shoot to the trunk, and from here descends to the almost ripe bunch of grapes. What then follows is the province of the winemakers: we are only interested in pinpointing the fact that it escaped (to our advantage, since we would not know how to put it in words) the alcoholic fermentation, and reached the wine without changing its nature.

It is the destiny of wine to be drunk, and it is the destiny of glucose to be oxidized. But it was not oxidized immediately: its drinker kept it in his liver for more than a week, well curled up and tranquil, as a reserve aliment for a sudden effort; an effort that he was forced to make the following Sunday, pursuing a bolting horse. Farewell to the hexagonal structure: in the space of a few instants the skein was unwound and became glucose again, and this was dragged by the bloodstream all the way to a minute muscle fiber in the thigh, and here brutally split into two molecules of lactic acid, the grim harbinger of fatigue: only later, some minutes after, the panting of the lungs was able to supply the oxygen necessary to quietly oxidize the latter. So a new molecule of carbon dioxide returned to the atmosphere, and a parcel of the energy that the sun had handed to the vine-shoot passed from the state of chemical energy to that of mechanical energy, and thereafter settled down in the slothful condition of heat, warming up imperceptibly the air moved by the running and the blood of the runner. 'Such is life', although rarely is it described in this manner: an inserting itself, a drawing off to its advantage, a parasitizing of the downward course of energy, from its noble solar form to the degraded one of low-temperature heat. In this downward course, which leads to equilibrium and thus death, life draws a bend and nests in it.

Our atom is again carbon dioxide, for which we apologize: this too is an obligatory passage; one can imagine and invent others, but on earth that's the way it is. Once again the wind, which this time travels far; sails over the Apennines and the Adriatic, Greece, the Aegean, and

Cyprus: we are over Lebanon, and the dance is repeated. The atom we are concerned with is now trapped in a structure that promises to last for a long time: it is the venerable trunk of a cedar, one of the last; it is passed again through the stages we have already described, and the glucose of which it is a part belongs, like the bead of a rosary, to a long chain of cellulose. This is no longer the hallucinatory and geological fixity of rock, this is no longer millions of years, but we can easily speak of centuries because the cedar is a tree of great longevity. It is our whim to abandon it for a year or five hundred years: let us say that after twenty years (we are in 1868) a wood worm has taken an interest in it. It has dug its tunnel between the trunk and the bark, with the obstinate and blind voracity of its race; as it drills it grows, and its tunnel grows with it. There it has swallowed and provided a setting for the subject of this story; then it has formed a pupa, and in the spring it has come out in the shape of an ugly gray moth which is now drying in the sun, confused and dazzled by the splendor of the day. Our atom is in one of the insect's thousand eyes, contributing to the summary and crude vision with which it orients itself in space. The insect is fecundated, lays its eggs, and dies: the small cadaver lies in the undergrowth of the woods, it is emptied of its fluids, but the chitin carapace resists for a long time, almost indestructible. The snow and sun return above it without injuring it: it is buried by the dead leaves and the loam, it has become a slough, a 'thing', but the death of atoms, unlike ours, is never irrevocable. Here are at work the omnipresent, untiring, and invisible gravediggers of the undergrowth, the microorganisms of the humus. The carapace, with its eyes by now blind, has slowly disintegrated, and the ex-drinker, ex-cedar, ex-wood worm has once again taken wing.

We will let it fly three times around the world, until 1960, and in justification of so long an interval in respect to the human measure we will point out that it is, however, much shorter than the average: which, we understand, is two hundred years. Every two hundred years, every atom of carbon that is not congealed in materials by now stable (such as, precisely, limestone, or coal, or diamond, or certain plastics) enters and re-enters the cycle of life, through the narrow door of photosynthesis. Do other doors exist? Yes, some syntheses created by man; they are a title of nobility for man-the-maker, but until now their quantitative importance is negligible. They are doors still much narrower than that of the vegetal greenery; knowingly or not, man has not tried until now to compete with nature on this terrain, that is, he has not striven to draw from the carbon dioxide in the air the carbon that is necessary to nourish him, clothe him, warm him, and for the hundred other more sophisticated needs of modern life. He has not done it because he has not needed to: he has found, and is still finding (but for how many more decades?) gigantic reserves of carbon already organicized, or at least reduced. Besides the vegetable and animal worlds, these reserves are constituted by deposits of coal and petroleum: but these too are the inheritance of photosynthetic activity carried out in distant epochs, so that one can well affirm that photosynthesis is not only the sole path by which carbon becomes living matter, but also the sole path by which the sun's energy becomes chemically usable.

It is possible to demonstrate that this completely arbitrary story is nevertheless true. I could tell innumerable other stories, and they would all be true: all literally true, in the nature of the transitions, in their order and data. The number of atoms is so great that one could always be found whose story coincides with any capriciously invented story. I could recount an endless number of stories about carbon atoms that become colors or perfumes in flowers; of others which, from tiny algae to small crustaceans to fish, gradually return as carbon dioxide to the waters of the sea, in a perpetual, frightening round-dance of life and death, in which every devourer is immediately devoured; of others which instead attain a decorous semi-eternity in the yellowed pages of some archival document, or the canvas of a famous painter; or those to which fell the privilege of forming part of a grain of pollen and left their fossil imprint in the rocks for our curiosity; of others still that descended to become part of the mysterious shape-messengers of the human seed, and participated in the subtle process of division, duplication, and fusion from which each of us is born. Instead, I will tell just one more story, the most secret, and I will tell it with the humility and restraint of him who knows from the start that his theme is desperate, his means feeble, and the trade of clothing facts in words is bound by its very nature to fail.

It is again among us, in a glass of milk. It is inserted in a very complex, long chain, yet such that almost all of its links are acceptable to the human body. It is swallowed; and since every living structure harbors a savage distrust toward every contribution of any material of living origin, the chain is meticulously broken apart and the fragments, one by one, are accepted or rejected. One, the one that concerns us, crosses the intestinal threshold and enters the bloodstream: it migrates, knocks at the door of a nerve cell, enters, and supplants the carbon which was part of it. This cell belongs to a brain, and it is my brain, the brain of the *me* who is writing; and the cell in question, and within it the atom in question, is in charge of my writing, in a gigantic minuscule game which nobody has yet described. It is that which at this instant, issuing out of a labyrinthine tangle of yeses and nos, makes my hand run along a certain path on the paper, mark it with these volutes that are signs: a double snap, up and down, between two levels of energy, guides this hand of mine to impress on the paper this dot, here, this one.

Richard Fortey

from LIFE: AN UNAUTHORIZED BIOGRAPHY

◼ Levi reminds us of the long history of our planet. As we draw to a close, three pieces now from writers we have already met, reflecting on life's long, slow march through deep time, to the point where evolved brains can contemplate the universe with curiosity and wonder. Richard Fortey, George Gaylord Simpson, and Loren Eiseley are all palaeontologists, and you can't be a good palaeontologist unless, for you, the poetry of earth is never dead. As Loren Eiseley said, in a line of his poetry, which adorns the headstone that he shares with his wife, 'We loved the earth, but could not stay'. ◼

I can imagine standing upon a Cambrian shore in the evening, much as I stood on the shore at Spitsbergen and wondered about the biography of life for the first time. The sea lapping at my feet would look and feel much the same. Where the sea meets the land there is a patch of slightly sticky, rounded stromatolite pillows, survivors from the vast groves of the Precambrian. The wind is whistling across the red plains behind me, where nothing visible lives, and I can feel the sharp sting of wind-blown sand on the back of my legs. But in the muddy sand at my feet I can see worm casts, little curled wiggles that look familiar. I can see trails of dimpled impressions left by the scuttling of crustacean-like animals. On the strand line a whole range of shells glistens—washed up by the last storm, I suppose—some of them mother-of-pearl, others darkly shining, made of calcium phosphate. At the edge of the sea a dead sponge washes back and forth in the waves, tumbling over and over in the foam. There are heaps of seaweed, red and brown, and several stranded jellyfish, one, partly submerged, still feebly pulsing. Apart from the whistle of the breeze and the crash and suck of the breakers, it is completely silent, and nothing cries in the wind. I wade out into a rock pool. In the clear water I can see several creatures which could fit into the palm of my hand crawling or gliding very slowly along the bottom. Some of them carry an armour of plates on their backs. I can recognize a chiton, but others are unfamiliar. In the sand there are shy tube-worms. A trilobite the size of a crab has caught one of them and is shredding it with its limbs. Another one crawls across my foot, and I can feel the tickle of its numerous legs on my bare flesh—but wait, it is not a trilobite, but a different kind of arthropod with eyes on stalks at the front and delicate grasping 'hands'. Now that I look out to sea I can see a swarm of similar arthropods sculling together in the bright surface water—and can that dark shape with glistening eyes be *Anomalocaris* in pursuit? Yes, for the top of its body briefly breaks the surface, and I can glimpse its fierce arms for an instant. Where the water breaks it shines luminously for a while in the dying light—the seawater must be full of light-producing plankton—and I have to imagine millions more microscopic organisms in the shimmering sea.

George Gaylord Simpson

from THE MEANING OF EVOLUTION

In preceding pages evidence was given, thoroughly conclusive evidence, as I believe, that organic evolution is a process entirely materialistic in its origin and operation, although no explicit conclusion was made or considered possible as to the origin of the laws and properties of matter in general under which organic evolution operates. Life is materialistic in nature, but it has properties unique to itself which reside in its organization, not in its materials or mechanics. Man arose as a result of the operation of organic evolution and his being and activities are also materialistic, but the human species has properties unique to itself among all forms of life, superadded to the properties unique to life among all forms of matter and of action. Man's intellectual, social, and spiritual natures are altogether exceptional among animals in degree, but they arose by organic evolution. They usher in a new phase of evolution, and not a new phase merely but also a new kind, which is thus also a product of organic evolution and can be no less materialistic in its essence even though its organization and activities are essentially different from those in the process that brought it into being.

It has also been shown that purpose and plan are not characteristic of organic evolution and are not a key to any of its operations. But purpose and plan are characteristic in the new evolution, because man has purposes and he makes plans. Here purpose and plan do definitely enter into evolution, as a result and not as a cause of the processes seen in the long history of life. The purposes and plans are ours, not those of the universe, which displays convincing evidence of their absence.

Man was certainly not the goal of evolution, which evidently had no goal. He was not planned, in an operation wholly planless. He is not the ultimate in a single constant trend toward higher things, in a history of life with innumerable trends, none of them constant, and some toward the lower rather than the higher. Is his place in nature, then, that of a mere accident, without significance? The affirmative answer that some have felt constrained to give is another example of the 'nothing but' fallacy. The situation is as badly misrepresented and the lesson as poorly learned when man is considered nothing but an accident as when he is considered as the destined crown of creation. His rise was neither insignificant nor inevitable. Man *did* originate after a tremendously long sequence of events in which both chance and orientation played a part. Not all the chance favored his appearance, none *might* have, but enough did. Not all the orientation was in his direction, it did not lead unerringly human-ward, but some of it came this way. The result *is* the most highly endowed organization of matter that has yet appeared on the earth—and we certainly have no good reason to believe there is any higher in the universe. To think that this result is insignificant would be unworthy of that high endowment, which includes among its riches a sense of values.

Loren Eiseley

from 'LITTLE MEN AND FLYING SAUCERS'

Darwin saw clearly that the succession of life on this planet was not a formal pattern imposed from without, or moving exclusively in one direction. Whatever else life might be, it was adjustable and not fixed. It worked its way through difficult environments. It modified and then, if necessary, it modified again, along roads which would never be retraced. Every creature alive is the product of a unique history. The statistical probability of its precise reduplication on another planet is so small as to be meaningless. Life, even cellular life, may exist out yonder in the dark. But high or low in nature, it will not wear the shape of man. That shape is the evolutionary product of a strange, long wandering through the attics of the forest roof, and so great are the chances of failure, that nothing precisely and identically human is likely ever to come that way again.

[...]

In a universe whose size is beyond human imagining, where our world floats like a dust mote in the void of night, men have grown inconceivably lonely. We scan the time scale and the mechanisms of life itself for portents and signs of the invisible. As the only thinking mammals on the planet— perhaps the only thinking animals in the entire sidereal universe—the burden of consciousness has grown heavy upon us. We watch the stars, but the signs are uncertain. We uncover the bones of the past and seek for our origins. There is a path there, but it appears to wander. The vagaries of the road may have a meaning however; it is thus we torture ourselves.

Lights come and go in the night sky. Men, troubled at last by the things they build, may toss in their sleep and dream bad dreams, or lie awake while the meteors whisper greenly overhead. But nowhere in all space or on a thousand worlds will there be men to share our loneliness. There may be wisdom; there may be power; somewhere across space great instruments, handled by strange, manipulative organs, may stare vainly at our floating cloud wrack, their owners yearning as we yearn. Nevertheless, in the nature of life and in the principles of evolution we have had our answer. Of men elsewhere, and beyond, there will be none forever.

Carl Sagan

from PALE BLUE DOT

■ Unlike the palaeontologists, Carl Sagan's poetry of earth came from viewing our planet from the outside, as the pale blue dot that would be the last thing any of us would see of it if we could ever leave our native parish and travel outbound through the eternal cold. Read Sagan's words. Read them again. Read them for that special kind of humility which only science can give, the special kind of humility with which this book began, and which we cannot afford to forget. ■

Look again at that dot. That's here. That's home. That's us. On it everyone you love, everyone you know, everyone you ever heard of, every human being who ever was, lived out their lives. The aggregate of our joy and suffering, thousands of confident religions, ideologies, and economic doctrines, every hunter and forager, every hero and coward, every creator and destroyer of civilization, every king and peasant, every young couple in love, every mother and father, hopeful child, inventor and explorer, every teacher of morals, every corrupt politician, every 'superstar', every 'supreme leader', every saint and sinner in the history of our species lived there—on a mote of dust suspended in a sunbeam.

The Earth is a very small stage in a vast cosmic arena. Think of the rivers of blood spilled by all those generals and emperors so that, in glory and triumph, they could become the momentary masters of a fraction of a dot. Think of the endless cruelties visited by the inhabitants of one corner of this pixel on the scarcely distinguishable inhabitants of some other corner, how frequent their misunderstandings, how eager they are to kill one another, how fervent their hatreds.

Our posturings, our imagined self-importance, the delusion that we have some privileged position in the Universe, are challenged by this point of pale light. Our planet is a lonely speck in the great enveloping cosmic dark. In our obscurity, in all this vastness, there is no hint that help will come from elsewhere to save us from ourselves.

The Earth is the only world known so far to harbor life. There is nowhere else, at least in the near future, to which our species could migrate. Visit, yes. Settle, not yet. Like it or not, for the moment the Earth is where we make our stand.

It has been said that astronomy is a humbling and character-building experience. There is perhaps no better demonstration of the folly of human conceits than this distant image of our tiny world. To me, it underscores our responsibility to deal more kindly with one another, and to preserve and cherish the pale blue dot, the only home we've ever known.

ACKNOWLEDGEMENTS

■ **Peter Atkins:** from *Creation Revisited* (Penguin, 1992), copyright © Peter Atkins 1992, by permission of Penguin Books Ltd. ■ **Per Bak:** from *How Nature Works: The Science of Self-Organized Criticality* (OUP, 1997), by permission of Oxford University Press. ■ **Colin Blakemore:** from *The Mind Machine* (BBC Books, 1988), by permission of the author. ■ **John Tyler Bonner:** from *Life Cycles: Reflections of an Evolutionary Biologist* (Princeton University Press, 1993), copyright © 1993, by permission of Princeton University Press. ■ **Sydney Brenner:** 'Theoretical Biology in the Third Millennium' in *Philosophical Transactions: Biological Sciences*, Vol. 354, No 1392, Millennium Issue (Dec 29, 1999), by permission of the author and The Royal Society. ■ **Jacob Bronowski:** from *The Identity of Man* (Prometheus Books, 2002), copyright © Jacob Bronowski 1965, 1966, by permission of Doubleday, a division of Random House, Inc. ■ **Rachel Carson:** from *The Sea Around Us* (Oxford University Press, 1989), copyright © Rachel Carson 1989, by permission of Pollinger Ltd and the proprietor. ■ **S Chandrasekhar:** from The Nora and Edward Ryerson Lecture in *Truth and Beauty: Aesthetics and Motivations in Science* (University of Chicago Press, 1987), copyright © The University of Chicago 1987, by permission of the publisher. ■ **Francis Crick:** from *Life Itself: Its Origin and Nature* (Simon & Schuster, 1981), copyright © Francis Crick 1981, by permission of Felicity Bryan Literary Agency; from *What Mad Pursuit: A Personal View of Scientific Discovery* (Basic Books, 1988), copyright © Francis Crick 1988, by permission of the publishers, Basic Books, a member of the Perseus Books Group and Weidenfeld and Nicolson, an imprint of The Orion Publishing Group. ■ **Helena Cronin:** from *The Ant and the Peacock* (Cambridge, 1991), copyright © Cambridge University Press 1991, by permission of the author and the publisher. ■ **Paul Davies:** from *The Goldilocks Enigma: Why is the Universe Just Right for Life?* (Penguin, 2006), copyright © Paul Davies 2006, published in the USA as *Cosmic Jackpot: Why Our Universe is Just Right for Life* (Houghton Mifflin, 2007), copyright © Paul Davies 2007, by permission of the publishers, Penguin Books Ltd and Houghton Mifflin Company. All rights reserved. ■ **Daniel C Dennett:** from *Darwin's Dangerous Idea: Evolution and the Meanings of Life* (Penguin Press/Simon & Schuster, 1995), copyright © Daniel C Dennett 1995, by permission of the publishers Penguin Books Ltd and Simon & Schuster Adult Publishing Group; from *Consciousness Explained* (Viking, 1991), copyright © 1991 by Daniel C Dennett, by permission of Little Brown & Company, Hachette Book Group USA. ■ **David Deutsch:** 'Virtual Reality' from *The Fabric of Reality* (Viking 1997/Allen Lane, 1998), copyright © David Deutsch 1997, 1998, by permission of Viking Penguin, a division of Penguin Group (USA) Inc and Penguin Books Ltd. ■ **Jared Diamond:** from *The Rise & Fall of the Third Chimpanzee* (Hutchinson/Vintage, 1991), by permission of the Random House Group Ltd; published in the USA as *The Third Chimpanzee* (HarperCollins, 1992), copyright © Jared Diamond 1992, by permission of HarperCollins Publishers. ■ **Theodosius Dobzhansky:** from *Mankind Evolving* (Yale University Press, 1962), copyright © Yale University Press 1962, by permission of the publisher. ■ **Freeman Dyson:** from *Disturbing the Universe* (Pan, 1979), copyright © Freeman J Dyson 1979, by permission of Basic Books, a member of Perseus Books Group. ■ **Arthur Eddington:** from *The Expanding Universe* (Pelican Books, 1940), by permission of the Master and Fellows of Trinity College Cambridge. ■ **Albert Einstein:** from 'What is the Theory of Relativity' first published in the *London Times*, 1919, and

'Religion and Science' first published in English in the *New York Times Magazine*, 1930, from *The Collected Papers of Albert Einstein* (Princeton University Press), copyright © 1987–2007 Hebrew University and Princeton University Press, by permission of Princeton University Press. ■ **Loren Eiseley:** from *The Immense Journey* (1946, Vintage 1959), copyright © Loren Eiseley 1946, 1950, 1951, 1953, renewed 1956, 1957, by permission of Random House, Inc. ■ **Richard P Feynman:** from *The Character of Physical Law* (BBC, 1965/ Penguin, 1992), copyright © Richard P Feynman 1965, copyright © 1967 Massachusetts Institute of Technology, by permission of the publishers, Penguin Books Ltd and MIT Press. ■ **R A Fisher:** from *The Genetical Theory of Natural Selection* (Clarendon Press, 1930/OUP, 1999), by permission of Oxford University Press. ■ **Richard Fortey:** from *Trilobite! Eyewitness to Evolution* (HarperCollins, 2000), copyright © Richard Fortey 2000; and from *Life: An Unauthorized Biography* (HarperCollins, 1997), copyright © Richard Fortey 1997, both by permission of HarperCollins Publishers Ltd and Alfred A Knopf, a division of Random House Inc. ■ **George Gamow:** from *Mr Tompkins in Paperback* (Cambridge, 2007), copyright © Cambridge University Press 1965, 1993, by permission of the author's Estate and the publisher. ■ **Martin Gardner:** 'Mathematical Games', *Scientific American* 223, October 1970, copyright © Scientific American Inc 1979, by permission of Scientific American Inc. All rights reserved. ■ **Stephen Jay Gould:** 'Worm for a Century, and All Seasons' from *Hen's Teeth & Horse's Toes: Further Reflections in Natural History* (Norton, 1983), copyright © Stephen Jay Gould 1983, by permission of W W Norton & Company, Inc. ■ **Brian Greene:** from *The Elegant Universe: Superstrings, Hidden Dimensions, and the Quest for the Ultimate Theory* (Jonathan Cape, 1999), copyright © Brian R Greene 1999, by permission of the publishers, The Random House Group Ltd and W W Norton & Company, Inc. ■ **Richard Gregory:** from *Mirrors in Mind* (W H Freeman, 1997), copyright © Richard Gregory 1997; copyright holder not traced. ■ **J B S Haldane:** 'On Being the Right Size', first published in *Possible Worlds and Other Essays* (Chatto & Windus, 1927), from *On Being the Right Size and Other Essays* (OUP, 1985); 'Cancer's a Funny Thing', first published in the *New Statesman*, London, 21 February 1964, by permission of the heirs of J B S Haldane. ■ **W D Hamilton:** from 'Geometry for the Selfish Herd' first published in *Journal of Theoretical Biology* 31, 295 (1971), copyright © Elsevier 1971, by permission of Elsevier Ltd; and from *Narrow Roads of Gene Land: The Collected Papers of W D Hamilton* (W H Freeman, 1996) Volume 1 *Evolution of Social Behaviour*, copyright © W D Hamilton 1996; copyright holder not traced. ■ **Garrett Hardin:** 'The Tragedy of the Commons', *Science* 162: 1243 (1968), by permission of American Association for the Advancement of Science (AAAS). ■ **Alister Hardy:** from *The Open Sea: Its Natural History* (Collins, 1956), by permission of the heirs of Sir Alister Hardy. ■ **G H Hardy:** from *A Mathematician's Apology* (Cambridge, 1969), copyright © Cambridge University Press 1969, by permission of the publisher. ■ **Stephen Hawking:** from *A Brief History of Time: From the Big Bang to Black Holes* (Bantam, 1988), by permission of The Random House Group Ltd and the author. ■ **Douglas R Hofstadter:** from *Godel, Escher, Bach: An Eternal Golden Braid* (Penguin, 1980), copyright © Basic Books Inc, 1979, by permission of Basic Books, a member of Perseus Books Group. ■ **Lancelot Hogben:** from *Mathematics for the Millions: How to Master the Magic of Numbers* (4e, W W Norton, 1971), copyright © 1937, 1940, 1943, 1951 by W W Norton & Co, Inc, copyright renewed 1964 by Lancelot Hogben, copyright © 1967, 1968 by Lancelot Hogben, copyright renewed 1971 by Lancelot Hogben, by permission of W W Norton & Company, Inc, and HarperCollins Publishers Ltd. ■ **Fred Hoyle:** from *Man in the Universe* (Columbia University Press, 1966), copyright © Columbia University Press 1966, by permission of the publisher. ■ **Nicholas Humphrey:** from *The Mind Made Flesh* (OUP, 2002), by permission of Oxford University Press. ■ **Julian Huxley:** 'God and Man' from *The Captive Shrew and other Poems of a Biologist* (Basil Blackwell, 1932), by permission of PFD (www.pfd.co.uk) on behalf of the Estate of Julian Huxley. ■ **James Jeans:** from *The Mysterious Universe* (Cambridge, 1930), copyright © Cambridge University Press, by permission of the publisher. ■ **Donald C Johanson and**

Maitland A Edey: from *Lucy: The Beginnings of Mankind* (Penguin, 1990), copyright © Donald C Johanson and Maitland A Edey 1981, by permission of SLL/Sterling Lord Literistic, Inc. ■ **Steve Jones:** from *The Language of Genes* (HarperCollins, 1993), copyright © Steve Jones 1993, by permission of HarperCollins Publishers Ltd. ■ **Jonathan Kingdon:** from *Self-Made Man & His Undoing* (Simon & Schuster, 1993), copyright © Jonathan Kingdon 1993, by permission of Simon & Schuster, Inc. ■ **David Lack:** from *The Life of the Robin* (Fontana, 1970); copyright holder not traced. ■ **Richard Leakey and Roger Levin:** from 'A Giant Lake' in *Origins Reconsidered* (Little Brown/Doubleday, 1992), copyright © B V Sherma 1992, by permission of the publishers, Little, Brown Books and Doubleday, a division of Random House, Inc. ■ **Primo Levi:** from *The Periodic Table* translated by Raymond Rosenthal (Michael Joseph, 1985), translation copyright © Schocken Books Inc 1984, by permission of Penguin Books Ltd and Schocken Books, a division of Random House, Inc. ■ **John Maynard Smith:** from 'The Importance of the Nervous System in the Evolution of Animal Flight' in *On Evolution* (Edinburgh University Press, 1972), by permission of the publishers, www.eup.ed. ac.uk ■ **Ernst Mayr:** from 'The Place of Biology in the Sciences and its Conceptual Structure' in *The Growth of Biological Thought: Diversity, Evolution, and Inheritance* (The Belknap Press of Harvard University Press, 1982), copyright © Ernst Mayr 1982, by permission of the publisher. ■ **Peter B Medawar:** 'Science and Literature' from *The Hope of Progress* (Methuen, 1972), by permission of the Medawar Estate. ■ **J Robert Oppenheimer:** from *The Flying Trapeze: Three Crises for Physicists* (OUP, 1964); copyright holder not traced. ■ **Roger Penrose:** from *The Emperor's New Mind: Concerning Computers, Minds and The Laws of Physics* (OUP, 1989), by permission of Oxford University Press. ■ **Max F Perutz:** 'A Passion for Crystals' from *I Wish I'd Made You Angry Earlier: Essays on Science, Scientists, and Humanity* (Cold Spring Harbor Laboratory Press/OUP, 1998), first published as 'Professor Dorothy Hodgkin', in *The Independent*, Obituaries, 1 August 1986, copyright © Vivien and Robin Perutz, by permission of the publishers. ■ **Steven Pinker:** from *How the Mind Works* (Norton, 1997/Allen Lane 1998), copyright © Steven Pinker 1997, 1998, by permission of W W Norton & Company, Inc and Penguin Books Ltd; from *The Language Instinct: The New Science of Language and Mind* (HarperCollins/Allen Lane, 1994), copyright © Steven Pinker 1994, by permission of the publishers, HarperCollins Publishers and Penguin Books Ltd. ■ **Mark Ridley:** from *Explanation of Organic Diversity: The Comparative Method and Adaptations for Mating* (Clarendon Press, 1983), by permission of Oxford University Press. ■ **Matt Ridley:** from *Genome* (Fourth Estate, 1999), copyright © Matt Ridley 1999, by permission of Felicity Bryan Literary Agency and the author and HarperCollins Publishers, USA. ■ **Martin Rees:** from *Just Six Numbers: The Deep Forces that Shape the Universe* (Weidenfeld & Nicolson, 1999), by permission of the author. ■ **Oliver Sacks:** from *Uncle Tungsten: Memories of a Chemical Boyhood* (Macmillan, 2001), copyright © Oliver Sacks 2001, by permission of Pan Macmillan, London ■ **Carl Sagan:** from *The Demon-Haunted World: Science as a Candle in the Dark* (Headline, 1996), copyright © Carl Sagan 1996; and from *Pale Blue Dot: A Vision of the Human Future in Space* (Random House, 1994), copyright © Carl Sagan 1994, both by permission of Democritus Properties for The Estate of Carl Sagan. ■ **Erwin Schrödinger:** from *What is Life?* (Cambridge, 1944, 1967), copyright © Cambridge University Press 1967, by permission of the author's Estate and the publisher. ■ **Claude E Shannon and Warren Weaver:** from *The Mathematical Theory of Communication* (University of Illinois Press, 1949), copyright © 1949, 1998 by Board of Trustees of the University of Illinois, by permission of the authors and the University of Illinois Press. ■ **George Gaylord Simpson:** from *The Meaning of Evolution: A Study of the History of Life and of Its Significance for Man* (Yale University Press, 1949), copyright © Yale University Press 1949, by permission of the publisher. ■ **Lee Smolin:** from *The Life of the Cosmos* (Weidenfeld & Nicolson, 1997), by permission of the publishers, Weidenfeld & Nicolson, a division of The Orion Publishing Group, and Oxford University Press. ■ **C P Snow:** foreword to G H Hardy *A Mathematician's Apology* (Cambridge, 1967, 2007), copyright © C P Snow 1967, by permission of Curtis Brown Group Ltd, London on behalf of the Estate of C P Snow. ■ **Russell Stannard**: extract from 'The Light Beam that Got Away' from *The Time and Space of Uncle Albert* (Faber, 1989), copyright © Russell Stannard 1989, by permission of the publishers, Faber & Faber Ltd. ■ **Ian Stewart**: from *From Here to Infinity* (OUP, 1996), first published as *The Problems of Mathematics* (1987), by permission of Oxford University Press. ■ **Lewis Thomas**: 'Seven Wonders' from *Late Night Thoughts* (Viking, 1983), copyright © Lewis Thomas 1983, by permission of Viking Penguin, a division of Penguin Group (USA) Inc. ■ **D'Arcy Wentworth Thompson**: from *On Growth and Form*, abridged by John Tyler Bonner (Cambridge, 1961), copyright © Cambridge University Press 1961, by permission of Professor J T Bonner and the publisher. ■ **Niko Tinbergen**: from *Curious Naturalists* (Country Life Ltd, 1958) by permission of the Executor of the Estate of Niko Tinbergen. ■ **Robert Trivers**: from *Social Evolution* (Benjamin/Cummings Publishing Co, 1985), by permission of the author. ■ **Alan Turing**: from 'Computing Machinery and Intelligence', *Mind* (1950) Vol. 59, by permission of Oxford University Press/The Mind Association. ■ **James Watson**: from *Avoid Boring People: Remembered Lessons* (OUP, 2007), by permission of Oxford University Press. ■ **James Watson and Francis Crick**: from 'Molecular Structure of Nucleic Acids', *Nature* 171: 737, 25 April 1953. ■ **Steven Weinberg**: from *Dreams of a Final Theory: The Search for the Fundamental Laws of Nature* (Hutchinson, 1993), copyright © Steven Weinberg 1992, by permission of the publishers, The Random House Group Ltd and Pantheon Books, a division of Random House, Inc. ■ **John Archibald Wheeler with Kenneth Ford**: from *Geons, Black Holes, and Quantum Foam: A Life in Physics* (W W Norton & Co, 1998), copyright ©1998 by John Archibald Wheeler and Kenneth Ford, by permission of W W Norton & Company, Inc. ■ **George C Williams**: from *Adaptation and Natural Selection* (Princeton University Press, 1966), copyright © 1966, renewed 1994 Princeton University Press, by permission of Princeton University Press. ■ **Edward O Wilson**: from *The Diversity of Life* (The Belknap Press of Harvard University Press, 1992/Allen Lane, 1993), copyright © Edward O Wilson 1992, 1993, 2001, by permission of the publishers Harvard University Press and Penguin Books Ltd. ■ **Lewis Wolpert**: from *The Unnatural Nature of Science* (Faber & Faber Ltd/The Belknap Press of Harvard University Press, 1989), copyright © Lewis Wolpert 1989, by permission of the publishers.

Every effort has been made to trace and contact all copyright holders before publication. If notified, the publishers will be pleased to rectify any errors or omissions at the earliest opportunity.

INDEX